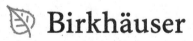

ISNM

International Series of Numerical Mathematics

Volume 167

More information about this series at www.birkhauser-science.com/series/4819

Svetlana M. Bauer · Sergei B. Filippov
Andrei L. Smirnov · Petr E. Tovstik
Rémi Vaillancourt

Asymptotic Methods
in Mechanics of Solids

Svetlana M. Bauer
Faculty of Mathematics and Mechanics
St. Petersburg State University
St. Petersburg
Russia

Sergei B. Filippov
Faculty of Mathematics and Mechanics
St. Petersburg State University
St. Petersburg
Russia

Andrei L. Smirnov
Faculty of Mathematics and Mechanics
St. Petersburg State University
St. Petersburg
Russia

Petr E. Tovstik
Faculty of Mathematics and Mechanics
St. Petersburg State University
St. Petersburg
Russia

Rémi Vaillancourt
Department of Mathematics and Statistics
University of Ottawa
Ottawa
Canada

Based on the Russian version: Асимптотические методы в механике твердого тела, ISBN 5-93972-475-2, © Regular and Chaotic Dynamics Publisher (R&C Dynamics) 2007. All rights reserved

ISSN 0373-3149   ISSN 2296-6072   (electronic)
International Series of Numerical Mathematics
ISBN 978-3-319-38682-9   ISBN 978-3-319-18311-4   (eBook)
DOI 10.1007/978-3-319-18311-4

Mathematics Subject Classification (2010): 34Exx, 74Kxx, 35Pxx

Springer Cham Heidelberg New York Dordrecht London
© Springer International Publishing Switzerland 2015
Softcover reprint of the hardcover 1st edition 2015

Printed on acid-free paper

Springer International Publishing AG Switzerland is part of Springer Science+Business Media (www.birkhauser-science.com)

Preface

The book is based on the special courses "Introduction to the asymptotic methods" and "Asymptotic method in mechanics" for postgraduate students at St. Petersburg State University and first read by Prof. P.E. Tovstik more than 40 years ago. The authors would like to underline the special role of Prof. P.E. Tovstik, who initiated the study of asymptotic methods applied to problems of solid mechanics at Saint-Petersburg (then Leningrad) State University and who is a teacher of the contributors.

The present book is a result of the scientific cooperation of researchers from the Departments of Theoretical and Applied Mechanics of the Faculty of Mathematics and Mechanics of St. Petersburg State University and the Department of Mathematics of the University of Ottawa.

Since in most of the papers in the collection on mechanics of solids published in 1993 [10] asymptotic ideas and methods were used the publisher proposed to supply the volume with survey by S.M. Bauer, S.B. Filippov, A.L. Smirnov, and P.E. Tovstik entitled "Asymptotic Methods in Mechanics with Applications to Thin Shells and Plate." Later this survey encouraged the authors to write a textbook on the application of the asymptotic method in mechanics. The present book is the elaborated version of the Russian edition published in 2007. The book is supplied with the Introduction containing a brief discussion of publications on asymptotic methods in mechanics of solids, especially those that are not referred to in the main text. The reference section is significantly enlarged.

The authors believe that studying the basics of asymptotic methods may be useful to advanced undergraduate, postgraduate, and Ph.D. students in Mathematics, Physics, and Engineering, to researchers and engineers working in the analysis and construction of thin-walled structures and continuous media, and to applied mathematicians who are interested in asymptotic methods in problems of mechanics.

This work was supported in part by the Russian Foundation for Basic Research through Grants #13-01-00523-a and #15-01-06311-a.

The authors are thankful to master's and Ph.D. students of the Department of Theoretical and Applied Mechanics of St. Petersburg State University for their dedicated help in reducing the number of errors in the solutions of the exercises.

Svetlana M. Bauer
Sergei B. Filippov
Andrei L. Smirnov
Petr E. Tovstik
Rémi Vaillancourt

Contents

Introduction

Asymptotic methods of various types have been successfully used since almost the birth of science itself. Transformation of the ideas of asymptotic analysis to a specific area in mathematics happened at the end of the nineteenth century when Henri Poincare introduced the idea of the asymptotic series and gave a rigorous definition of an asymptotic expansion. In the twentieth century asymptotic methods were widely used in different areas of applied mathematics. Now asymptotic methods based on the expansion of solutions in series in small or large parameters or coordinates hold a central place among approximate methods. Asymptotic methods give a qualitative characteristic of the behavior of solutions. Besides that, in some cases, asymptotic expansions have small errors for a rather wide parameter domain.

The number of textbooks, monographs, and journal papers devoted to the asymptotic methods is rather large and it grows constantly. The asymptotic expansions are also discussed in publications on general methods of solution of applied problems. For example, the book by Bender and Orszag [11] contains many interesting examples of application of the perturbation methods.

For introduction to the general principles of asymptotic analysis, the textbooks by Nayfeh [49, 50] may be recommended. In these books the definitions of the asymptotic series and simple operations with them are introduced together with methods of solution of algebraic and transcendent equations, methods of integrations (Laplace method, stationary phase method, and steepest descent method), and classical methods of solution of linear and nonlinear ordinary differential equations with a parameter, including the multiscale method. Some of these problems are discussed in detail in books by de Bruijn [14], Erdèlyi [20], Kevorkian and Cole [37], and Holms [34]. In addition, monograph [34] includes chapters on homogenization, discrete equations, wave propagation, and Lyapunov-Schmidt method. Books by Maslov [43], Maslov, and Nazaikinskii [44] contain descriptions of the most general methods of asymptotic integration of linear and nonlinear nonstationary partial differential equations.

In 1963, Martin David Kruskal coined the term Asymptology to describe the "art of dealing with applied mathematical systems in limiting cases." He tried to show that asymptology is a special branch of knowledge, intermediate, in some sense, between science and art. Kruskal's ideas were lately developed in works by Andrianov and Manevitch [3] and Barantsev [8], which contained the heuristic description of different asymptotic methods.

In the majority of cases in the listed books the authors limit themselves with construction of a few first terms of the asymptotic series without rigorous estimating the errors. These are so-called formal asymptotic expansions. The estimates for the asymptotic expansions are given by Murdock [48], Fröman and Fröman [27], Fedoruk [23], Evgrafov [21].

One of the main areas of application of asymptotic methods is the analysis of differential equations. In asymptotic integration of differential equations containing small parameter the cases of regular and singular perturbations are considered separately [33]. The perturbation is called regular if the orders of the differential equation or the system of equations do not change when the small parameter becomes equal to zero. For singular perturbation, when the small parameter is set equal to zero the order of the equation or system decreases since the small parameter is a multiplier at the higher derivatives. Asymptotic expansions of solutions of singularly perturbed equations are usually divergent series [50].

In the presented book special attention is devoted to the analysis of singular perturbed differential equations. The authors of this book use different asymptotic methods to solve applied problems not pretending to develop the general theory of singular perturbations. The systematic studies of asymptotic solutions of some singular problems may be found in monographs by Eckhaus [19] and Lomov [42].

Consider singularly perturbed linear differential equation of the nth order

$$\sum_{k=0}^{n} \mu^k a_k(x) \frac{d^k y}{dx^k} = 0, \tag{1}$$

where $\mu > 0$ is a small parameter. Solution of (1) we seek in the form

$$y(x, \mu) = \sum_{k=0}^{\infty} \mu^k u_k(x) \exp\left(\frac{1}{\mu} \int_{x_0}^{x} \lambda(x) dx\right). \tag{2}$$

Substituting (2) into (1) and equating the coefficients at μ^k to zero we get the system of equations to find the unknown functions $\lambda(x)$ and $u_k(x)$. For nontrivial solutions $\lambda(x)$ is a root of the characteristic equation

$$\sum_{k=0}^{n} a_k(x) \lambda^k = 0 \tag{3}$$

For $a_n(x) \neq 0$ Eq. (3) has n roots. Let $\lambda(x)$ be a simple root of Eq. (3). Then series (2) may be constructed with the coefficients $u_k(x)$, which makes Eq. (1) an identity. Such series is called formal asymptotic solution. Further, we limit ourselves to construction of such solutions leaving aside the question of existence of exact solutions, for which the obtained solutions are asymptotic expansions.

For the system of singularly perturbed linear differential equations

$$\mu \frac{dy}{dx} = A(x)y, \quad \mu > 0, \tag{4}$$

where y is the nth dimensional vector and A is the square matrix of the nth order, for formal asymptotic solution we seek the form

$$y(x,\mu) \simeq \sum_{k=0}^{\infty} U_k(x)\mu^k \exp\left(\frac{1}{\mu}\int_{x_0}^{x} \lambda(x)dx\right).$$

The function $\lambda(x)$ satisfies the characteristic equation

$$\det(A_0(x) - \lambda(x)I_n) = 0,$$

where I_n is the identity matrix of the order n.

If all n roots of the characteristic equation are simple, then we get n linearly independent asymptotic solutions of Eq. 1 or system (4), which may be used to solve the boundary value problems. Solutions, for which $\Re(\lambda) \neq 0$, increase or decrease exponentially are called the edge effect integrals in solid mechanics and boundary layer integrals in hydromechanics. For $\Re(\lambda) = 0$ the solution rapidly oscillates and for $\lambda = 0$ the solution changes slowly.

The difficulties arise when the characteristic equation has multiple roots. First consider the case of the zero root of the multiplicity m that is often met in applications. The linear differential equation of the order $n = l + m$

$$L_\mu y = \sum_{k=0}^{l} \mu^k a_{k+m}(x) \frac{d^{k+m}y}{dx^{k+m}} + \sum_{k=0}^{m-1} a_k(x) \frac{d^k y}{dx^k} = 0 \tag{5}$$

for $\mu = 0$ transforms to the equation of the order m

$$L_0 y = \sum_{k=0}^{m} a_k(x) \frac{d^k y}{dx^k} = 0. \tag{6}$$

If Eq. (5) is multiplied by μ^m, we get it in the form (1), for which the characteristic equation

$$\sum_{k=0}^{l} a_{k+m}(x)\lambda^{k+m} = 0$$

has zero root of multiplicity m. Let $a_n(x) \neq 0$, $a_m(x) \neq 0$ and all roots of equation

$$\sum_{k=0}^{l} a_{k+m}(x)\lambda^k = 0$$

are simple. Then Eq. (5) has l solutions of form (2). The remaining m solutions are slowly changing functions in x and have the expansions

$$y(x, \mu) = \sum_{k=0}^{\infty} \mu^k v_k(x), \qquad (7)$$

If the multiplicity of the roots changes with the argument x, then the points at which the changes happen are called the turning points. The first approximate studies of the behavior of solutions at the neighborhoods of the turning points were made by Wentzel et al. [11, 37], thus the methods of integration of equations with the turning point are sometimes called WKB-methods. The constructions of asymptotic solutions for system of differential equation of the second and higher orders under different assumptions on the character of the turning point are made in the fundamental monograph by Wasow [65]. Usually, for construction of the asymptotic solutions for equations with turning point the method of comparison equations is used [64]. These equations have the same singularities as the initial equations, but they are simpler than the last ones.

The equation of the second order

$$\mu^2 \frac{d^2 y}{dx^2} - q(x)y = 0$$

with the small parameter at the derivative has the turning point $x = x_*$, if $q(x_*) = 0$. For the simple turning point, for which $q'(x_*) \neq 0$, the asymptotic expansions of the solutions may be expressed in the Airy functions $\mathrm{Ai}(\eta)$ and $\mathrm{Bi}(\eta)$. These functions are the solutions of the comparison equation

$$\frac{d^2 v}{d\eta^2} - \eta v = 0.$$

In the general formulation the problem of asymptotic integration of equations with the turning points has not been solved yet. Only some special cases for the equations encountered in applications has been analyzed.

In the paper by Lin and Rabestein [41] the fourth order equation describing the stability of the laminar viscous flow is considered. Its characteristic has the form $\lambda^4 + x\lambda^2 = 0$, and its roots are quadruple for $x = 0$. The analysis of axisymmetric vibrations of noncylindrical shell of revolution may be reduced to study of the sixth order equation with the characteristic equation $\lambda^6 + f(x)\lambda^2 = 0$ and sextuple turning point. The asymptotic solutions for that equation were constructed by Goldenveizer et al. [30] with the help of the comparison equations method.

The solution of many problems in mechanics of solids may be reduced to solution of the boundary value problems for linear differential equations. The approximate solution of the boundary value problem may be obtained by substituting the asymptotic expansions of solutions into the boundary conditions. For singular perturbed ordinary differential equations such expansion may have the form (2). In this case the method to obtain the approximate solution of the boundary value problem depends on behavior of integrals (2), which is defined, in turn, by the values of the roots of the characteristic equation.

Consider boundary value problem for Eq. (5). Assume that its solution satisfies n homogeneous boundary conditions. When the small parameter μ vanishes Eq. (5) degenerates to Eq. (6) which has the order of $m < n$. Therefore, solution of (6) cannot satisfy all n boundary conditions of the initial boundary value problem. The questions are: (i) can the solution of unperturbed Eq. (6) be zero approximation to solution of the initial problem and (ii) which m out of n given boundary conditions should be selected for Eq. (6)? The answer to these questions is given in the classical paper by Vishik and Lyusternik [62]. In that paper the concept of regular degeneracy, for which the solution of the boundary value problem converges to solution of the unperturbed problem as $\mu \to 0$, is given.

For regular degeneracy it is necessary that all solutions of form (2) be the edge effect integrals and the number of decreasing and increasing integrals corresponds to the number of the boundary conditions on the left and right ends of the integration interval. In this case the solution of the initial boundary value problem is represented as the sum of solutions (2) and (7). In problems of mechanics of solids solution of form (7) is called the main state, the boundary conditions for the unperturbed problem are the main boundary conditions, and the other conditions are the auxiliary conditions.

The thin shell theory provides numerous problems to be solved by asymptotic methods. Two-dimensional differential equations of the eighth order in the theory of shells are singularly perturbed since they contain the natural small parameter h, the dimensionless relative thickness of the shell, which is a factor at the higher derivatives. Often the dimensionless parameter μ, which is proportional to $\sqrt{h}$ is used instead of h. For $\mu = 0$ we have the unperturbed system of equations of the fourth order, which is called membrane (or momentless).

The foundations of asymptotic analysis of linear equations of the theory of shells are formulated in the classical works by Gol'denveizer [28, 29]. In those works an important concept of the index of variation of solution is introduced. The index of variation for function F is a real number t, such that

$$\frac{\partial F}{\partial x} \sim \mu^{-t} F, \quad \text{as} \quad \mu \to 0.$$

For $t > 0$, the function F varies fast while for $t \leq 0$ it varies slowly. Solutions (2) have the index of variation $t = 1$, at the same time the index of variation for solutions (7) is zero.

    The analysis of possible solutions of shell equilibrium equations with different indices of variations in two space variables given in [29] permits to classify the main stress states: membrane (momentless) state, the edge effect, etc. Based on this analysis the approximate methods of solution of the problems of shell statics have been developed. For example, under some conditions the solution may be sought in the form of a sum of the main membrane state and the edge effect integrals.

    For the shells closed in the circumferential direction it is convenient to select as space coordinates on the shell mid-surface (neutral surface) the length of the meridian arc s and the angle in the circumferential direction φ. After separating the variables

$$y(s, \varphi) = y(s)e^{im\varphi}, \tag{8}$$

the equations of shell statics transform to ordinary differential equations, the coefficients of which depend on the wavenumber in the circumferential direction m. In equations describing vibrations or buckling of shells the additional dimensionless parameter Λ, which is proportional to the square of the natural frequency or critical loading, appears. The form of the asymptotic solutions depends on the relations between parameters μ, m and Λ.

    For $m = 0$ the shell deformation is axisymmetric and it is described by the system of differential equations of the sixth order. For low frequency vibrations of cylindrical and conical shells, which are of the great importance to the applications, $m \sim \mu^{-1/2}$ and $\Lambda \sim \mu^2$. In this case the degeneration of the initial system of the eighth order to the system of the fourth order is regular. The stress state described by the unperturbed system is called semi-momentless. For the cylindrical shell the unperturbed system has an explicit solution. The detailed asymptotic analysis of free vibrations of shells is given by Goldenveizer et al. [30].

    To solve the linear problems of buckling of momentless initial stress state, the same methods of asymptotic integration as for the problems of free vibrations are used. The only difference is that for buckling problems the lowest eigenvalue corresponding to the critical load is sought as a rule. Numerous methods and results on buckling of shells are included in the book by Tovstik and Smirnov [56]. In this book the main attention is devoted to the methods of construction of asymptotic expansions of localized buckling modes based on algorithm proposed by Maslov [43]. In the classical problems of shell buckling the radii of the curvature of the mid-surface, its thickness, and momentless initial stress resultants are usually constant. In this case the pits cover the entire surface of a shell under buckling. On the other hand, if the parameters of the shell and the initial stress state depend on the space coordinates, then the localization of the buckling pits may happen at the vicinities of some lines or points on the mid-surface, which are called the weakest lines (points).

    In the book by Tovstik and Smirnov [56] the buckling modes for the convex shells of revolution localized at the neighborhood of the weakest parallel have been constructed. Under nonhomogeneous axial compression of cylindrical shells the buckling modes are localized at the vicinity of the weakest generatrix. The convex shell and cylindrical shell may buckle under nonhomogeneous compression with the buckling mode localized at the weakest point. For these cases the asymptotic expansions for the buckling modes are found. The method of asymptotic separation of the variables is developed and applied to represent the total stress state of the shell as a sum of the semi-momentless state and edge effect. Simultaneously, the problem of separation of the boundary conditions at the shell edges for the main and auxiliary conditions is solved. This method is applied for cylindrical and conic shells, for which the problem may not be reduced to one dimensional by separating the variables in form (8).

In the book by Filippov [26] the method of asymptotic separation of the variables is applied for analysis of free vibrations and buckling under external pressure of the joint shells and shells reinforced by rings. For cylindrical and conic shells the boundary conditions on the shells joint lines and on the lines of the contact of the shell and rings are split into the main and auxiliary boundary conditions.

The solutions of the problems of shell theory by means of the Lyapunov-Schmidt procedure, multiscale method, homogenization, Padé approximants, and other asymptotic methods are included in the monograph by Andrianov et al. [2], which contains a vast bibliography.

The theoretical results in this book are supplemented with the analysis of problems and exercises. In the solution of many problems, asymptotic and numerical methods are used together. The combination of these two methods makes the results more reliable, permits to estimate the applicability domain for asymptotic formulas, and makes easier the numerical analysis of the problem. For example, when evaluating a root of an equation one should know the interval boundary for the root. This boundary may be found by means of asymptotic methods.

Asymptotic estimates of functions, solutions of algebraic and transcendental equations, and also systems of linear algebraic equations are considered in the first chapter. This part is traditional for many manuals on asymptotic methods. However, some of the questions are rarely discussed in textbooks. For example, the Newton polyhedron, which is a generalization of the Newton polygon for equations with two or more parameters, is considered in Chap. 1. Then the important concept of the index of variation for functions is introduced. Special attention is devoted to eigenvalue problems containing a small parameter.

Chapter 2 is dedicated to asymptotic methods for calculating integrals (integration by parts, Laplace transform, stationary phase, saddle point) which are used later in the book to construct asymptotic expansions for solutions of differential equations containing small parameters.

In Chap. 3 the construction of solutions of regularly perturbed ordinary differential equations is discussed. The traditional methods include Poincaré's averaging and multiscale methods. In addition, linear boundary value problems for differential equations with small parameters are considered. Problems for equations with fast oscillating coefficients are also analyzed.

The main part of the book is Chaps. 4 and 5, which deal with methods of asymptotic solutions of linear singularly perturbed boundary value and eigenvalue problems without or with turning points, respectively. In Chap. 4, the asymptotic expansions of linearly independent solutions of systems of linear ordinary differential equations with small parameters at the derivatives are constructed. These asymptotic expansions are later used in the book for approximating solution of nonhomogeneous boundary value and eigenvalue problems. The cases where the eigenfunctions are localized near the edge of the integration interval are also studied. As examples, one-dimensional equilibrium, dynamics, and stability problems for rigid bodies and solids are examined.

In Chap. 5, the singular perturbed problems are analyzed in the case where there exist turning points inside the interval of integration. At a turning point, the asymptotic expansions obtained in Chap. 4 are not valid, since in the expressions one of the functions in the denominator is zero. Approximate asymptotic solutions in a neighborhood of a turning point for linear differential equations of the second order with small parameters at the higher derivative are given in Chap. 5. Then the eigenvalue problems describing the vibration of circular plates and shells of revolution are examined. Asymptotic expansions for the eigenfunctions localized near the internal point of the interval of integration are also found.

Finally, in Chap. 6 the asymptotic integration of nonlinear differential equations is considered, where questions of singular perturbation and ramification of solutions are discussed.

Many of the problems of asymptotic integration are not discussed in this book. Among them there is the method of matching of asymptotic expansions, which is widely used in hydromechanics. Its description may be found, for example, in books by Van Dyke [59] and Hinch [33]. One of the versions of this method is the application of Padé approximants, numerous examples of which are given in the monograph by Baker and Graves-Morris [6].

In our book we analyze only stationary vibrations. We note that the considerable progress has been also made in study of the process of wave propagations by asymptotic methods. In the book by Mikhasev and Tovstik [47] the authors study both localized buckling modes and the motion of the wave packages running on the shell of revolution either in circumferential on axial directions. The book by Babich and Buldyrev [5] concerns the analysis of short-wave asymptotics for solution of Helmholtz equation for the wave propagation with constant or variable wave speed in two-dimensional or three-dimensional spaces. In the book by Kaplunov et al. [36] the asymptotic approach is used to describe the waves of different types in thin elastic solids and in particularly in shells of revolution.

One of the important areas of application of asymptotic methods, which is not included in this book due to its complexity, is the continuum mechanics in the narrow domains. These are the problems of thin-walled beams, plates, and shells theory and also the contact problems for solids on different dimensions. The asymptotic methods are applicable here since these problems contain the geometric small parameter, ratio of the minimum and maximal solid dimensions. In applications the equations are simplified as usual as a result of assumptions on distribution of the unknown functions in the thickness direction. In this case one of the goals of the asymptotic analysis is the verification of the hypotheses and the estimate of the errors happened under the assumptions.

The monograph by Nazarov [51] concerns these problems. It contains the asymptotic expansions of solutions of static problems and problems of free vibrations of beam, plates, and shells. The main attention is devoted to the error estimates, when only the main terms of the asymptotic expansions are considered and to the solvability of the boundary value problems, which appear in the process of the asymptotic integration.

In the book by Gol'denveizer [28] the method to derive the equations of the theory of shells from the three-dimensional equations of the theory of elasticity is proposed. The asymptotic approach developed by Kaplunov et al. in [36] is a dynamical generalization of Goldenveizer's method of asymptotic integration of partial differential equations in narrow domains. In the monograph by Ciarlet [16] two-dimensional equations of the theory of shells describing the membrane stress state are derived by means of the asymptotic method and strong error estimates are obtained. An ingenious sequence of the shell theories refining one another is given by Libai and Simmonds [40].

The authors do not claim the bibliography section to be complete. The references include mostly textbooks and monographs concerning the methods and problems considered in the book.

Chapter 1
Asymptotic Estimates

In this chapter, asymptotic estimates for functions, algebraic and transcendental equations are considered.

1.1 Estimates of Functions

1.1.1 Basic Definitions

Let the functions $f(z)$ and $g(z)$ be defined on a set S of complex or real numbers and let a be a limit point in S. The point a is a *limit point of the set S* if any neighborhood of a contains at least one point of S different from a. Consider $z \in S$. We recall the definitions of some symbols used to compare a function $f(z)$ with a known and, as a rule, simpler function $g(z)$.

The big "O" notation, $f(z) = O(g(z))$ as $z \to a$, means that there exists a neighborhood U of the point a and a constant C (depending on U) such that

$$|f(z)| \leq C|g(z)|, \quad \text{for} \quad z \in U \cap S. \tag{1.1.1}$$

The notation $f(z) = O(g(z))$ is also used if there exists a constant C such that (1.1.1) is valid for all z. It is clear that if this inequality is satisfied for some C, it is also satisfies for any larger C. The least upper bound of the ratio $|f(z)|/|g(z)|$ for $z \in U$ is called the exact upper bound or the boundary constant:

$$C_{\min}(U) = \sup_{z \in U} \frac{|f(z)|}{|g(z)|}.$$

© Springer International Publishing Switzerland 2015
S.M. Bauer et al., *Asymptotic Methods in Mechanics of Solids*,
International Series of Numerical Mathematics 167,
DOI 10.1007/978-3-319-18311-4_1

We list several examples where $S = U = \mathbb{R}$ is the set of real numbers:

$$\sin x = O(x), \quad x \in \mathbb{R}, \qquad C_{\min} = 1,$$
$$\sin x = O(1), \quad x \in \mathbb{R}, \qquad C_{\min} = 1,$$
$$(x+1)^2 = O(x^2), \quad x \in [1, \infty), \quad C_{\min} = 4.$$

If the functions $f(z)$ and $g(z)$ also depend on other variables or parameters but the neighborhood U and the constant C in (1.1.1) do not depend on them, then relation (1.1.1) is said to be uniform in those parameters. For example, if u is a parameter in the interval $[0, a]$, where a is a positive constant, then

$$e^{(x-u)^2} = O\left(e^{x^2}\right)$$

as $x \to \infty$ uniformly in u.

The small "o" notation, $f(z) = o(g(z))$ as $z \to a$, means that

$$\lim_{z \to a} \frac{f(z)}{g(z)} = 0 \quad \text{for} \quad z \in S. \tag{1.1.2}$$

For example, for $S = \mathbb{R}$,

$$\ln x = o(x^\alpha), \quad \text{as} \quad x \to \infty, \quad \alpha > 0,$$
$$x^\alpha = o(e^x), \quad \text{as} \quad x \to \infty, \quad \alpha > 0,$$
$$x^{\alpha_1} = o(x^{\alpha_2}), \quad \text{as} \quad x \to 0, \quad \alpha_1 > \alpha_2.$$

If $f(z)$ and $g(z)$ depend on parameters then relation (1.1.2) is called uniform in these parameters if $f(z)/g(z)$ converges to zero uniformly in these parameters, i.e. for any $\varepsilon > 0$ one can find a neighborhood U_a of the point a such that the inequality

$$\left| \frac{f(z)}{g(z)} \right| < \varepsilon$$

holds for any $z \in U_a \cap S$ simultaneously for all values of the parameters. For example, if $u \in [0, a]$, then $e^{-|z-u|} = o(|z|^{-b})$ as $|z| \to \infty$ uniformly in u. Here a and b are arbitrary real numbers.

Thus $f = O(g)$ means that the order of f is not greater than the order of g, and $f = o(g)$ means that the order of f is less than the order of g as $z \to a$.

For functions f and g whose orders are equal as $z \to a$ we use the notation $f \sim g$ as $z \to a$. In this case $f = O(g)$ and $g = O(f)$ simultaneously as $z \to a$. In many publications the symbol "$\sim$" is only used when the functions f and g are equivalent, i.e.

$$\lim_{z \to a} \frac{f(z)}{g(z)} = 1.$$

In the sequel, we do not assume that the last equality holds for $f \sim g$. For example, for $S = \mathbb{R}$,

$$\sin x \sim x, \quad \text{and} \quad \sin x \sim 2x, \quad x \to 0,$$
$$\ln(1 + x) \sim x, \quad \text{as} \quad x \to 0,$$
$$\sin x - x \sim x^3, \quad \text{as} \quad x \to 0.$$

The relation "$\sim$" similar to relations (1.1.1) and (1.1.2) may be uniform (or non-uniform) in other variables or parameters. For example, for small ε, the relation

$$\frac{1}{1 - x - \varepsilon} \sim \frac{1}{1 - x}$$

is non-uniform in a neighborhood of the point $x = 1$ but uniform in any domain that does not contain this neighborhood, since, as $\varepsilon \to 0$ and $x \approx 1$, the second and the following terms in the expansion have the same orders as the first term:

$$\frac{1}{1 - x - \varepsilon} \sim \frac{1}{1 - x} - \frac{\varepsilon}{(1 - x)^2} + \cdots$$

1.1.2 Operations with Symbols

The notation $o\,(g(z))$ and $O\,(g(z))$ characterizes the classes of functions f, satisfying relations (1.1.1) and (1.1.2). From this follow the rules for operating with these symbols:

$$o\,(g(z)) + o\,(g(z)) = o\,(g(z)), \qquad\qquad o\,(g(z)) + O\,(g(z)) = O\,(g(z)),$$
$$o\,(g(z)) \times o\,(f(z)) = o\,(g(z) \times f(z)), \qquad o\,(g(z)) \times O\,(f(z)) = o\,(g(z) \times f(z)),$$
$$O\,(o\,(g(z))) = o\,(g(z)), \qquad\qquad\qquad o\,(O\,(g(z))) = o\,(g(z)),$$
$$o\,(o\,(g(z))) = o\,(g(z)), \qquad\qquad\qquad o\,(g(z)) = O\,(g(z)),$$

as $z \to a$ and $z \in S$. We note that some of the relations of this type, for example, the last two are irreversible. For example, the equality $O\,(g(z)) = o\,(g(z))$ does not hold.

We prove the first formula in the list. The reader may prove the others in a similar way.

Let $f_1(z) = o\,(g(z))$ and $f_2(z) = o\,(g(z))$ as $z \to a$. Then

$$\lim_{z \to a} \frac{f_1(z) + f_2(z)}{g(z)} = \lim_{z \to a} \frac{f_1(z)}{g(z)} + \lim_{z \to a} \frac{f_2(z)}{g(z)} = 0,$$

i.e. $f_1(z) + f_2(z) = o\,(g(z))$, as was to be proved.

Asymptotic relations and order relations may be integrated under some evident conditions guaranteeing the convergence of the integrals.

Example

If $S = \mathbb{R}$, the function $|f(t)|$ is integrable and $f(x) = O(x^\alpha)$ as $x \to \infty$, then, for $\alpha > -1$,

$$\int_0^x f(t)dt = O(x^{\alpha+1}), \quad \text{as} \quad x \to \infty.$$

Indeed, there exists X such that for $x > X$ the relation $|f(t)| \leq C|x^\alpha|$ is satisfied and

$$\int_0^x f(t)dt \leq \int_0^x |f(t)|dt = \int_0^X |f(t)|dt + \int_X^x |f(t)|\,dt$$

$$\leq C_1 + C\,\frac{x^{\alpha+1}}{\alpha} - C\,\frac{X^{\alpha+1}}{\alpha} = C_2 + C\,\frac{x^{\alpha+1}}{\alpha} \leq Dx^{\alpha+1}$$

$$(1.1.3)$$

for

$$D \geq \frac{C}{\alpha} + \frac{C_2}{X^{\alpha+1}}.$$

Similarly, it may be proved that under the same assumptions for $\alpha < -1$

$$\int_x^\infty f(t)dt = O\left(x^{\alpha+1}\right), \quad \text{as} \quad x \to \infty.$$

For $\alpha = -1$, if the function $|f(t)|$ is locally integrable,

$$\int_0^x f(t)dt = O(\ln x), \quad \text{as} \quad x \to \infty.$$

Operations with the symbols O, o and $\sim$ and many examples can be found in [14, 20, 21, 49, 50, 53].

1.1.3 Exercises

1.1.1. Arrange the following relations in decreasing order for small ε.

$$\ln(1 + \varepsilon), \quad \frac{1 - \cos\varepsilon}{1 + \cos\varepsilon}, \quad \sqrt{\varepsilon(1 - \varepsilon)}.$$

1.1.2. Determine the boundary constants in the interval $[1, \infty)$ for the relations
(a) $\sqrt{(x^2 - 1)} = O(x)$ and **(b)** $x^n = O(e^x)$.

1.1.3. Let S be the sector $|\arg z| \le \frac{\pi}{2} - \varepsilon < \frac{\pi}{2}$ in the complex plane $z \in \mathbb{C}$. Show that there exists a real number $c_1 > 0$ such that $e^{-cz} = O(e^{-c_1|z|})$ as $z \to \infty$ for any $c > 0$.

1.1.4. Show that if f is integrable and $f(x) = o(g(x))$ as $x \to \infty$, where $g(x)$ is a positive non-decreasing differentiable function x, then

$$\int_a^x f(t)\, dt = o(xg(x)).$$

1.1.5. Show the following order relations:

(a) $\dfrac{1 - \sqrt{\cos x}}{1 - \cos(\sqrt{x})} \sim x,$ as $x \to 0,$

(b) $\sin \sqrt{x+1} - \sin \sqrt{x} = O\left(\dfrac{1}{\sqrt{x}}\right),$ as $x \to \infty,$

(c) $\sin \ln(x+1) - \sin \ln x = O\left(\dfrac{1}{x}\right),$ as $x \to \infty,$

(d) For $n \ne 1$, $\ln \dfrac{nx + \sqrt{1 - n^2 x^2}}{x + \sqrt{1 - x^2}} \sim x,$ as $x \to 0.$

1.2 Asymptotic Series

1.2.1 Definitions

Consider the sequence of functions

$$\varphi_n(z), \quad n = 0, 1, 2, \ldots, \tag{1.2.1}$$

defined on a set S that has a limit point a. The sequence (1.2.1) is called *asymptotic* as $z \to a$ if, for any integer $n \ge 0$,

$$\varphi_{n+1}(z) = o\left(\varphi_n(z)\right), \quad \text{as}\ \ z \to a. \tag{1.2.2}$$

We give some examples of asymptotic series under the assumption that $S \subset \mathbb{R}$ is a set of real numbers:

$$(x - a)^n, \quad x \to a; \qquad x^{-n}, \quad x \to \infty;$$

$$e^{\lambda_n x}, \quad x \to \infty, \quad \Re(\lambda_{n+1} - \lambda_n) < 0; \tag{1.2.3}$$

$$\Phi(x)(x-a)^{\alpha_n}, \quad x \to a, \quad \alpha_{n+1} > \alpha_n,$$

where $\Phi(x)$ is an arbitrary function defined on S. The first two sequences in (1.2.3) are called *power sequences* .

Let the function $f(z)$ be defined on the same set S. The series

$$f(z) \simeq \sum_{n=0}^{\infty} a_n \varphi_n(z), \quad \text{as} \quad z \to a, \tag{1.2.4}$$

is called *Poincaré asymptotic expansion* of $f(z)$ in terms of the asymptotic sequence (1.2.2) if a_n are constants and, for any integer $N \geq 0$,

$$R_N(z) = f(z) - \sum_{n=0}^{N} a_n \varphi_n(z) = o(\varphi_N(z)), \quad \text{as} \quad z \to a, \tag{1.2.5}$$

or, what is same,

$$R_N(z) = O(\varphi_{N+1}(z)), \quad \text{as} \quad z \to a. \tag{1.2.6}$$

For asymptotic expansion here and later we use the approximation symbol $\simeq$ instead of the equality sign $=$.

Series (1.2.4) may diverge. Relation (1.2.5) produces a sequence of approximate formulas,

$$f(z) \approx \sum_{n=0}^{N-1} a_n \varphi_n(z), \quad N = 1, 2, \ldots,$$

the errors of which, $R_N(z)$, has the order of the first deleted term in series (1.2.4), $R_N(z) \sim \varphi_{N+1}(z)$. The sum of the first N terms of series (1.2.4) is called the N-terms approximation to the function $f(z)$.

In the books [1, 21, 35, 49, 50, 53] one can find many asymptotic expansions for special functions. For example, the integral exponential function has the asymptotic expansion [1]

$$\text{Ei}(z) = \int_z^{\infty} t^{-1} e^{-t} dt \simeq e^{-z} \sum_{n=0}^{\infty} \frac{(-1)^n n!}{z^{n+1}}, \quad \text{as} \quad |z| \to \infty, \tag{1.2.7}$$

and series (1.2.7) diverges for almost all z. At the same time, the error $R_N(z)$ (see formula (1.2.5)) is not larger than the first deleted term in expansion (1.2.7).

1.2.2 Properties of Asymptotic Series

Properties of asymptotic series and operations on them are discussed in [14, 20, 21, 35]. Here we list only the main ones.

If the function $f(z)$ is expanded in the asymptotic series (1.2.4) in terms of the sequence (1.2.1), then this expansion is unique, i.e. the coefficients a_n in (1.2.4) are defined in a unique way:

$$a_n = \lim_{z \to a} \frac{f(z) - \sum_{k=0}^{n-1} a_k \varphi_k(z)}{\varphi_n(z)}.$$

However $f(z)$ is not defined in a unique way by the asymptotic series. Indeed, the function $f(z) + f_1(z)$ has the same asymptotic expansion (1.2.4) if, for all $n \geq 0$,

$$f_1(z) = o(\varphi_n(z)), \quad \text{as} \quad z \to a.$$

For example, the functions $\mathrm{Ei}(z)$ and $\mathrm{Ei}(z) + e^{-2z}$ have the same expansion (1.2.7).

From the above remarks, it is clear that in the choice of sequence (1.2.1) for the asymptotic expansion of $f(z)$ one should take the behavior of $f(z)$ as $z \to a$ into account. For a bad choice of $\varphi_n(z)$, expansion (1.2.4) either does not exist or produces the trivial result $f(z) \simeq 0$. For example, expanding e^{-z} in the sequence z^{-n}, one gets $e^{-z} \simeq 0$ as $z \to \infty$, i.e. in (1.2.4) $a_n = 0$ for all n.

Asymptotic expansions may be summed, multiplied by functions, differentiated and integrated under special assumptions. The power asymptotic expansions

$$f(z) \simeq \sum_{n=0}^{\infty} a_n z^n, \quad g(z) \simeq \sum_{n=0}^{\infty} b_n z^n, \quad \text{as} \quad z \to 0 \tag{1.2.8}$$

may be multiplied: $f(z)g(z)$, and, if $b_0 \neq 0$, divided: $f(z)/g(z)$. If $b_0 = 0$ we may substitute the function $g(z)$ into $f(z)$: $f[g(z)]$. In this case, the asymptotic expansions for the resulting functions are obtained following the rules for convergent Maclaurin series (1.2.8). If the series

$$\sum_{n=0}^{\infty} a_n z^n \tag{1.2.9}$$

converges in the open disk $S = \{z : |z| < R\}$, then it defines an analytic function $f(z)$ in this domain, the asymptotic expansion of which coincides with (1.2.9) as $z \to 0$. It is not difficult to give an example where series (1.2.9) diverges for all $z \neq 0$ (for example, when $a_n = n$).

It should be noted that any coefficients a_n of series (1.2.9) gives the asymptotic expansion as $z \to 0$ of some function $f(z)$, analytic in the sector

Fig. 1.1 Errors for asymptotic expansions of Bi(x)

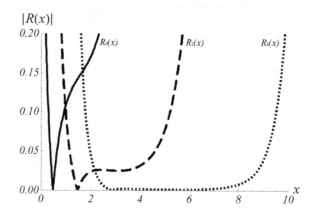

$$S = \{z : |z| < R, \quad \alpha_1 < \arg z < \alpha_2\}$$

for any α_1, α_2.

We underline the significant difference from the computational point of view between convergent series and divergent asymptotic series. If series (1.2.4) converges for some z, this means that for given z

$$\lim_{N \to \infty} R_N(z) = 0,$$

i.e. the value of the function $f(z)$ may be (in principle) found to any accuracy. If series (1.2.4) diverges, then the error $R_N(z)$, keeping the $(N+1)$th term of the series, attains its maximum value for some $N = N_0(z)$ and the accuracy of the computation may not be higher than $R_{N_0}(z)$. In Fig. 1.1 the errors $R_0(x)$, $R_2(x)$ and $R_6(x)$ are plotted for the Airy function Bi(x) for $x > 0$ (see Sect. 5.1).

In the simplest cases, for example, special functions, one can construct the entire asymptotic series. In other cases, in particular, for the integration of differential equations, we must limit ourselves to first or the first two terms of the series because of awkward calculations (see the following chapters).

If the function $f(z)$ (and, perhaps, $\varphi_n(z)$) depends on the parameter u, and the terms with symbols O and o in (1.2.2), (1.2.5) and (1.2.6) are uniform (non-uniform) in u in some set U then the asymptotic expansions is called uniform (non-uniform) in u in U.

For example, the expansion

$$\frac{1}{x - 1 + \varepsilon} = \sum_{n=0}^{\infty} \frac{(-1)^n \varepsilon^n}{(x - 1)^{n+1}}, \quad x > 1, \quad \text{as} \quad \varepsilon \to 0,$$

is not uniform for $(x - 1)/\varepsilon = O(1)$. In that domain the error due to the deletion of the terms beyond the Nth term does not have order $O(\varepsilon^N)$.

1.2.3 Exercises

1.2.1. Find the first three terms of the asymptotic expansions of the following functions for small ε:

(a) $\sqrt{1 - \dfrac{\varepsilon}{2} + 2\varepsilon^2}$, (b) $\sin\left(1 + \varepsilon - \varepsilon^2\right)$, (c) $\left(1 - a\varepsilon + a^4\varepsilon^2\right)^{-1}$,

(d) $\sin^{-1}\left(\dfrac{\varepsilon}{\sqrt{1+\varepsilon}}\right)$, (e) $\ln \dfrac{1 + 2\varepsilon - \varepsilon^2}{\sqrt[3]{1+\varepsilon}}$, (f) $\ln\left[1 + \ln(1+\varepsilon)\right]$.

1.2.2. Which of the following expansions are not uniform for all x as $\varepsilon \to 0$? In which domains these expansions are not uniform?

(a) $f = \displaystyle\sum_{n=0}^{\infty} (-1)^n (\varepsilon x)^n$, (b) $g = \displaystyle\sum_{n=1}^{\infty} \varepsilon^{n-1} \cos nx$, (c) $q = \displaystyle\sum_{n=0}^{\infty} \left(\dfrac{\varepsilon}{x}\right)^n$.

1.2.3. Show that the asymptotic expansion of the function $f(x) = e^{-x} \sin e^x$ in terms of the sequence x^{-n} as $x \to \infty$ gives the trivial result $f(z) \simeq 0$.

1.2.4. Find a series

$$\sum_{n=0}^{\infty} a_n x^n,$$

that formally satisfies the differential equation

$$x^3 u'' + (x^2 + x)u' - u = 0.$$

Show that this series diverges.

1.3 Newton Polygons

1.3.1 Introduction

We consider the problem of finding a solution, $x(\mu)$, of the implicit equation

$$F(\mu, x) = 0,$$

where μ and x are real variables and $F(\mu_0, x_0) = 0$.

Expanding $F(\mu, x)$ in a power series of $(\mu - \mu_0)$ and $(x - x_0)$, we seek a solution of the form

$$x = x_0 + \gamma(\mu - \mu_0)^\beta + \gamma'(\mu - \mu_0)^{\beta'} + \cdots,$$

where $\beta, \beta', \ldots$ is an increasing sequence of positive integers.

Presumably, Isaac Newton was the first who studied this problem without the assumption that $F'_x(\mu_0, x_0) \neq 0$ and proposed a geometrical method for evaluating $\gamma, \beta, \gamma', \beta', \ldots$, which was later called the Newton diagram (or Newton polygon), the description of which may be found in [58].

1.3.2 Statement of the Problem

Consider the case where $F(\mu, x)$ is a polynomial in x:

$$\sum_{k=0}^{n} a_k(\mu) x^k = 0. \tag{1.3.1}$$

Let the coefficients a_k have the form

$$a_k(\mu) = \sum_{j=0}^{m_k} a_{kj} \mu^{\alpha_{kj}}, \tag{1.3.2}$$

where

$$a_{k0} \neq 0, \quad \alpha_{k,j+1} > \alpha_{kj}, \quad 0 \leqslant m_k \leqslant \infty, \quad k = 0, 1, \ldots, n.$$

Thus, $a_k(\mu) \sim \mu^{\alpha_{k0}}$ as $\mu \to 0$. We assume that $m_p = 0$ and $\alpha_{p0} = \infty$ for $a_p(\mu) \equiv 0$, and let $a_n(\mu) \neq 0$ and $a_0(\mu) \neq 0$.

We seek asymptotic expansions for the roots x_q of equation (1.3.1) in the form

$$x_q \simeq \sum_{j=0}^{\infty} x_{qj} \mu^{\beta_{qj}}, \quad \mu \to 0, \quad \beta_{q,j+1} > \beta_{qj}, \quad q = 1, \ldots, n, \tag{1.3.3}$$

or

$$x_q \simeq x_{q0} \mu^{\beta_{q0}} + o\left(\mu^{\beta_{q0}}\right). \tag{1.3.4}$$

To find the values of x_{q0} and β_{q0} we substitute relation (1.3.4) in (1.3.1), collect terms with the lowest power in μ and set to zero the coefficient of that power. Until

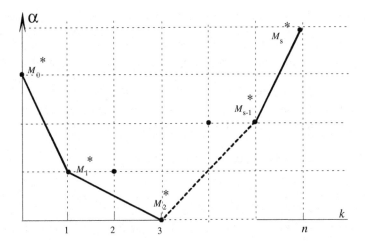

Fig. 1.2 Newton polygon

the value of β_{q0} is determined, we do not know which terms have the lowest power in μ. It is only clear that these terms are among the followings:

$$a_{00}\mu^{\alpha_{00}}, \quad a_{10}x_{q0}\mu^{\alpha_{10}+\beta_{q0}}, \quad a_{20}x_{q0}^2\mu^{\alpha_{20}+2\beta_{q0}}, \quad \ldots, \quad a_{n0}x_{q0}^n\mu^{\alpha_{n0}+n\beta_{q0}}.$$

For cancelling the terms of lowest order in μ (the main terms) two of the exponents

$$\alpha_{00}, \quad \alpha_{10}+\beta_{q0}, \quad \alpha_{20}+2\beta_{q0}, \quad \ldots, \quad \alpha_{n0}+n\beta_{q0},$$

must coincide and the rest should be not less than them. Equating the powers we find all the values for β_{q0}, and then obtain x_{q0}.

To find the values of the parameter β_{q0} by means of the Newton polygon, draw $n+1$ points $M_k = \{k, \alpha_{k0}\}$ in the (k, α) plane with integer abscissas (Fig. 1.2). Then draw the segment determined by the points M_0 and M_1. The tangent of the angle between the segment and the axis of abscissas, k, is equal to the value of β_{q0}, for which the orders of the first and second terms coincide. It is not hard to check that the points which lie above the line passing through M_0 and M_1 correspond to the terms with highest order in μ (the smaller terms).

We are interested only in the main terms; this is why one should join the points M_k with segments in such a way that the points M_k not belonging to this segment lie above the obtained broken line.

To construct the broken line we rotate anticlockwise the ray going vertically down from the point $M_0^* = M_0$. We denote by M_1^* the first of the points M_k that is touched by the ray. Then we rotate anticlockwise the vertical ray going from the point M_1^* until it touches the next point M_2^*. And so on until the final ray touches the point $M_s^* = M_n$. If the ray contains several points M_k then for M_i^* we take the rightmost point (with maximal k).

The broken line connecting the points M_i^* is called the *Newton polygon* which is the lower part of a convex hull for a point set. The slope ratio for the segment determined by the points M_i^* and M_{i+1}^* gives the order β_{q0} for the root x_q. The length of the projection of the segment on the k axis is equal to the number of roots x_q of such power and the number of points M_k through which the segment passes is equal to the number of terms in the equation for evaluating x_{q0}.

We note that the Newton polygon usually permits to estimate the order of the correction (the order of the next term) in the expression for the root,

$$x \sim \mu^{\beta_{q0}} \left[1 + O\left(\mu^{\delta/\varkappa}\right)\right] = \mu^{\beta_{q0}} + O\left(\mu^{\delta/\varkappa+\beta_{q0}}\right),$$

where the value of δ is defined as the minimal length of the segments going vertically from points M_k ($M_k \neq M_i^*$ and $M_k \neq M_{i+1}^*$) till the straight line containing the segment determined by the points M_i^* and M_{i+1}^*, and $\varkappa$ is the multiplicity of the root x_{q0}.

The equations

$$-\mu x^3 + x^2 - \mu^2 = 0$$

and

$$-\mu x^3 + x^2 - 2\mu x + \mu^2 = 0$$

have similar Newton polygons (Fig. 1.3), and also, in both cases, $\delta = 2$ for the roots $x_{1,2} = O(\mu)$.

However, the multiplicity of the root ($\varkappa = 2$) for the second equation increases the error in the expansion for a root. So, for the first equation $x_{1,2} = \pm\mu + O(\mu^3)$ and for the second $x_{1,2} = \mu + O(\mu^2)$.

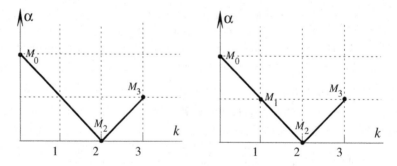

Fig. 1.3 Effect of the root multiplicity on the error of the expansion

Example 1

Consider the equation

$$\mu^3 x^5 + x^4 - 2x^3 + x^2 - \mu = 0 \qquad (1.3.5)$$

and find initially the first term in expansion (1.3.3), omitting the index q. Substituting $x = x_0 \mu^{\beta_0}$ into (1.3.5) we get

$$\mu^{3+5\beta_0} x_0^5 + \mu^{4\beta_0} x_0^4 - 2\mu^{3\beta_0} x_0^3 + \mu^{2\beta_0} x_0^2 - \mu = 0. \qquad (1.3.6)$$

Now we must find the main terms in (1.3.6) and the values of β_0 for which at least two main terms have equal orders. For $\beta_0 < 0$, the main terms are the first two terms. The value of $\beta_0 = -3$ is found from the condition for equal orders of the main terms $3+5\beta_0 = 4\beta_0$, and $x_0 = -1$ is obtained from the equation $x_0^5 + x_0^4 = 0$. The Newton polygon for Eq. (1.3.5) is plotted in Fig. 1.4. It consists of three parts: descending, constant and ascending. Moreover,

$$\beta_{10} = -3, \quad x_1 = -\mu^{-3}\left[1 + O\left(\mu^3\right)\right], \quad \text{since } |K_1 M_3| = 3;$$

$$\beta_{20} = \beta_{30} = 0, \quad \beta_{40} = \beta_{50} = 1/2, \quad \text{since } |K M_3| = 1/2.$$

Therefore $x_{2,3} = 1 + O\left(\mu^{1/2}\right)$ and $x_{4,5} = \pm\mu^{1/2}\left[1 + O\left(\mu^{1/2}\right)\right]$.

To obtain the following terms of series (1.3.3) we substitute $x_q = \mu^{\beta_{q0}}(1 + R_q)$ in (1.3.2). Then for R_q we get an algebraic equation of type (1.3.2). However, now we seek only the roots that satisfy the relation $R_q = o(1)$. The number of such roots is equal to the multiplicity of the root x_{q0}.

Fig. 1.4 Newton polygon for (1.3.5)

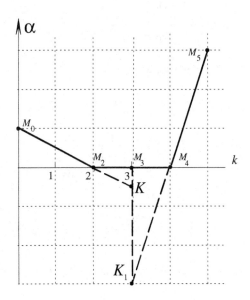

Fig. 1.5 Newton polygon
for (1.3.7)

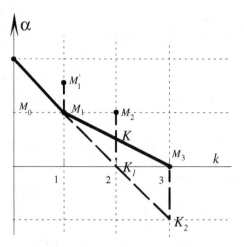

The two-term approximations for the roots of equation (1.3.6) are the followings:

$$x_1 = -\mu^{-3} - 2 + o(1), \quad x_{2,3} = 1 \pm \mu^{1/2} + o\left(\mu^{1/2}\right), \quad x_{4,5} = \pm\mu^{1/2} + \mu + o(\mu).$$

Example 2

Consider the equation

$$x^3 + 3\mu x^2 - \left(\mu + \mu^{3/2}\right) x + 2\mu^2 = 0. \tag{1.3.7}$$

The Newton polygon for this equation is shown in Fig. 1.5.

It is clear that

$$\beta_{10} = \beta_{20} = 1/2, \quad |KM_2| = 1/2, \quad x_{1,2} = \pm\mu^{1/2}\left[1 + O\left(\mu^{1/2}\right)\right].$$

For the third root, we get

$$\beta_{30} = 1, \quad |K_1 M_2| = |K_2 M_3| = 1, \quad x_3 = 2\mu\left(1 + O\left(\mu^{1/2}\right)\right),$$

since $|M_1' M_1| = 1/2$. Here $M_1' = \{k, \alpha_{k1}\}$. The second term in the coefficient $a_1(\mu) = \mu + \mu^{3/2}$ affects the order of the correction.

The approximate values for the roots of equation (1.3.7) are

$$x_1 = -\mu^{1/2} - 3\mu + O\left(\mu^{3/2}\right),$$

$$x_2 = \mu^{1/2} - 2\mu + O\left(\mu^{3/2}\right),$$

$$x_3 = 2\mu - 2\mu^{3/2} + O\left(\mu^2\right).$$

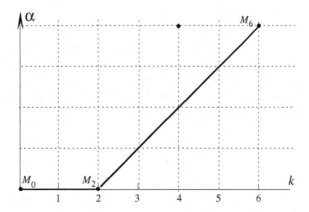

Fig. 1.6 Newton polygon for (1.3.8)

Example 3

We consider one more equation similar to that for the problem of the spectrum of axisymmetric free vibrations of a circular cylindrical thin shell (see Sect. 4.4 in Chap. 4),

$$\mu^4 x^6 + \left(1 - \nu^2\right) \lambda \mu^4 x^4 + (1 - \lambda) x^2 + \lambda \left[1 - \left(1 - \nu^2\right) \lambda\right] = 0. \qquad (1.3.8)$$

Here λ is the frequency parameter, where λ, ν, $(1 - \nu^2) \sim 1$ and also $(1 - \lambda) \sim 1$. The Newton polygon for Eq. (1.3.8) is plotted in Fig. 1.6. Clearly,

$$\beta_{10} = \beta_{20} = 0 \quad \text{and} \quad x_{1,2} = \pm ib, \quad b^2 = \frac{\lambda \left[1 - \left(1 - \nu^2\right) \lambda\right]}{(1 - \lambda)}.$$

For the other four roots,

$$\beta_{k0} = -1, \quad \text{where} \quad k = 3, 4, 5, 6, \quad x_k = \frac{c}{\mu}, \quad c^4 = (\lambda - 1)^{1/4}.$$

We note that considering a linear differential equation of order n with constant coefficients depending in the parameter μ,

$$\sum_{k=0}^{n} a_k(\mu) \frac{d^k z}{dt^k} = 0,$$

after the substitution $z = e^{xt}$ we obtain the algebraic equation (1.3.1).

1.3.3 Newton Polyhedra

A more difficult problem is to find asymptotics for the roots of equations containing two or more parameters. In this case, the shape of the Newton polygon depends on the relation between the two parameters.

Consider an equation containing two parameters, the main small parameter μ and an arbitrary parameter λ,

$$P(x; \mu, \lambda) = \sum_{k=0}^{n} \sum_{i} a_{ki} \mu^{\alpha_{ki}} \lambda^{\beta_{ki}} x^k = 0, \qquad a_{ki} \geqslant 0, \quad a_{ki} = O(1).$$

We draw points $\{k, \alpha_{ki}, \beta_{ki}\}$ in 3D space and construct the convex hull of such point set. Since one of the parameters, μ, is small, one should construct only the "lower" facets of the hull, i.e. only the facets visible from the point $(0, -\infty, 0)$ in the $\{k, \alpha, \beta\}$ space. The facets of the convex hull determine the relations for the parameters for which the Newton polygon changes shape. To get these relations (abridged equations) one should equate the orders of the terms corresponding to the nodal points of such facets.

Remark The above method may be used also when the coefficients a_{ki} have arbitrary signs. But in this case, the order of the main term in the abridged equation, which is the difference of two terms of the same order, may go down. This is a specific situation that must be considered separately (see Example 4).

Example 4

Find the main terms of the equation

$$\mu^4 x^6 + \left(1 - \nu^2\right) \lambda \mu^4 x^4 + (1 - \lambda) x^2 + \lambda \left[1 - \left(1 - \nu^2\right) \lambda\right] = 0$$

for $\mu \ll 1$. As noted before, an equation of this type describes the spectrum of free axisymmetric vibrations of a circular cylindrical thin shell. Now the order of the frequency parameter λ may be arbitrary (compare with Example 3 where λ has order 1).

The 3D convex hull for the above equation consists of the three facets:

1. (M_1, M_2, M_3, M_4),
2. (M_3, M_4, M_6),
3. (M_2, M_4, M_5, M_6),

where the points M_i have the following coordinates:

$$M_1 = \{0, 0, 1\}, \quad M_2 = \{0, 0, 2\}, \quad M_3 = \{2, 0, 0\},$$
$$M_4 = \{2, 0, 1\}, \quad M_5 = \{4, 4, 1\}, \quad M_6 = \{6, 4, 0\},$$

(see Fig. 1.7).

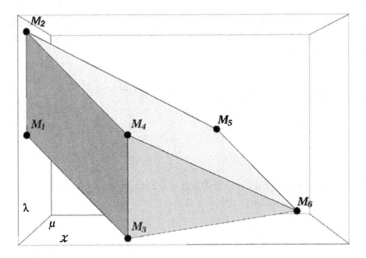

Fig. 1.7 Newton polyhedron

Equating the orders of the terms which define the facets we get

$$\lambda \sim \lambda^2 \sim x^2 \sim \lambda x^2,$$
$$\lambda x^2 \sim x^2 \sim \mu^4 x^6,$$
$$\lambda^2 \sim \lambda x^2 \sim \lambda \mu^4 x^4 \sim \mu^4 x^6.$$

Hence, for the first and second relations, $\lambda \sim 1$, and for the third, $\lambda \sim \mu^{-4}$, and the entire domain of the parameter λ splits into three subdomains, where the Newton polygon has a similar structure.

In the subdomain $\lambda \ll 1$, the Newton polygon has two segments determined by the points M_1, M_3, and M_3, M_6, respectively, which correspond to the abridged equations $\lambda + x^2 = 0$ and $1 + \mu^4 x^4 = 0$.

In the subdomain $1 \ll \lambda \ll \mu^{-4}$, the Newton polygon has two segments determined by the points M_2, M_4, and M_4, M_6, respectively, which correspond to the abridged equations $(1 - \nu^2)\lambda + x^2$ and $-\lambda + \mu^4 x^4 = 0$.

In the subdomain $\lambda \gg \mu^{-4}$, the Newton polygon has two segments determined by the points M_2, M_5, and M_5, M_6, respectively, which correspond to the abridged equations $-\lambda + \mu^4 x^4 = 0$ and $(1 - \nu^2)\lambda + x^2 = 0$.

On the boundary of the subdomains for $\lambda \sim 1$, the Newton polygon has two segments determined by the points M_1, M_2, M_3, M_4, and M_3, M_4, M_6, respectively, which correspond to the abridged equations $(1 - \lambda)x^2 + \lambda(1 - (1 - \nu^2)\lambda) = 0$ and $\mu^4 x^4 + (1 - \lambda) = 0$ (see Example 3). The situation becomes more difficult if $1 - \lambda \sim \mu^\alpha$ for $\alpha > 0$. Here the order of the coefficient at the main term in the abridged equation goes down and this requires a special consideration (see Sect. 4.4.5).

On the boundary of the domains for $\lambda \sim \mu^{-4}$, the Newton polygon has one segment determined by the points M_2, M_4, M_5 and M_6, which corresponds to the abridged equation

$$\mu^4 x^6 + \left(1 - \nu^2\right) \lambda \mu^4 x^4 - \lambda x^2 - \left(1 - \nu^2\right) \lambda^2 = 0.$$

1.3.4 Exercises

Use Newton polygons to find the first and second terms in the expansions for the roots of the following equations for $\mu \ll 1$.

1.3.1. $x^3 - 3x\mu + \mu^3 = 0$.
1.3.2. $\mu^4 x^4 - x^2 + x - \mu = 0$.
1.3.3. $\mu^{-3} x^3 + \mu^{-1} x^2 - \mu^{-2} x + 1 = 0$.
1.3.4. $\mu^5 x^5 - \mu^2 x^3 + x - \mu^3 = 0$.

1.3.5. Find the main terms in the expansion for the roots of the following equation for $\mu \ll 1$:

$$\mu^4 x^4 + \lambda^2 \mu^2 x^4 + \mu \lambda x^2 + \lambda = 0.$$

1.4 Variation Index of a Function

1.4.1 Definitions

When integrating ordinary differential equations with parameter $\lambda > 0$, we obtain functions $F(x, \lambda)$ of two variables. For the asymptotic integration and qualitative study of solutions as $\lambda \to \infty$, one should have a way to describe the rate of change of $F(x, \lambda)$ as a function of x, i.e. the derivative of $\partial F/\partial x$, as $\lambda \to \infty$. Gol'denveizer in [28] introduced the concept of the variation index of a function.

The variation index of a function $F(x, \lambda)$ as $\lambda \to \infty$ is the value t such that

$$\frac{\partial F}{\partial x} \sim \lambda^t F, \quad \text{as} \quad \lambda \to \infty. \tag{1.4.1}$$

For $t > 0$, the function F varies fast and for $t < 0$ it varies slowly.
    As an example consider the function

$$F(x, \lambda) = g(x) e^z \tag{1.4.2}$$

where

$$z = \lambda^t f(x) \quad \text{as} \quad \lambda \to \infty, \quad t \geq 0, \quad f'(x) \neq 0 \tag{1.4.3}$$

(f and g do not depend on λ). Since

$$\frac{\partial F}{\partial x} = \lambda^t e^{\lambda^t f(x)} \left[g(x) f'(x) + \frac{g'(x)}{\lambda^t} \right] \sim \lambda^t f'(x) F, \quad \text{as} \quad \lambda \to \infty,$$

then, the variation index of function (1.4.2) is equal to t. Similarly, the variation index of the function

$$F(x, \lambda) = g(x) \sin z \tag{1.4.4}$$

under condition (1.4.3) is also equal to t since, for $\lambda \to \infty$,

$$\frac{\partial F}{\partial x} = \lambda^t \left[g(x) f'(x) \cos(\lambda^t f(x)) + g'(x) \sin(\lambda^t f(x)) \right] \sim \lambda^t f'(x) F \frac{\cos z}{\sin z}.$$

In fact, for function (1.4.4) the estimate (1.4.1) is not valid in neighborhoods of the points where $\cos z$ or $\sin z$ are equal to zero, but these cases used to be ignored when introducing the variation index.

Example
   The variation index may be different in different domains of variation of x. As an example consider the equation

$$F'' - \lambda q(x) F = 0, \quad ()' = \partial/\partial x, \tag{1.4.5}$$

with holomorphic coefficient $q(x)$. For $q(x) \neq 0$, the variation index of the solutions $F(x, \lambda)$ for Eq. (1.4.5) is $t = 1/2$ since, in this case, for $q > 0$ (see Chap. 4)

$$F(x, \lambda) \simeq \frac{C_1 \exp\left(\int \sqrt{\lambda q(x)}\, dx \right) + C_2 \exp\left(- \int \sqrt{\lambda q(x)}\, dx \right)}{\sqrt[4]{q(x)}},$$

and for $q < 0$

$$F(x, \lambda) \simeq \frac{C_1 \cos\left(\int \sqrt{-\lambda q(x)}\, dx \right) + C_2 \sin\left(\int (\sqrt{-\lambda q(x)}\, dx \right)}{\sqrt[4]{-q(x)}}.$$

If $q(x_0) = 0$ and $q'(x_0) = a \neq 0$, then, for $x - x_0 \sim \lambda^{-1/3}$, the variation index is equal to $t = 1/3$. To prove this we use the substitution $x = x_0 + (\lambda a)^{-1/3} \eta$. Then Eq. (1.4.5) in the given domain may be represented in the form

$$\frac{\partial^2 F}{\partial \eta^2} - \eta F = 0, \tag{1.4.6}$$

which does not contain the parameter λ, and, therefore, we may consider $\partial F/\partial \eta \sim F$ as $\lambda \to \infty$; consequently $\partial F/\partial x \sim \lambda^{1/3} F$.

Equation (1.4.6) is called *Airy's equation*.

The *characteristic length of the deformation profile*, $l \sim \lambda^{-t}$, is connected with the variation index t (this concept is used to describe the field of deformation of solids).

1.4.2 Auxiliary Definitions

The variation index may also be introduced for functions of a larger number of variables. Consider the function $F(x, y, \lambda)$. The *general variation index* of F is the number t, such that

$$\max \left\{ \left| \frac{\partial F}{\partial x} \right|, \left| \frac{\partial F}{\partial y} \right| \right\} \sim \lambda^t F, \quad \text{as} \quad \lambda \to \infty.$$

If arbitrary functions F have different orders in different variables or different directions in the (x, y) plane, then we introduce *partial variation indexes*, $t_i < t$. For example, the function $F(x, y, \lambda) = A \sin(\lambda^{t_1} x) \sin(\lambda^{t_2} y)$ has variation indexes t_1 and t_2 in x and y, respectively.

To compare the orders of several functions, we use the *indexes of intensity*. Consider two functions $F_1(x, \lambda)$ and $F_2(x, \lambda)$ which can be represented in the form

$$F_k = g_k(x, \lambda) H_k(x, \lambda), \quad k = 1, 2,$$

where

$$g_k \sim \lambda^{p_k}, \quad k = 1, 2; \quad H_1 \sim H_2, \quad \text{as} \quad \lambda \to \infty.$$

Then the numbers p_k are called the indexes of intensity of the functions F_k.

For example, if the functions F_k have the form (1.4.4), $F_k = g_k \sin(z + \alpha_k)$, then we may consider $H_1 \sim H_2 \sim 1$, and the variation indexes coincide with the orders of the functions F_k, i.e.

$$F_k \sim \lambda^{p_k}. \tag{1.4.7}$$

For functions $F_k = g_k e^z$ of type (1.4.3), the estimate (1.4.7) is not valid.

Let the function F have variation index t and index of intensity p. Then by (1.4.3), for $t \geq 0$ the index of intensity p_1 for its derivative $\partial F/\partial x$ is equal to $p_1 = p + t$.

Clearly the variation index cannot be introduced for every function. Often solutions of differential equations with a parameter may be represented as a sum of several functions with different variation indexes.

1.4.3 Exercises

1.4.1. Find the variation index of the function $F(x, \lambda) = g(x) \sinh z$ for $z = \lambda^t f(x)$, $\lambda \to \infty$, $t \geq 0$, $f'(x) \neq 0$.

1.4.2. For the following functions, find the general variation index. Show the directions in the (x, y) plane, in which the partial variation indexes attain their minima as $\lambda \to \infty$ for $n \geq 0$.

(a) $F(x, y, \lambda) = \sin(\lambda^n (x - ay))$,
(b) $F(x, y, \lambda) = e^{\lambda^n x / y}$.

1.4.3. Find the indexes of variation of the solutions of the differential equations:

(a) $y''' + 4\lambda^2 y'' - \lambda y' - \lambda^2 y = 0$.
(b) $y^{(5)} - y^{(4)} - \lambda^2 y''' + \lambda^2 y'' + \lambda y' - \lambda y = 0$.

1.5 Asymptotic Solution of Transcendental Equations

Asymptotic methods are also used to solve transcendental equations, though, naturally, apart from algebraic equations there does not exist a unified approach.

Example 1

We want to solve the equation

$$\tan x = \frac{x}{a}, \tag{1.5.1}$$

for large positive values of x. Here a is a positive number. For example, such equations occur in the solution of problems of high frequency transverse vibrations of a clamped beam with variable cross-section area.

It is evident that for large values of x the solution of equation (1.5.1) has the form $x = \left(\frac{1}{2} + n\right)\pi + o(1)$, where n is a positive integer. Denote $\lambda = \left(\frac{1}{2} + n\right)\pi$ and seek the correction δ for this root. Then, Eq. (1.5.1) may be represented as:

$$\tan(\lambda + \delta) = -\frac{1}{\tan \delta} = \frac{\lambda + \delta}{a}$$

or

$$\tan \delta = -\frac{a}{\lambda + \delta} = -\frac{a}{\lambda} + O\left(\frac{\delta}{\lambda^2}\right).$$

Thus, it follows that

$$\delta = -a\lambda^{-1} + O\left(\lambda^{-3}\right) \quad \text{or} \quad x = \lambda - a\lambda^{-1} + O\left(\lambda^{-3}\right).$$

Further, the main part of the correction δ_1 is

$$x = \lambda - a\lambda^{-1} + \delta_1, \quad \delta_1 = O\left(\lambda^{-3}\right).$$

Then, Eq. (1.5.1) becomes

$$\frac{1}{\tan\left(a\lambda^{-1} - \delta_1\right)} = \frac{\lambda - a\lambda^{-1} + \delta_1}{a},$$

or

$$a\left(\lambda - a\lambda^{-1} + \delta_1\right)^{-1} = \tan\left(a\lambda^{-1} - \delta_1\right). \tag{1.5.2}$$

Expanding both sides of equality (1.5.2) in series, we obtain

$$\sum_{k=0}^{\infty} \frac{b_k}{\lambda^{2k+1}} = \frac{a}{\lambda} + \frac{a^2}{\lambda^3} + o\left(\lambda^{-3}\right)$$

$$= \left(a\lambda^{-1} - \delta_1\right) + \frac{\left(a\lambda^{-1} - \delta_1\right)^3}{3} + O\left(\lambda^{-5}\right)$$

$$= \frac{a}{\lambda} - \delta_1 + \frac{a^3}{3\lambda^3} + o\left(\lambda^{-3}\right),$$

which, in turn, gives

$$\delta_1 = a^2\left(\frac{a}{3} - 1\right)\lambda^{-3} + O\left(\lambda^{-5}\right),$$

i.e.

$$x = \lambda - a\lambda^{-1} + a^2\left(\frac{a}{3} - 1\right)\lambda^{-3} + O(\lambda^{-5}).$$

Similarly, one can find the next approximations. Hence,

$$x = \lambda + \sum_{k=0}^{\infty} \frac{c_k}{\lambda^{2k+1}}, \tag{1.5.3}$$

where

$$c_0 = -a, \quad c_1 = a^2\left(\frac{a}{3} - 1\right), \quad c_2 = a^3\left(\frac{4a}{3} - 2 - \frac{a^2}{5}\right), \quad \ldots.$$

The asymptotic order of the error steadily goes down with the number of approximation. Note that, for $a = 1$, the first root of equation (1.5.1) with an accuracy of five decimal places is equal to $x_1 = 4.49341$, and the asymptotic approximations for the first three roots are

$$x_{n,i} = \lambda_n + \sum_{k=0}^{k=i} \frac{c_k}{\lambda_n^{2k+1}}, \quad \text{where} \quad \lambda_n = \frac{\pi}{2} + \pi n.$$

Results are in the next table.

i	$x_{1,i}$	$x_{2,i}$	$x_{3,i}$
0	4.71239	7.85398	10.99557
1	4.50018	7.72666	10.90463
2	4.49381	7.72528	10.90413
3	4.49344	7.72525	10.90412
∞	4.49341	7.72525	10.90412

Example 2

Consider the similar equation

$$\tan \frac{b}{x} = \sum_{k=0}^{\infty} a_k x^{2k+1} \tag{1.5.4}$$

for small x. Here a_k and b are real numbers. The problem of free transverse vibrations of a string with variable density with fixed ends leads to such equation.

Clearly, the solution of equation (1.5.4) has the form

$$\frac{b}{x} = \pi n \left[1 + o\left(\frac{1}{n}\right) \right] \quad \text{or} \quad x = \frac{b}{\pi n} + o\left(\frac{1}{n^2}\right),$$

where n is a positive integer and $x \to 0$.

Similar to Example 1, the asymptotic expansion for the root is

$$x = \sum_{k=0}^{\infty} \frac{c_k}{(\pi n)^{2k+1}}, \quad c_0 = b. \tag{1.5.5}$$

Substituting this series into Eq. (1.5.4) and equating the coefficients of the same power of the small parameter $1/(\pi n)$ to zero, we can find the coefficients c_k, $(k > 0)$ for series (1.5.5):

$$c_1 = -a_0 b^2, \quad c_2 = b^3 \left(2a_0^2 - a_1 b + \frac{a_0^3}{3} \right), \quad \ldots$$

Example 3

Consider the equation,

$$x^2 - \ln x = u, \tag{1.5.6}$$

where u is a large positive parameter.

It is evident that $x^2 - \ln x \sim x^2$ as $x \to \infty$. Then Eq. (1.5.6) may be written in the form

$$u = [1 + o(1)]x^2, \quad \text{as} \quad x \to \infty,$$

whence

$$x = u^{1/2}[1 + o(1)], \quad \text{as} \quad u \to \infty.$$

Substituting this approximation in equation $x^2 = u + \ln x$ and taking $\ln[1 + o(1)] = o(1)$ into account, we find

$$x^2 = u + \frac{1}{2}\ln u + o(1),$$

i.e.

$$x = u^{1/2}\left[1 + \frac{\ln u}{4u} + o\left(\frac{1}{u}\right)\right]. \tag{1.5.7}$$

As in Example 1, one can iterate this substitution and find the asymptotic approximation to any order.

For $u = 10$, Eq. (1.5.6) has the solution $x = 3.3478$, while relation (1.5.7) gives $x \approx 3.3443$.

For $u = 100$ the exact solution is $x = 10.11504$ and relation (1.5.7) gives $x \approx 10.11512$.

Example 4

Construct the asymptotic solution of the equation

$$\sin z + z = 0 \tag{1.5.8}$$

in the complex plane for large values of $|z|$.

Note that on the real axis this equation has only the root $z = 0$. Substituting $z = x + iy$, where i is the imaginary unit, into Eq. (1.5.8) and setting the real and imaginary parts equal to zero, we obtain the system:

$$\begin{aligned} \frac{e^y + e^{-y}}{2}\sin x + x &= 0, \\ \frac{e^y - e^{-y}}{2}\cos x + y &= 0. \end{aligned} \tag{1.5.9}$$

One can see from system (1.5.9) that $|y|$ increases as $|x|$ goes up. It is also clear that if the pair of real numbers x, y is a solution of system (1.5.9) then the pair x, $-y$ is also a solution. That is why one can consider $y > 0$ and for large values of y system (1.5.9) can be approximately represented in the form

$$\frac{e^y}{2} \sin x + x = 0,$$

$$\frac{e^y}{2} \cos x + y = 0. \tag{1.5.10}$$

From the second equation in (1.5.10) we get

$$x = \pm \arccos \left(-2y\, e^{-y} \right) + 2\pi n, \tag{1.5.11}$$

and, therefore, $x \sim n$ as $n \to \infty$. Since $|-2y\, e^{-y}| \to 0$ as $y \to \infty$, then

$$x = 2\pi n \pm \frac{\pi}{2} + o(1). \tag{1.5.12}$$

Thus, Eq. (1.5.10) leads to

$$y = \ln \left(\frac{-2x}{\sin x} \right) \simeq \ln(4\pi n - \pi) \simeq \ln 4\pi n + O\left(n^{-1} \right). \tag{1.5.13}$$

Note that the "+" sign is not acceptable in equality (1.5.12), since the argument of the logarithm must be positive. Substituting expression (1.5.13) into (1.5.11) one gets the next approximation for x:

$$x \approx 2\pi n - \frac{\pi}{2} - \frac{\ln 4\pi n}{2\pi n}$$

Therefore,

$$z \simeq 2\pi n - \frac{\pi}{2} - \frac{\ln 4\pi n}{2\pi n} + i \ln 4\pi n + O\left(n^{-1} \right) \quad \text{as} \quad n \to \infty.$$

Example 5

Many asymptotic solutions of transcendental equations can be obtained by the direct use of the Lagrange–Bürmann inversion formula. This formula is defined for functions of a complex variable, but it can also be used for functions of a real variable.

Let the function $f(z)$ be analytic in a neighborhood of the point $z = 0$ and $f(0) \neq 0$. Consider the equation

$$\mu = \frac{z}{f(z)}, \tag{1.5.14}$$

where z is unknown. Then, there exists $a > 0$ such that, for $|\mu| < a$, Eq. (1.5.14) has only one solution in a neighborhood of $z = 0$ and that solution is an analytic function of μ:

$$z = \sum_{n=1}^{\infty} c_n \mu^n, \quad c_n = \frac{1}{n!}\left[\left(\frac{d}{dz}\right)^{n-1}(f(z))^n\right]\Big|_{z=0}, \qquad (1.5.15)$$

The generalized formula determines the value of $g(z)$, if the function $g(z)$ is analytic in a neighborhood of $z = 0$:

$$g(z(\mu)) = g(0) + \sum_{n=1}^{\infty} d_n \mu^n, \quad d_n = \frac{1}{n!}\left[\left(\frac{d}{dz}\right)^{n-1} g'(z)(f(z))^n\right]\Big|_{z=0}.$$

For example, solution (1.5.3) can be obtained by formula (1.5.14) since it is obvious that each segment $\pi(n + 1/2) < x < \pi(n + 3/2)$ contains only the root x_* and the difference $x_* - \pi(n + 1/2) \to 0$ as $n \to \infty$.

Assuming that $x = \pi(n+1/2)+z = \lambda+z$ and $\mu = \lambda^{-1}$, Eq. (1.5.1) is transformed into an equation of type (1.5.14), where

$$f(z) = \frac{-z}{\sin z}(a\cos z + z\sin z), \quad f(0) = -a.$$

Note that Eq. (1.5.4) can be transformed into Lagrange–Bürmann equation in a similar manner, but this method of solution is more cumbersome than the above direct solution. To transform equation (1.5.6) into an equation of type (1.5.14) is much more difficult than for Example 1.

1.5.1 Exercises

1.5.1. Find the first three terms of the asymptotic expansion for the roots of the equation $xe^{1/x} = e^u$, where u is a large positive number.

1.5.2. Find the first two terms of the asymptotic expansion of the equation $\cos x \cosh x = -1$ for large positive roots. The problem of buckling under a distributive loading of a beam with one clamped edge and one free edge is reduced to this equation.

1.5.3. Find the first three terms of the asymptotic expansion for large positive roots of the equation $x \sin x = 1$.

1.5.4. Find first three terms of the asymptotic expansion for large positive roots of the equation $x \tan x = 1$.

1.5.5. Find the first three terms of the asymptotic expansion for large positive roots of the equation $\tan x = x^k$, where k is a positive integer.

1.5.6. Find the first two terms of the asymptotic expansion for a root of the equation $x \ln x = u$, where u is a large positive number.

1.5.7. Find first three terms of the asymptotic expansion for a root of the equation $x + \tanh x = u$, where u is a large positive number.

1.5.8. Find the first three terms of the asymptotic expansion for a root in the interval $(0, \pi/2)$ for the equation $x \tan x = u$, where u is a large positive number.

1.5.9. Find the asymptotic solution of the equation $\cos z + z = 0$ in the complex plane for large values of $|z|$ ($\Re z > 0$, $\Im z > 0$).

1.6 Solution of Systems of Linear Algebraic Equations

1.6.1 Regular Unperturbed Systems

Consider the system of linear algebraic equations

$$A(\mu)x = b(\mu), \quad x = (x_1, \ldots x_N)^T, \quad b = (b_1, \ldots b_N)^T, \tag{1.6.1}$$

where

$$A(\mu) = \sum_{n=0}^{\infty} A_n \mu^n, \quad b(\mu) = \sum_{n=0}^{\infty} b_n \mu^n. \tag{1.6.2}$$

As $\mu \to 0$, system (1.6.1) degenerates into the system

$$A_0 x = b_0. \tag{1.6.3}$$

If for small μ the determinant $\det A(\mu) \neq 0$, then Eq. (1.6.1) has a solution $x(\mu)$. We shall discuss shortly the problem of expanding the solution $x(\mu)$ in powers of μ and the relation between that solution and the solution of the confluent equation (1.6.3).

In the simple case where $\det A_0 \neq 0$ and A_0^{-1} exists, the solution of system (1.6.1) is obtained in the form of a power series in μ:

$$x = x_0 + \mu x_1 + \mu^2 x_2 + \cdots. \tag{1.6.4}$$

Substituting the asymptotic expansion (1.6.4) into Eq. (1.6.1) and equating the terms with the same power of μ we get

$$
\begin{aligned}
\mu^0: &\quad A_0 x_0 = b_0, \\
\mu^1: &\quad A_0 x_1 = b_1 - A_1 x_0, \\
\mu^2: &\quad A_0 x_2 = b_2 - A_1 x_1 - A_2 x_0, \\
&\quad \vdots
\end{aligned} \tag{1.6.5}
$$

System (1.6.5) is an iterative process for defining the expansion coefficients of (1.6.4).

1.6.2 Singular Unperturbed Systems in Special Cases

In the case $\det A_0 = 0$, problem (1.6.3) is, generally speaking, unsolvable and the solution of problem (1.6.1), $x(\mu)$, if it exists, is singular at the point $\mu = 0$. If we assume that $\det(A_0 + \mu A_1) \neq 0$ for small μ, then Eq. (1.6.1) has solution $x(\mu)$. A detailed discussion of this issue may be found in [63].

In the sequel, we only quote some results.

If $\det A_0 = 0$, then the confluent equation (1.6.3) is not solvable for all b_0 and the corresponding homogeneous equation

$$A_0 x = 0 \tag{1.6.6}$$

has a limited number, $r < N$, of linearly independent solutions, x_{i0} ($i = 1, 2, \ldots, r$). The nonhomogeneous equation (1.6.3) admits a solution if and only if its right side, b_0, is orthogonal to all solutions $y^{(k)}$ ($k = 1, 2, \ldots, r$) of the adjoint homogeneous equation

$$A_0^* y^{(k)} = 0 \quad (k = 1, 2, \ldots, r), \tag{1.6.7}$$

i.e.

$$\left(b_0, \bar{y}^{(k)}\right) \equiv \sum_{j=1}^{N} b_j \bar{y}_j = 0.$$

Here A_0^* denotes the adjoint transpose of matrix A_0 ($a_{ij}^* = \bar{a}_{ji}$, where the bar means complex conjugation) and $(\cdot, \cdot)$ is the vector scalar product.

For simplicity assume that $r = 1$. Then the homogeneous equation (1.6.6) has only one solution, x_{10}, up to a constant factor. In this case, the adjoint equation (1.6.7) also has just one solution, $y^{(1)}$. We expand the solution $x(\mu)$ of Eq. (1.6.1) in terms of μ:

$$x = \frac{C_0 x_{10}}{\mu} + (x_0 + C_1 x_{10}) + \mu(x_1 + C_2 x_{10}) + \cdots + \mu^n(x_n + C_{n+1} x_{10}) + \cdots \tag{1.6.8}$$

Here the unknowns are the coefficients C_i and the vectors x_i. To evaluate these we substitute expression (1.6.8) into Eq. (1.6.1) and equate coefficients of equal powers of μ:

$$\begin{aligned}
\mu^{-1}: \quad & C_0 A_0 x_{10} = 0, \\
\mu^0: \quad & A_0 x_0 + C_0 A_1 x_{10} = b_0, \\
\mu^1: \quad & A_0 x_1 + A_1 x_0 + C_1 A_1 x_{10} + C_0 A_2 x_{10} = b_1, \\
& \vdots
\end{aligned} \tag{1.6.9}$$

The first equality in (1.6.9) holds for any C_0. The second equation is solvable for x_0 if and only if

$$\left(b_0 - C_0 A_1 x_{10}, y^{(1)} \right) = 0.$$

From this equation the constant C_0 may be found:

$$C_0 = \frac{\left(b_0, y^{(1)} \right)}{\left(A_1 x_{10}, y^{(1)} \right)}.$$

Once C_0 is calculated, one can obtain x_0 as a partial solution of the second equation in (1.6.9). Hence, the first two equations in (1.6.9) provide C_0 and x_0. We write the third equation in the form

$$A_0 x_1 = b_1 - A_1 x_0 - C_1 A_1 x_{10} - C_0 A_2 x_{10}.$$

From the solvability conditions for that equation,

$$\left(b_1 - A_1 x_0, -C_1 A_1 x_{10} - C_0 A_2 x_{10}, y^{(1)} \right) = 0$$

we find the next coefficient C_1 and then the next partial solution x_1. Continuing this process, one can find C_2 and x_2, etc.

1.6.3 Singular Unperturbed Systems in General Cases

To construct a solution in the general case where $r > 1$ we must recall some definitions of linear algebra.

Let A be a matrix of order N. If the vector u_0 satisfies the equation $Au_0 = 0$, and the nonzero vectors $u_1, \ldots, u_{k-1}$ satisfy the equations $Au_i = u_{i-1}$ for $i \geq 1$ but the equation $Ax = u_{k-1}$ is unsolvable, then the vectors $u_1, \ldots, u_{k-1}$ form a *Jordan chain* of length k, and the vector u_i is called *a generalized eigenvector* of the ith order.

Let $A_0, \ldots, A_l$ be matrices of order N from (1.6.2) and consider the generalized characteristic equation

$$P(\lambda) \equiv \det \left(A_0 + \lambda A_1 + \lambda^2 A_2 + \cdots + \lambda^l A_l \right) = 0.$$

If $\lambda = 0$ is a root of this equation (a generalized eigenvalue), then there exists an eigenvector u_0 ($Au_0 = 0$). In this case the so-called generalized eigenvectors $u_0, u_1, \ldots, u_{k-1}$ satisfy the equations

$$
\begin{aligned}
A_0 u_1 &= -A_1 u_0 = z_1, \\
A_0 u_2 &= -A_1 u_1 - A_2 u_0 = z_2, \\
&\vdots = \vdots \\
A_0 u_{k-1} &= -A_1 u_{k-2} - A_2 u_{k-3} - \cdots - A_l u_{k-l} = z_{k-1},
\end{aligned}
\tag{1.6.10}
$$

where $z_1, z_2, \ldots, z_{k-1}$ are orthogonal to all eigenvectors of the adjoint matrix A_0^* corresponding to the eigenvalue $\lambda = 0$ and z_k is not orthogonal to all eigenvectors.

Thus, suppose Eq. (1.6.6) has $r > 1$ linearly independent solutions x_{i0} ($i = 1, 2, \ldots, r$). Let x_{ij} ($j = 0, 1, \ldots, n_i - 1$) be a complete system of eigenvectors and generalized eigenvectors with respect to the matrices $A_0, A_1, \ldots$. Then we want to expand the solution of equation (1.6.1) in the form

$$
\begin{aligned}
x(\mu) = \sum_{i=1}^{r} & \left[\frac{C_{i0} x_{i0}}{\mu^{n_i}} + \frac{C_{i0} x_{i1} + C_{i1} x_{i0}}{\mu^{n_i - 1}} + \cdots \right. \\
& \left. + \frac{C_{i0} x_{i,n_i-1} + C_{i1} x_{i,n_i-2} + \cdots + C_{i,n_i-1} x_{i0}}{\mu} \right] \\
& + x_0 + \sum_{i=1}^{r} (C_{i1} x_{i,n_i-1} + \cdots + C_{in_i} x_{i0}) + \cdots \\
& + \mu^s \left[x_s + \sum_{i=1}^{r} (C_{i,s+1} x_{i,n_i-1} + \cdots + C_{i,s+n_i} x_{i0}) \right] + \cdots
\end{aligned}
\tag{1.6.11}
$$

The coefficients C_{im} and the vectors x_s are the unknowns. As before, to obtain these unknowns we substitute (1.6.11) into Eq. (1.6.1) and equate the coefficients with equal powers of μ. This case is considered in greater detail in [63].

1.6.4 Exercises

Find the first three terms of the asymptotic expansion for the solution of the following systems of equations.

1.6.1.

$$
\begin{aligned}
(1 + \mu)x + \mu y + z &= a_1, \\
\mu x + y + \mu z &= a_2, \\
x + \mu y + (1 + \mu)z &= a_3.
\end{aligned}
$$

1.6.2.

$$
\begin{aligned}
(6 + \mu)x + \mu y + (-2 + \mu)z &= 16 + 26\mu, \\
(-3 + 2\mu)y + (2 + \mu)z &= 8\mu, \\
(1 + \mu)x + 4y + (-3 - 2\mu)z &= 6 + 2\mu.
\end{aligned}
$$

1.6.3.

$$(1 + \mu)x + y + z = b_1,$$
$$x + (1 + \mu)y + z = b_2,$$
$$x + y + (1 + \mu)z = b_3.$$

1.6.4.

$$(-4 - \mu)x + y + 2z = d_1,$$
$$-8x + (2 - \mu)y + 4z = d_2,$$
$$-4x + y + (2 - \mu + \alpha\mu^2)z = d_3.$$

Note: In Exercise 1.6.4, the form of the expansion depends on the value of α.

1.7 Eigenvalue Problems

Now, we consider eigenvalue problems for systems of linear algebraic equations:

$$(A(\mu) - \lambda I)x = 0, \tag{1.7.1}$$

where, as before,

$$A(\mu) = \sum_{n=0}^{\infty} A_n \mu^n.$$

Here I is the identity matrix. Let A_0 be a Hermitian matrix, that is $A_0 = A_0^*$. If A is an $N \times N$ Hermitian matrix, then for any N vectors x, y we have

$$(Ax, y) = (x, Ay).$$

We recall that Hermitian matrices have only real eigenvalues and a full set of eigenvectors which can be chosen to be orthonormal.

Given a solution of an eigenvalue problem, we expand the eigenvalue and the eigenvector in power series in μ:

$$\lambda = \lambda_0 + \mu\lambda_1 + \mu^2\lambda_2 + \cdots, \quad x = x_0 + \mu x_1 + \mu^2 x_2 + \cdots \tag{1.7.2}$$

Substituting the asymptotic expansions (1.7.2) into (1.7.1) and equating the terms of equal power of μ we get:

$$\mu^0 : (A_0 - \lambda_0 I) x_0 = 0,$$
$$\mu^1 : (A_0 - \lambda_0 I) x_1 = \lambda_1 x_0 - A_1 x_0,$$
$$\vdots$$
$$\mu^k : (A_0 - \lambda_0 I) x_k = \sum_{l=0}^{k-1} (\lambda_{k-l} x_l - A_{k-l} x_l), \qquad (1.7.3)$$
$$\vdots$$

The first equation in (1.7.3) has N eigenvalues λ_{0i} and N eigenvectors x_{0i}. First, consider the case where all eigenvalues are simple. We seek the corrections to the eigenvalues λ_{0i}. Normalize the eigenvectors x_{0i}: $(x_{0i}, x_{0i}) = 1$. Since $\det(A_0 - \lambda_{0i} I) = 0$, for the second equation in (1.7.3) to have a solution it is necessary that its right side be orthogonal to the eigenvectors of the matrix on the left side:

$$((A_0 - \lambda_{0i} I) x_{1i}, x_{0i}) = \lambda_{1i} |x_{0i}|^2 - (A_1 x_{0i}, x_{0i}).$$

Since the matrix $(A_0 - \lambda_{0i} I)$ is Hermitian, then

$$((A_0 - \lambda_{0i} I) x_{1i}, x_{0i}) = (x_{1i}, (A_0 - \lambda_{0i} I) x_{0i}) = 0,$$

and the correction to the eigenvalue is

$$\lambda_{1i} = \frac{(A_1 x_{0i}, x_{0i})}{|x_{0i}|^2} = (A_1 x_{0i}, x_{0i}).$$

After obtaining the value of λ_{1i}, we can find the correction x_{1i} to the eigenvector x_{0i} from the second equation in (1.7.3). Continuing this process one can get the next terms in the expansions (1.7.2).

1.7.1 Multiple Eigenvalues

Now consider the more complex case of multiple eigenvalues. Let the unperturbed problem (the first equation in (1.7.3)) have an eigenvalue λ_0 of multiplicity r with associated pairwise orthonormal eigenvectors $e_1, e_2, \ldots, e_r$. For the perturbed problem, a bifurcation of the r-fold eigenvalue occurs.

As before, represent the perturbed eigenvalues and eigenvectors in power series $(j = 1, 2, \ldots, r)$

$$\lambda_j = \lambda_0 + \mu \lambda_{1j} + \mu^2 \lambda_{2j} + \cdots, \quad x = x_{0j} + \mu x_{1j} + \mu^2 x_{2j} + \cdots.$$

As opposed to the case of simple eigenvalues, it is not clear now how to find the vectors x_{0j}. It is only known that these vectors are linear combinations of the vectors $e_1, e_2, \ldots, e_r$. Write

$$x_{0j} = a^0_{j1}e_1 + a^0_{j2}e_2 + \cdots + a^0_{jr}e_r. \tag{1.7.4}$$

Comparing the coefficients of equal powers of μ we get:

$$\mu^1: \quad A_1 x_{0j} + A_0 x_{1j} = \lambda_0 x_{1j} + \lambda_{1j} x_{0j}. \tag{1.7.5}$$

One should find the scalar coefficients $a^0_{j1}, \ldots, a^0_{jr}$ and the values of λ_{1j}. Taking the scalar product of both sides of equality (1.7.5) with e_p we get

$$\left(A_1 x_{0j}, e_p\right) + \left(A_0 x_{1j}, e_p\right) = \lambda_0 \left(x_{1,j}, e_p\right) + \lambda_{1j} \left(x_{0j}, e_p\right),$$

or taking

$$\left(A_0 x_{1j}, e_p\right) = \left(x_{1j}, A_0 e_p\right) = \lambda_0 \left(x_{1,j}, e_p\right)$$

into account, we have

$$\left(A_1 x_{0j}, e_p\right) = \lambda_{1j} \left(x_{0j}, e_p\right).$$

Substituting (1.7.4) in the above equality, we find

$$\sum_{i=1}^{r} \left(A_1 e_i, e_p\right) a^0_{ji} = \lambda_{1j} a^0_{jp},$$

or

$$\sum_{i=1}^{r} c_{ip} a^0_{ji} = \lambda_{1j} a^0_{jp}, \quad \text{where} \quad c_{ip} = \left(A_1 e_i, e_p\right), \tag{1.7.6}$$

i.e. the quantities λ_{1j} are the eigenvalues of the matrix $C = (c_{ip}), i, p = 1, 2, \ldots, r$, and may be determined from the equation

$$\det |C - \lambda_{1j} I| = 0, \quad j = 1, 2, \ldots, r.$$

The coefficients a^0_{jk} are obtained from Eq. (1.7.6).

1.7.2 Generalized Eigenvalue Problems

A similar approach is used to solve the generalized eigenvalue problem

$$Ax = \lambda Bx, \quad \text{where} \quad A = \sum_{n=1}^{\infty} A_n \mu^n, \quad B = \sum_{n=1}^{\infty} B_n \mu^n,$$

where the matrices A_0 and B_0 are symmetric.

One seeks the eigenvalues and eigenvectors as series in small parameters. The eigenvectors for the unperturbed problem, e_j, possess the following properties

$$\left(A_0 e_j, e_i\right) = \delta_{ij}\lambda_{0j}, \quad \left(B_0 e_j, e_i\right) = \delta_{ij}.$$

Applying the same algorithm and taking these properties into account, one gets formulas for a_{ji}^k and the next approximations to λ_{kj}.

1.7.3 Spectrum of a Bundle of Operators

Now we come to the problem of non-selfadjoint perturbation, i.e. of the spectrum of a bundle of operators. Consider the problem

$$Ax = B(\lambda)x, \tag{1.7.7}$$

where the matrices A and B are represented in the form

$$A = \sum_{n=0}^{\infty} A_n \mu^n, \quad B(\lambda) = \sum_{i=1}^{\infty} B_i \lambda^i, \quad B_i = \sum_{n=0}^{\infty} B_{n,i}\mu^n.$$

As before, we seek a solution as a series in the small parameter μ. Consider a particular case of the problem where

$$A = A_0 + A_2\mu^2, \quad B = B_{1,1}\mu\lambda + I\lambda^2. \tag{1.7.8}$$

Such problems arise in the study of free vibrations of rotating solids. In this case the term A_0 is related to the elastic strain energy of the non-rotating solid, A_2 is related to the energy of initial stresses and centrifugal forces, and $B_{1,1}$ is related to the energy of the Coriolis force, x is the displacement vector, λ is the natural frequency, and μ is the angular velocity. For convenience, we denote $B_{1,1} = A_1$. Now, representing eigenvectors and eigenvalues for the matrix A in the form (1.7.2), substituting (1.7.2) into (1.7.7) and equating the coefficients of equal powers of μ we get

$$\mu^0 : \left(A_0 - \lambda_0^2 I\right)x_0 = 0,$$
$$\mu^1 : \left(A_0 - \lambda_0^2 I\right)x_1 = \lambda_0 A_1 x_0 + 2\lambda_0\lambda_1 x_0,$$
$$\vdots$$

$$\mu^k : \left(A_0 - \lambda_0^2 I\right)x_k = \sum_{l=1}^{k-1}\left(\lambda_{k-l-1}A_1 x_l - A_2 x_{k-2} + \sum_{p=0}^{k-l}\lambda_p\lambda_{k-p-l}x_l\right),$$

$$\tag{1.7.9}$$

The spectrum of the unperturbed problem (the first equation in (1.7.9)) consists of N pairs of frequencies $+\lambda_{0j}$, $-\lambda_{0j}$, both of which corresponding to the eigenvector x_{0j}. Consider the case where all eigenvalues λ_0 are simple and nonzero. Then the eigenvectors form an orthogonal basis, the normalized elements of which we denote as e_i. Representing the vectors in a form similar to (1.7.2) we have

$$x_{kj} = \sum_{i=1}^{N} a_{ji}^{k} e_i, \qquad j = 1, \ldots, N, \tag{1.7.10}$$

and substituting them into (1.7.9) we obtain

$$\mu^1 : \sum_{i=1}^{N} \left(\lambda_{0i}^2 - \lambda_{0j}^2 \right) a_{ji}^1 e_i = \left(\lambda_{0j} A_1 + 2\lambda_{0j}\lambda_{1j} I \right) e_j,$$

$$\vdots$$

$$\mu^k : \sum_{i=1}^{N} \left(\lambda_{0i}^2 - \lambda_{0j}^2 \right) a_{ji}^k e_i =$$

$$\sum_{i=1}^{N} \sum_{l=0}^{k-1} \left(a_{j,i}^l \lambda_{j,k-l-1} A_1 e_i - a_{j,i}^{k-2} A_2 e_i + \sum_{p=0}^{k-l} \lambda_{j,p} \lambda_{j,k-p-l} a_{j,i}^l e_i \right). \tag{1.7.11}$$

Multiplying (1.7.11) by e_p and assuming that $p = j$, we obtain an expression for the correction to the frequency λ_{jk}, and for $p \neq j$ we get the corrections to the eigenvector x_{kj}. Here we took $a_{ij}^0 = \delta_{ij}$ and $a_{ii}^k = 0$ into account. We list the first corrections to the eigenvalue and eigenvector.

$$\lambda_{1j} = -\frac{(A_1 e_j, e_j)}{2}, \qquad j = 1, \ldots, N,$$

$$a_{ij}^1 = \frac{\lambda_{0j}(A_1 e_j, e_i)}{\lambda_{0i}^2 - \lambda_{0j}^2}, \qquad i, j = 1, \ldots, N, \quad j \neq i. \tag{1.7.12}$$

One can see that the first corrections to the eigenvalues from the pair $\pm\lambda_j$ are equal to each other and the corrections to the corresponding eigenvectors have opposite signs. For the second correction the situation is reversed.

$$\pm\lambda_j = \pm\lambda_{0j} + \mu\lambda_{1j} \pm \mu^2\lambda_{2j}\cdots, \quad x_j = x_{0j} \pm \mu x_{1j} + \mu^2 x_{2j} \pm \cdots.$$

Perturbations in the problem under consideration lead to the bifurcation of eigenvalues and eigenvectors and their shift. In Fig. 1.8, three lower frequencies of free vibrations of a rotating cylindrical shell versus the angular velocity μ are plotted. Here the solid line is an exact solution, and the dashed line is a two-term approximation.

Fig. 1.8 Three lower
frequencies of free vibrations
versus the angular velocity μ

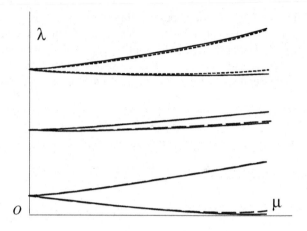

1.7.4 Exercises

Find the eigenvalues and eigenvectors of the matrix $A(\mu)$ to accuracy $O(\mu)$.

1.7.1.

$$A(\mu) = \begin{bmatrix} \mu & 1 & 1+\mu \\ 1 & \mu & -1 \\ 1+\mu & -1 & \mu \end{bmatrix}.$$

1.7.2.

$$A(\mu) = \begin{bmatrix} 1+2\mu & 3\mu & 0 \\ \mu & 1+4\mu & 0 \\ 0 & 0 & 1+2\mu \end{bmatrix}.$$

1.7.3.

$$A(\mu) = \begin{bmatrix} 1+\mu & 0 & \mu \\ 0 & 1+\mu & \mu^2 \\ -\mu & 0 & 1+3\mu \end{bmatrix}.$$

1.7.4.

$$A(\mu) = \begin{bmatrix} 1+\mu & 0 & \mu \\ 0 & 1+\mu & \mu^2 \\ 0 & 0 & 1+3\mu \end{bmatrix}.$$

1.7.5.
$$A(\mu)x = \left(\lambda\mu A_1 + \lambda^2 I\right)x,$$

where

$$A(\mu) = \begin{bmatrix} 1 & 0 & \mu^2 \\ 0 & 4+\mu^2 & 0 \\ 0 & 0 & 9 \end{bmatrix}, \quad A_1 = \begin{bmatrix} 1 & 0 & 0 \\ 0 & 1 & -1 \\ 0 & 1 & 1 \end{bmatrix}.$$

1.7.6.
$$A(\mu)x = \left(\lambda\mu A_1 + \lambda^2 I\right)x,$$

where

$$A(\mu) = \begin{bmatrix} 1 & 0 & -\mu^2 \\ 0 & 4+\mu^2 & 0 \\ \mu^2 & 0 & 9 \end{bmatrix}, \quad A_1 = \begin{bmatrix} 0 & -1 & 0 \\ 1 & 0 & -1 \\ 0 & 1 & 0 \end{bmatrix}.$$

1.8 Answers and Solutions

1.1.1. For small ε we have

$$\ln(1+\varepsilon) \sim \varepsilon, \quad \frac{1-\cos\varepsilon}{1+\cos\varepsilon} \sim \varepsilon^2, \quad \sqrt{\varepsilon(1-\varepsilon)} \sim \sqrt{\varepsilon}.$$

1.1.2a. $\dfrac{\sqrt{x^2-1}}{|x|} \leq 1$; therefore, $C_{\min} = 1$.

1.1.2b. $|x^n| \leq C|e^x|$, $C \geq |x^n e^{-x}|$. The maximal value of $x^n e^{-x}$ is attained at $x = n$ and is equal to $C_{\min} = n^n e^{-n}$.

1.1.3. Since $\Re z \geq |z|/\sin\varepsilon$ as $z \in S$, then $|e^{-cz}| \leq e^{-c_1|z|}$, $c_1 = c/\sin\varepsilon$.

1.1.4. Show that $\displaystyle\lim_{x\to\infty}\left(\int_a^x f(t)dt\right)\Big/ xg(x) = 0$. Use l'Hôpital's rule:

$$\lim_{x\to\infty}\frac{\int_a^x f(t)dt}{xg(x)} = \lim_{x\to\infty}\frac{f(x)}{g(x)+x\cdot g'(x)}, \quad \left|\frac{f(x)}{g(x)+x\cdot g'(x)}\right| \leq \left|\frac{f(x)}{g(x)}\right|,$$

since $g'(x) \geq 0$, and $|f(x)|/|g(x)| \to 0$ as $x \to \infty$.

Fig. 1.9 Graphs of the
function in Exercise 1.1.5b
and the function $\frac{1}{2\sqrt{x}}$

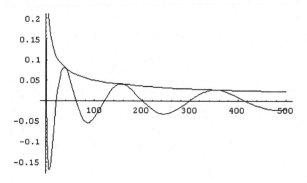

1.1.5a. As $x \to 0$

$$\frac{1 - \sqrt{\cos x}}{1 - \cos(\sqrt{x})} \simeq \frac{1 - \sqrt{1 - x^2/2}}{1 - (1 - x/2)} \simeq \frac{1 - (1 - x^2/4)}{x/2} = \frac{x}{2} \sim x.$$

1.1.5b.

$$\sin\sqrt{x+1} - \sin\sqrt{x} = 2\sin\frac{\sqrt{x+1} - \sqrt{x}}{2}\cos\frac{\sqrt{x+1} + \sqrt{x}}{2}$$

$$= 2\sin\left[\frac{\sqrt{x}}{2}\left(\sqrt{1 + 1/x} - 1\right)\right]\cos\left[\frac{\sqrt{x}}{2}\left(\sqrt{1 + 1/x} + 1\right)\right].$$

In the last expression, the second factor is bounded and the first one has order

$$2\sin\left(\frac{\sqrt{x}}{2}\frac{1}{2x}\right) \simeq \frac{1}{2\sqrt{x}} = O\left(\frac{1}{\sqrt{x}}\right) \quad \text{as} \quad x \to \infty.$$

Figure 1.9 shows graphs of the given function and the function $\frac{1}{2\sqrt{x}}$.

1.1.5c.

$$\sin\ln(x+1) - \sin\ln x = 2\sin\left(\frac{1}{2}\ln\frac{1+x}{x}\right)\cos\left(\frac{1}{2}\ln x(1+x)\right),$$

$$\left|\cos\left(\frac{1}{2}\ln x(1+x)\right)\right| < 1,$$

and

$$\sin\left(\frac{1}{2}\ln\left(1 + \frac{1}{x}\right)\right) \simeq \sin\frac{1}{2x} \quad \text{as} \quad x \to \infty.$$

Hence $\sin(\ln(x+1)) - \sin(\ln x) = O(1/x)$. Figure 1.10 shows the graphs of the given function and the function $1/x$.

Fig. 1.10 Graphs of the
function in exercise 1.1.5c
and the function $1/x$

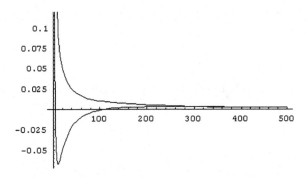

1.1.5d.

$$\ln \frac{nx + \sqrt{1 - n^2x^2}}{x + \sqrt{1 - x^2}} \simeq \ln \frac{nx + 1 - n^2x^2/2}{x + 1 - x^2/2}$$

$$\simeq \ln \left[1 + (n-1)x\right] \simeq (n-1)x \sim x, \quad \text{as} \quad x \to 0.$$

1.2.1a. $\sqrt{1 - \dfrac{\varepsilon}{2} + 2\varepsilon^2} = 1 - \dfrac{\varepsilon}{4} + \dfrac{31\varepsilon^2}{32} + O\left(\varepsilon^3\right).$

1.2.1b. $\sin\left(1 + \varepsilon - \varepsilon^2\right) = \sin(1) + \cos(1)\varepsilon - \left[\cos(1) + \dfrac{\sin(1)}{2}\right]\varepsilon^2 + O\left(\varepsilon^3\right).$

1.2.1c. $(1 - a\varepsilon + a^4\varepsilon^2)^{-1} = 1 + a\varepsilon + a^2\varepsilon^2(1 - a^2) + O(\varepsilon^3).$

1.2.1d. $\sin^{-1}\left(\dfrac{\varepsilon}{\sqrt{1 + \varepsilon}}\right) = \dfrac{1}{\varepsilon} + \dfrac{1}{2} + \dfrac{\varepsilon}{24} + O(\varepsilon^2).$

1.2.1e. $\ln \dfrac{1 + 2\varepsilon - \varepsilon^2}{\sqrt[3]{1 + \varepsilon}} = \dfrac{5\varepsilon}{3} - \dfrac{17\varepsilon^2}{6} + \dfrac{41\varepsilon^3}{9} + O(\varepsilon^4).$

1.2.1f. $\ln\left[1 + \ln(1 + \varepsilon)\right] = \varepsilon - \varepsilon^2 + \dfrac{7\varepsilon^3}{6} + O(\varepsilon^4).$

1.2.2a. The expansion does not hold as $x \to \infty$.

1.2.2b. Uniformly.

1.2.2c. The expansion does not hold for $x = O(\varepsilon)$.

1.2.3. $a_n = \lim\limits_{x \to \infty} x^n e^{-x} \sin e^x = 0$ for all n.

1.2.4. Let $u = \sum\limits_{n=0}^{\infty} a_n x^n$. Substituting this series into the initial equation

$$\sum_{n=3}^{\infty} \left[a_{n-1}(n-1)^2 + a_n(n-1)\right] x^n + (a_1 + a_2)x^2 - a_0 = 0$$

we get

$$a_0 = 0, \quad a_2 = -a_1, \quad a_n = -(n-1)a_{n-1} \quad \text{or} \quad a_n = (-1)^{n-1}(n-1)!a_1,$$

i.e. $u = a_1 \sum_{n=1}^{\infty} (n-1)!(-1)^{n-1}x^n$. The series diverges by d'Alembert criterion since

$$\lim_{n\to\infty} \frac{a_n x^n}{a_{n-1}x^{n-1}} = -\lim_{n\to\infty}(n-1)x.$$

One should note that the given equation of the second order has two independent solutions. The second solution, which is singular at the point $x = 0$, may be represented in the form

$$u = \sum_{n=0}^{\infty} \frac{b_0}{n!x^n}.$$

1.3.1. $x_1 = \frac{1}{3}\mu^2 + \frac{1}{81}\mu^5 + O\left(\mu^8\right), \quad x_{2,3} = \pm\sqrt{3}\mu^{1/2} - \frac{1}{6}\mu^2 + O\left(\mu^{7/2}\right).$

1.3.2. $x_1 \simeq \mu + \mu^2 + O\left(\mu^3\right), x_2 \simeq 1 - \mu + O\left(\mu^2\right), x_{3,4} \simeq \pm\mu^{-2} - 1/2 + O\left(\mu^2\right).$

1.3.3. $x_1 \simeq \mu^2 - 2\mu^5 + O\left(\mu^8\right), \quad x_{2,3} \simeq \pm\mu^{1/2} - \mu^2 + O\left(\mu^{7/2}\right).$

1.3.4. $x_1 = \mu^3 + \mu^{11} + o(\mu^{11}), \quad x_{2,3} = \pm\left(\frac{1}{\mu} + \frac{1}{2}\right) + o(1),$

$x_{4,5} = \pm\left(\frac{1}{\mu^{3/2}} - \frac{1}{2\mu^{1/2}}\right) + o(1/\mu^{1/2}).$

1.3.5. We find 3 boundary points corresponding to the lower (in μ) facets of the Newton polyhedron which, in this case, is a tetrahedron with vertices $M_1 = \{0, 0, 1\}$, $M_2 = \{4, 4, 0\}$, $M_3 = \{4, 2, 2\}$, and $M_4 = \{2, 1, 1\}$ and, four intervals, or four ranges, for λ with a specific structure of the Newton polygon: (I) $\lambda \ll \mu^2$, (II) $\mu^2 \ll \lambda \ll \mu$, (III) $\mu \ll \lambda \ll 1$, and (IV) $\lambda \gg 1$ (see Fig. 1.11). For example, in the closed domains I and IV the Newton polygon consists of one segment and in the remaining domains it consists of two segments.

Fig. 1.11 Newton polyhedron for equation in Exercise 1.3.5

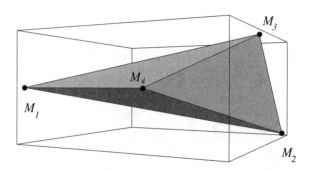

1.4.1. $\dfrac{\partial F}{\partial x} = g'(x)\sinh z + g(x)\lambda^t f'(x)\cosh z \sim \lambda^t F(x,\lambda)$. The variation index is equal to t.

1.4.2a. The general variation index is equal to n. On the line $ay = x$, the partial variation index is equal to zero.

1.4.2b. The general variation index is equal to n. On the line $y = x$, the partial variation index is equal to $-\infty$.

1.4.3. We seek a solution of the differential equation in the form $y = e^z$. The characteristic equation is $z^3 + 4z^2\lambda^2 - z\lambda - \lambda^2 = 0$. Upon substituting $\lambda = 1/\mu$, we get $\mu^2 z^3 + 4z^2 - \mu z - 1 = 0$. Using the Newton polygon, we can get the roots of the characteristic equation:

$$z_{1,2} \simeq \pm\frac{1}{2} + \frac{\mu}{8} + O\left(\mu^2\right), \quad \text{or } z_{1,2} \simeq \pm\frac{1}{2} + \frac{1}{8\lambda} + O\left(\frac{1}{\lambda^2}\right),$$

$$z_3 \simeq -\frac{4}{\mu^2} - \frac{\mu}{4} + O\left(\mu^2\right), \quad \text{or } z_3 \simeq -4\lambda^2 - \frac{1}{4\lambda} + O\left(\frac{1}{\lambda^2}\right).$$

Hence, the given differential equation has two roots with variation index equal to 0: $y(x,\lambda) \simeq e^{\pm\frac{x}{2}}$ and $\frac{\partial y}{\partial x} \simeq \pm\frac{1}{2} y(x,\lambda)$, and one solution with variation index equal to 2: $y(x,\lambda) \simeq e^{-4\lambda^2 x}$ and $\frac{\partial y}{\partial x} \simeq -4\lambda^2 y(x,\lambda)$.

1.4.4. After the substitution $\lambda = 1/\mu$, the characteristic equation $z^5 - z^4 - \lambda^2 z^3 + \lambda^2 z^2 + \lambda z - \lambda = 0$ becomes $\mu^2 z^5 - \mu^2 z^4 - z^3 + z^2 + \mu z - \mu = 0$ with roots $z_{1,2} = \pm\mu^{1/2} + O(\mu)$, or $z_{1,2} = \pm\lambda^{-1/2} + O(\lambda^{-1})$, $z_3 = 1$, $z_{4,5} = \pm 1/\mu + O(1)$, or $z_{4,5} = \pm\lambda + O(1)$.

Hence, the initial differential equation has two solutions with variation index equal to $-1/2$:

$$y(x,\lambda) \simeq e^{\pm x/\lambda^{1/2}}, \quad \frac{\partial y}{\partial x} \simeq \pm\frac{1}{\lambda^{1/2}} y(x,\lambda),$$

One solution with variation index equal to 0

$$y(x,\lambda) \simeq e^x, \quad \frac{\partial y}{\partial x} \simeq y(x,\lambda),$$

and two solutions with variation index equal to 1:

$$y(x,\lambda) \simeq e^{\pm\lambda x}, \quad \frac{\partial y}{\partial x} \simeq \pm\lambda y(x,\lambda).$$

1.5.1. Equation $xe^{1/x} = e^u$ is equivalent to the equation

$$\frac{1}{x} + \ln x = u. \tag{1.8.1}$$

For large u, this equation has two roots. One root corresponds to small values of x and for this case the main term in the left side of the equation is equal to $1/x$. The other root corresponds to large values of x and then the main term in the left side of equation is $\ln x$. So, in the first case, to first approximation $x = 1/u + o(1/u)$. We seek a correction to this root. Substituting $x = 1/u + \delta$ into Eq. (1.8.1) and, expanding the left side of the equation, we obtain

$$\frac{1}{1/u + \delta} + \ln\left(\frac{1 + \delta u}{u}\right) \approx u - u^2\delta + \delta u - \ln u = u,$$

and $\delta = -\ln u/u^2$. In a similar way, one can get the next correction:

$$x = \frac{1}{u} - \frac{\ln u}{u^2} + \frac{\ln^2 u}{u^3} + O\left(\frac{\ln u}{u^3}\right).$$

For the second root, in the first approximation, one has $x = e^u$. In the next approximation,

$$x = e^u - \sum_{i=1}^{\infty} \frac{i^i}{(i+1)!} e^{-iu}.$$

1.5.2. The equation $\cos x \cosh x = -1$ may be represented in the form $\cos x = -1/\cosh x$. For large x, $\cos x = -\frac{2}{e^x}$. The right side of this equation converges to zero as $x \to \infty$; this allows us to write the roots of the equation as $x = \pi/2 + \pi n + o(1) = \lambda_n + o(1)$. Taking $\cos(\lambda_n + \delta) = (-1)^{n+1}\sin\delta$ into account and expanding $\sin\delta$ in a series, one gets $\delta = 2(-1)^n/e^{\lambda_n}$. Continuing this process, one can find the next terms of the series $x = \lambda_n + \sum_{k=1}^{\infty} \frac{c_k}{e^{k\lambda_n}}$.

1.5.3. Equation $x \sin x = 1$ may be written in the form $\sin x = 1/x$. Then for large x the roots of the equation are $x = \pi n + O(1/\pi n)$. Taking $\sin(\pi n + \delta) = (-1)^n \sin\delta$ into account and expanding $\sin\delta$ in a series, one can obtain the next terms of the series

$$x = \sum_{k=-1}^{\infty} a_k(\pi n)^{-2k-1} : \quad x = \pi n + \frac{(-1)^n}{\pi n} - \frac{1}{\pi^3 n^3}\left[1 - \frac{(-1)^n}{6}\right] + O\left(n^{-5}\right).$$

1.5.4. Equation $x \tan x = 1$ may be written in the form $\tan x = 1/x$. It follows that, for large x, the roots of the equation may be written as $x = \pi n + O(1/\pi n)$. Taking into account that $\tan(\pi n + \delta) = \tan\delta$ and expanding $\tan\delta$ in a series one can find the next terms in the expansion:

$$x = \sum_{k=-1}^{\infty} a_k(\pi n)^{-2k-1}, \quad x = \pi n + \frac{1}{\pi n} - \frac{4}{3\pi^3 n^3} + O\left(n^{-5}\right).$$

1.5.5. Similar to Example 1 of Sect. 1.5, for large x the solution of the equation has the form $x = (\frac{1}{2} + n)\pi + o(1)$, where n is a positive integer. We denote $\lambda = (\frac{1}{2} + n)\pi$ and seek the correction δ to this root. We obtain $\tan(\lambda + \delta) = -1/\tan\delta = (\lambda+\delta)^k$; whence $\delta = -1/\lambda^k + o(\lambda^{-k})$. Evaluating the next correction, we get $\delta_1 \approx 1/(3\lambda^{3k}) - k/\lambda^{2k+1}$. It is clear that for $k = 1$ both terms have equal orders and $\delta_1 = -2/3\lambda^{-3}$. For $k > 1$, the correction is $\delta_1 = -k/\lambda^{2k+1}$ or $x = \lambda - \lambda^{-k} - k/\lambda^{2k+1}$.

1.5.6. We show that the first approximation to the root of the equation $x = u/\ln x$ has the form $x = u/\ln u$. Substituting this expression into the initial equation we obtain

$$\frac{u}{\ln u} \ln\left(\frac{u}{\ln u}\right) = \frac{u}{\ln u}\left(\ln u - \ln\ln u\right) = u - u\frac{\ln\ln u}{\ln u} = u + o(u).$$

Further, we seek the correction δ. Substitute $x = u/\ln u + \delta$ into the given equation and keep only the main terms. Then $\delta = \dfrac{u\ln\ln u}{\ln^2 u}$ or $x \approx \dfrac{u}{\ln u} + \dfrac{u\ln\ln u}{\ln^2 u}$.

1.5.7. We write the initial equation in the form $x = u - \tanh x$. Then taking $\tanh x = 1 + o(1)$ for large x into account, we obtain $x = u - 1 + o(1)$. For the next approximation, expand $\tanh x$ in a series for large x:

$$\tanh x = \sum_{k=0}^{\infty} a_k e^{-2kx} = 1 - 2e^{-2x} + 2e^{-4x} - \cdots .$$

This gives us $x = u-1+O(e^{-2u})$ and at the next step $x = u-1+2e^{-2u+2}+O(e^{-4u})$.

1.5.8. For large u, the root of the equation $x\tan x = u$ in the interval $(0, \pi/2)$ may be written as $x = \frac{\pi}{2} - \delta$, where $\delta = o(1)$. Since $\tan\left(\frac{\pi}{2} - \delta\right) = 1/\tan\delta$ then $\left(\frac{\pi}{2} - \delta\right) u^{-1} = \tan\delta \approx \delta$ and $\delta = \pi/2u$. For the next approximations $x = \frac{\pi}{2}\left(1 - \frac{1}{u} + \frac{1}{u^2}\right) + O(u^{-3})$.

1.5.9. As in Example 3 in Sect. 1.5, the initial equation may be represented in the form of the system

$$(e^y + e^{-y})\cos x + 2x = 0,$$
$$(e^{-y} - e^y)\sin x + 2y = 0, \qquad z = x + iy.$$

Arguing as in Sect. 1.5.3, one gets

$$z \simeq (2n + 1)\pi - \frac{\ln(4\pi n)}{2\pi n} + i\ln(4\pi n) + O(n^{-1}) \quad \text{as } n \to \infty.$$

1.6.1. We write the initial system of equations in the form

$$(A_0 + \mu A_1)X = a, \quad \text{where} \quad X = [x, y, z]^T, \quad a = [a_1, a_2, a_3]^T,$$

$$A_0 = \begin{bmatrix} 1 & 0 & 1 \\ 0 & 1 & 0 \\ 1 & 0 & 1 \end{bmatrix}, \quad A_1 = \begin{bmatrix} 1 & 1 & 0 \\ 1 & 0 & 1 \\ 0 & 1 & 1 \end{bmatrix}.$$

We note that $\det(A_0 + \mu A_1) = \mu(2 + \mu - 2\mu^2) > 0$ for $0 < \mu \ll 1$, but $\det A_0 = 0$ and $r = 1$. The matrix A_0 is symmetric. Homogeneous equation (1.6.6) has a solution $X_{10} = [1, 0, -1]^T$. Expanding the solution of the initial system $X(\mu)$ in μ we seek a solution in the form (1.6.8). From the solvability condition with respect to x_0 for the second equation in (1.6.9) we obtain $C_0 = (a, X_{10})/(A_1 X_{10}, X_{10}) = (a_1 - a_3)/2$. One of the partial solutions of this equation is $x_0 = [a_1/2, a_2, a_3/2]^T$.

Continuing this process, one finds

$$C_1 = -C_0/2, \quad x_1 = -[(a_1 + a_3)/4, (a_1 + a_3)/2, a_2]^T, \quad C_2 = (a_1 + a_3 - 4a_2)/8,$$

or

$$x = \frac{1}{2\mu}(a_1 - a_3) + \frac{1}{4}(a_1 + a_3) - \frac{\mu}{8}(a_1 + a_3 + 4a_2) + O\left(\mu^2\right),$$

$$y = a_2 - \frac{\mu}{2}(a_2 + a_3) + O\left(\mu^2\right),$$

$$z = \frac{1}{2\mu}(a_3 - a_1) + \frac{1}{4}(a_1 + a_3) - \frac{\mu}{8}(a_1 + a_3 + 4a_2) + O\left(\mu^2\right).$$

1.6.2. We write the initial system of equations in the form

$$(A_0 + \mu A_1)X = b_0 + \mu b_1,$$

where

$$X = [x, y, z]^T, \quad b_0 = [16, 0, 6]^T, \quad b_1 = [26, 8, 2]^T.$$

$$A_0 = \begin{bmatrix} 6 & 0 & -2 \\ 0 & -3 & 2 \\ 1 & 4 & -3 \end{bmatrix}, \quad A_0^* = \begin{bmatrix} 6 & 0 & 1 \\ 0 & -3 & 4 \\ -2 & 2 & -3 \end{bmatrix}, \quad A_1 = \begin{bmatrix} 1 & 1 & 1 \\ 0 & 2 & 1 \\ 1 & 0 & -2 \end{bmatrix},$$

and $\det A_0 = 0$ and $r = 1$. The homogeneous equation (1.6.6) has the solution $X_{10} = [1, 2, 3]^T$. The adjoint homogeneous equation $A_0^* y = 0$ has the solution $y = [1, -8, -6]^T$. From the solvability condition with respect to x_0 for the second equation in (1.6.9) we get $C_0 = 1$, and one of the partial solutions of this equation is $x_0 = [1, 1, -2]^T$.

Subsequently, one can find

$$x = \frac{1}{\mu} + 2 + \frac{11}{4}\mu + O\left(\mu^2\right), \quad y = \frac{2}{\mu} + 3 - \frac{3}{2}\mu + O\left(\mu^2\right), \quad z = \frac{3}{\mu} + 1 - \frac{7}{4}\mu + O\left(\mu^2\right).$$

1.6.3. The initial system of equations may be written as

$$(A_0 + \mu A_1)X = b, \quad \text{where} \quad X = [x, y, z]^T, \quad b = [b_1, b_2, b_3]^T.$$

$$A_0 = \begin{bmatrix} 1 & 1 & 1 \\ 1 & 1 & 1 \\ 1 & 1 & 1 \end{bmatrix}, \quad A_1 = \begin{bmatrix} 1 & 0 & 0 \\ 0 & 1 & 0 \\ 0 & 0 & 1 \end{bmatrix}.$$

The matrix A_0 is symmetric and $\det A_0 = 0$. The solution of the homogeneous equation (1.6.6) may be represented in the form $X_0 = a\,[1, 0 - 1]^T + b\,[0, 1, -1]^T$, i.e. $r = 2$. It may be shown that for any value of the constants a and b there exist no solutions of system (1.6.10), i.e. there exist no generalized adjoint vectors. Hence, one may seek a solution of the given system in the form (1.6.11) for $n_1 = n_2 = 1$, i.e.

$$x = \frac{C_{10}x_{10} + C_{20}x_{20}}{\mu} + (x_0 + C_{11}x_{10} + C_{21}x_{20}) + \mu\,(x_1 + C_{12}x_{10} + C_{22}x_{20}) + \cdots.$$

As vectors x_{10} and x_{20}, one may chose any two linear independent vectors, for example, $x_{10} = [1, 0, -1]^T$ and $x_{20} = [1, -2, 1]^T$. Substituting asymptotic expansion (1.6.11) into the initial equation and equating the terms of equal powers of μ we get

$$\begin{aligned}
\mu^{-1}: \quad & C_{10}A_0x_{10} + C_{20}A_0x_{20} = 0, \\
\mu^0: \quad & A_0x_0 + C_{11}A_0x_{10} + C_{21}A_0x_{20} + C_{10}A_1x_{10} + C_{20}A_1x_{20} = b, \\
\mu^1: \quad & A_0x_1 + A_1x_0 + C_{11}A_1x_{10} + C_{21}A_1x_{20} = 0, \\
& \cdots
\end{aligned}$$

The first of these equalities holds for any C_{10} and C_{20}. For solvability of the second equation with respect to x_0 it is necessary and sufficient that

$$b - C_{10}x_{10} - C_{20}x_{20}, x_{i0} = 0, \quad i = 1, 2.$$

From that we find the constants

$$C_{10} = (b_1 - b_3)\,/2, \qquad C_{20} = (b_1 - 2b_2 + b_3)\,/6.$$

With these constants one can obtain the partial solution x_0 and from the solvability condition for the next equation $A_0x_1 = -A_1x_0 - C_{11}A_1x_{10} - C_{21}A_1x_{20}$ we find the next constants C_{11} and C_{21}. Continuing this process we get

$$X = \frac{1}{3\mu} \begin{bmatrix} 2b_1 - b_2 - b_3 \\ 2b_2 - b_1 - b_3 \\ 2b_3 - b_1 - b_2 \end{bmatrix} + \frac{b_1 + b_2 + b_3}{9} \begin{bmatrix} 1 \\ 1 \\ 1 \end{bmatrix} - \mu \frac{b_1 + b_2 + b_3}{27} \begin{bmatrix} 1 \\ 1 \\ 1 \end{bmatrix} + O\left(\mu^2\right).$$

1.6.4. The initial system of equations is written as

$$(A_0 + \mu A_1 + \mu^2 A_2) X = d, \quad \text{where} \quad X = [x, y, z]^T, \quad d = [d_1, d_2, d_3]^T,$$

and

$$A_0 = \begin{bmatrix} -4 & 1 & 2 \\ -8 & 2 & 4 \\ -4 & 1 & 2 \end{bmatrix}, \quad A_1 = \begin{bmatrix} -1 & 0 & 0 \\ 0 & -1 & 0 \\ 0 & 0 & -1 \end{bmatrix}, \quad A_2 = \begin{bmatrix} 0 & 0 & 0 \\ 0 & 0 & 0 \\ 0 & 0 & \alpha \end{bmatrix}.$$

The matrix A_0 is symmetric, $\det A_0 = 0$ and $r = 2$. The solution of the homogeneous equation (1.6.6) may be represented as $X_0 = a[1, 4, 0]^T + b[0, -2, 1]^T$, i.e. $r = 2$. The first equation in (1.6.10) has a solution if $u_0 = X_0$ for $a = b$, i.e. if $u_0 = a[1, 2, 1]^T$. Then the general solution depending on the two arbitrary constants a_1 and b_1 has the form $u_1 = [a_1, a + 4a_1 - 2b_1, b_1]^T$. The second equation in (1.6.10), $A_0 u_2 = -A_1 u_1 - A_2 u_0$, has a solution only if $\alpha = 1/2$. Then, for $b_1 = a_1 + a/2$, this equation is transformed into $A_0 u_2 = a_1[1, 2, 1]^T$. Therefore, one should seek the solution of the given system in the form (1.6.11), and if $\alpha \neq 1/2$, then $n_1 = 2$ and $n_2 = 1$, i.e.

$$x = \frac{C_{10}x_{10}}{\mu^2} + \frac{C_{10}x_{11} + C_{11}x_{10} + C_{20}x_{20}}{\mu} + (x_0 + C_{11}x_{11} + C_{12}x_{10} + C_{21}x_{20}) + \cdots$$

The vector x_{10} must have the form $x_{10} = a[1, 2, 1]^T$ for any a and the vector $x_{11} = [a_1, a + 4a_1 - 2b_1, b_1]^T$ for any a_1 and b_1. For example, we may assume

$$x_{10} = [4, 8, 4]^T, \quad x_{11} = [-1, 0, 0]^T, \quad x_{20} = [1, -2, 3]^T.$$

If $\alpha = 1/2$, then $n_1 = 3$ and $n_2 = 1$, i.e.

$$x = \frac{C_{10}x_{10}}{\mu^3} + \frac{C_{10}x_{11} + C_{11}x_{10}}{\mu^2} + \frac{C_{10}x_{12} + C_{11}x_{11} + C_{12}x_{10} + C_{20}x_{20}}{\mu}$$
$$+ (x_0 + C_{11}x_{12} + C_{12}x_{11} + C_{13}x_{10} + C_{21}x_{20}) + \cdots$$

and

$$x_{10} = [4, 8, 4]^T, \quad x_{11} = [-1, 0, 0]^T, \quad x_{12} = [0, -1, 0]^T, \quad x_{20} = [1, -2, 3]^T.$$

Substituting this asymptotic expansion into the initial equation and equating the terms with equal powers of μ one get the coefficients C_{10}, C_{20}, etc. successively.

For $\alpha = 1/2$,

$$X = \frac{4d_3 + 2d_2 - 8d_1}{\mu^3} \begin{bmatrix} 1 \\ 2 \\ 1 \end{bmatrix} + \frac{1}{\mu^2} \begin{bmatrix} 4d_1 - d_2 \\ 8d_1 - 2d_2 \\ 2d_3 \end{bmatrix} - \frac{1}{\mu} \begin{bmatrix} d_1 \\ d_2 \\ 0 \end{bmatrix}.$$

For $\alpha \neq 1/2$,

$$X = \frac{2d_3 + d_2 - 4d_1}{(2\alpha - 1)\mu^2} \begin{bmatrix} 1 \\ 2 \\ 1 \end{bmatrix} + O\left(\frac{1}{\mu}\right).$$

1.7.1. For this problem

$$A_0 = \begin{bmatrix} 0 & 1 & 1 \\ 1 & 0 & -1 \\ 1 & -1 & 0 \end{bmatrix}, \quad A_1 = \begin{bmatrix} 1 & 0 & 1 \\ 0 & 1 & 0 \\ 1 & 0 & 1 \end{bmatrix}.$$

The first equation in (1.7.3) (the unperturbed problem) has a double eigenvalue $\lambda_{0(1,2)} = 1$ and a simple eigenvalue $\lambda_{01} = -2$. Orthonormal eigenvectors corresponding to the first eigenvalue are

$$e_1 = \left[1/\sqrt{2}, 1/\sqrt{2}, 0\right]^T, \quad e_2 = \left[1/\sqrt{6}, -1/\sqrt{6}, \sqrt{2/3}\right]^T.$$

By relations (1.7.6) one can obtain the corrections to the eigenvalue $\lambda_{0(1,2)} = 1$ and find the coefficients a_j to determine the corresponding eigenvectors:

$$c_{11} = (A_1 e_1, e_1) = 1, \qquad c_{12} = (A_1 e_1, e_2) = 1/\sqrt{3},$$
$$c_{21} = (A_1 e_2, e_1) = 1/\sqrt{3}, \qquad c_{22} = (A_1 e_2, e_2) = 5/3.$$

The eigenvalues for the matrix c_{ij} are found from the equation $\tilde{\lambda}^2 - \frac{8}{3}\tilde{\lambda} + \frac{4}{3} = 0$, i.e. $\tilde{\lambda}_1 = 2$ and $\tilde{\lambda}_2 = 2/3$. The coefficients a_k, corresponding to $\tilde{\lambda}_1 = 2$, can be obtained from relations (1.7.5): $a_1 = 1$ and $a_2 = \sqrt{3}$. Similarly, for $\tilde{\lambda}_2 = 2/3$ one can get $a_1 = -\sqrt{3}$ and $a_2 = 1$. Therefore for the matrix A we have

$$\lambda_1 = 1 + 2\mu + O\left(\mu^2\right), \quad \lambda_2 = 1 + \frac{2}{3}\mu + O\left(\mu^2\right).$$

The eigenvector is $x_{01} = \left[\sqrt{2}, 0, \sqrt{2}\right]^T$ and, for convenience, we assume that $x_{01} = [1, 0, 1]^T$ and $x_{02} = \frac{1}{\sqrt{6}}[-2, -4, 2]^T$ or consider $x_{02} = [-1, -2, 1]^T$. By (1.7.4) the correction to the third eigenvalue may be found as $\lambda_3 = -2 + \frac{\mu}{3} + O\left(\mu^2\right)$ and taking (1.7.3) into account we finally find $x_{13} = [1/3, 0, -1/3]^T$, i.e. $x_3 = [1 + \mu/3, -1, -1 - \mu/3]^T + O\left(\mu^2\right)$.

1.7.2. For this problem

$$A_0 = \begin{bmatrix} 1 & 0 & 0 \\ 0 & 1 & 0 \\ 0 & 0 & 1 \end{bmatrix}, \quad A_1 = \begin{bmatrix} 2 & 3 & 0 \\ 1 & 4 & 0 \\ 0 & 0 & 2 \end{bmatrix}.$$

The unperturbed problem has a triple eigenvalue $\lambda_{0(1,2,3)} = 1$ with corresponding orthonormal eigenvectors $e_1 = [1, 0, 0)]^T$, $e_2 = [0, 1, 0]^T$, and $e_3 = [0, 0, 1]^T$.

Using relations (1.7.6) one can find the corrections to the eigenvalue $\lambda_{0(1,2,3)} = 1$ and the coefficients a_j to determine the corresponding eigenvectors which are the columns of the matrix C,

$$C = \begin{bmatrix} 2 & 3 & 0 \\ 1 & 4 & 0 \\ 0 & 0 & 2 \end{bmatrix}.$$

The eigenvalues of C are $\tilde{\lambda}_1 = 2$, $\tilde{\lambda}_2 = 1$, $\tilde{\lambda}_3 = 5$.

The coefficients a_k corresponding to $\tilde{\lambda}_1 = 2$ are $a_1 = a_2 = 0$, and $a_3 = 1$, for $\tilde{\lambda}_2 = 1$ they are $a_1 = -3$, $a_2 = 1$, and $a_3 = 0$, and for $\tilde{\lambda}_2 = 5$ they are $a_1 = a_2 = 1$, and $a_3 = 0$. Therefore for the matrix A

$$\lambda_1 = 1 + 2\mu + O\left(\mu^2\right), \quad \lambda_2 = 1 + \mu + O\left(\mu^2\right), \quad \lambda_3 = 1 + 5\mu + O\left(\mu^2\right),$$

$$x_{01} = [0, 0, 1]^T, \quad x_{02} = [-3, 1, 0]^T, \quad x_{03} = [1, 1, 0]^T.$$

1.7.3.

$$\lambda_1 = 1 + \mu, \quad \lambda_{2,3} = 1 + 2\mu,$$

$$x_1 = [0, 1, 0]^T, \quad x_{2,3} = [1, \mu, 1].$$

1.7.4.

$$\lambda_{1,2} = 1 + \mu, \quad \lambda_3 = 1 + 3\mu,$$

$$x_1 = [0, 1, 0]^T, \quad x_2 = [1, 0, 0]^T, \quad x_3 = [1, \mu, 2].$$

1.7.5. The unperturbed problem $\left(A_0 - \lambda^2 I\right) x = 0$ has eigenvalues $\lambda_{01} = \pm 1$, $\lambda_{02} = \pm 2$, $\lambda_{03} = \pm 3$ and the eigenvectors are the columns of the identity matrix. The corrections to the eigenvalues λ_{1i} and coefficients a_{ij}^1 can be obtained from (1.7.12):

$$\lambda_{11} = \lambda_{12} = \lambda_{13} = -\frac{1}{2}, \quad a_{23}^1 = \frac{\pm 2}{5}, \quad a_{32}^1 = \frac{\pm 3}{5}, \quad a_{12}^1 = a_{21}^1 = a_{13}^1 = a_{31}^1 = 0.$$

Then, from (1.7.10) one can get the eigenvectors

$$x_{01} = [1, 0, 0]^T + O\left(\mu^2\right),$$
$$x_{02} = [0, 1, \pm 2\mu/5]^T + O\left(\mu^2\right),$$
$$x_{03} = [0, \pm 3\mu/5, 1]^T + O\left(\mu^2\right).$$

1.7.6. The unperturbed problem has the same eigenvalues and eigenvectors as in 1.7.4. However, in this case $\lambda_{11} = \lambda_{12} = \lambda_{13} = 0$ and the eigenvectors are

$$x_{01} = [1, \pm \mu/3, 0]^T + O\left(\mu^2\right),$$
$$x_{02} = [\pm 2\mu/3, 1, \pm 2\mu/5]^T + O\left(\mu^2\right),$$
$$x_{03} = [0, \pm 3\mu/5, 1]^T + O\left(\mu^2\right).$$

Chapter 2
Asymptotic Estimates for Integrals

Mechanical problems can be described by differential equations, the solutions of which often cannot be expressed by elementary functions, but have an integral representation. In this chapter, we discuss asymptotic estimates of integrals of the form

$$F(\mu) = \int_{a(\mu)}^{b(\mu)} f(z, \mu)\, dz, \tag{2.1}$$

where μ is a small parameter and $z \in \mathbb{C}$. Sometimes, it is more convenient to use a large parameter λ and consider the integral

$$F(\lambda) = \int_{a(\lambda)}^{b(\lambda)} f(z, \lambda)\, dz. \tag{2.2}$$

With obvious adjustments, all statements which are valid for integrals hold also for (2.2). In the general case, asymptotic estimates are given as $\mu \to 0$ and $\lambda \to \infty$.

2.1 Series Expansion of Integrands

Suppose the integrand in (2.1) can be expanded in an asymptotic series of the form

$$f(z, \mu) = \sum_{n=0}^{N} a_n(z)\mu^{p_n} + O\left(\mu^{p_{n+1}}\right), \quad p_0 < p_1 < \cdots < p_n < \cdots < +\infty,$$

$$\tag{2.1.1}$$

uniformly in z over the interval $[a(\mu), b(\mu)]$. Then this series can be integrated term-by-term and

© Springer International Publishing Switzerland 2015
S.M. Bauer et al., *Asymptotic Methods in Mechanics of Solids*,
International Series of Numerical Mathematics 167,
DOI 10.1007/978-3-319-18311-4_2

$$F(\mu) = \sum_{n=0}^{N} b_n(\mu)\mu^{P_n} + O\left([b(\mu) - a(\mu)]\mu^{P_{n+1}}\right), \quad b_n(\mu) = \int_{a(\mu)}^{b(\mu)} a_n(z)\,dz.$$

$$(2.1.2)$$

The form of expansion (2.1.2) is simplest if the integrand can be expanded in Taylor series.

Example 1

Find the asymptotic expansion of the function

$$F(\mu) = \int_0^1 \frac{\sin \mu x}{x}\,dx, \quad x \in \mathbb{R}.$$

The function $\sin \mu x$ has the asymptotic expansion

$$\sin \mu x = \sum_{n=0}^{N} \frac{(-1)^n x^{2n+1}}{(2n+1)!}\,\mu^{2n+1} + O\left(\mu^{2N+3}\right),$$

which is uniform over the interval of integration [0, 1]. Therefore,

$$F(\mu) = \sum_{n=0}^{N} \int_0^1 \left[\frac{(-1)^n x^{2n}}{(2n+1)!}\,\mu^{2n+1} + O\left(\mu^{2N+3}\right)\right] dx$$

$$= \sum_{n=0}^{N} \frac{(-1)^n}{(2n+1)!(2n+1)}\,\mu^{2n+1} + O\left(\mu^{2N+3}\right).$$

When the limits of the interval of integration depend on a small parameter, the change of variable

$$t = \frac{z - a(\mu)}{b(\mu) - a(\mu)}$$

$$(2.1.3)$$

transforms integral (2.1) to an integral of the form

$$F(\mu) = \int_0^1 f_1(t, \mu)\,dt.$$

A similar problem is solved for integrals with large parameters.

Example 2

Find an asymptotic expansion for the function

$$F(\mu) = \int_0^{\mu} x^{-3/4} \exp(-x)\,dx.$$

Substitution (2.1.3) transforms the given integral into

$$F(\mu) = \mu^{1/4} \int_0^1 t^{-3/4} \exp(-\mu t)\, dt.$$

The function $\exp(-\mu t)$ has the asymptotic representation

$$\exp(-\mu t) = \sum_{n=0}^{N} \frac{(-1)^n}{n!} \mu^n t^n + O\left(\mu^{N+1}\right),$$

that is uniform for all t in the interval of integration $[0, 1]$. Therefore,

$$F(\mu) = \mu^{1/4} \sum_{n=0}^{N} \int_0^1 \left[\frac{(-1)^n}{n!} t^{n-3/4} \mu^n + O\left(\mu^{N+1}\right) \right] dt$$

$$= \sum_{n=0}^{N} \frac{(-1)^n}{n!(n+1/4)} \mu^{n+1/4} + O\left(x^{N+5/4}\right)$$

$$= 4\mu^{1/4} - \frac{4}{5}\mu^{5/4} + \frac{2}{9}\mu^{9/4} + O\left(\mu^{13/4}\right).$$

In some cases, the integrand does not have a uniform asymptotic representation on the entire interval of integration. In this case one should try to split the initial interval into parts such that, in some of them, the function has a uniform asymptotic expansion and in others the integral could be calculated by means of quadratures or estimated.

Example 3

Find an asymptotic expansion for the function

$$F(\mu) = \int_\mu^\infty \exp\left(-x^2\right) dx.$$

The function $\exp(-x^2)$ has the uniform asymptotic expansion

$$\exp\left(-x^2\right) = \sum_{n=0}^{N} (-1)^n \frac{x^{2n}}{n!} + O\left(x^{2N+2}\right)$$

in any segment $[0, x_0]$. Therefore, we write the given integral as

$$\int_\mu^\infty \exp\left(-x^2\right) dx = \int_0^\infty \exp\left(-x^2\right) dx - \int_0^\mu \exp\left(-x^2\right) dx.$$

The first integral in the right side can be easily calculated after a change of Cartesian coordinates to polar coordinates [49] and it is equal to $\sqrt{\pi}/2$. For the second integral, the expansion of the integrand is uniform in the interval of integration $[0, 1]$. Therefore,

$$F(\mu) = \sqrt{\pi}/2 - \sum_{n=0}^{N} \mu \int_0^1 \left[\frac{(-1)^n t^{2n}}{n!} \mu^{2n} + O\left(\mu^{2N+2}\right) \right] dt$$

$$= \sqrt{\pi}/2 - \sum_{n=0}^{N} \frac{(-1)^n \mu^{2n+1}}{n!(2n+1)} + O\left(\mu^{2N+2}\right).$$

2.1.1 Exercises

2.1.1. Find an asymptotic expansion of the function

$$F(\mu) = \int_0^{\pi/2} \sqrt{1 - \mu \sin^2 x}\, dx.$$

2.1.2. The function F is defined by the integral

$$F(\mu) = \int_0^{\mu} \frac{dx}{\sqrt{1 - m \sin^2 x}}, \quad m \in \mathbb{N}.$$

The Jacobi elliptic functions are defined as

$$sn(x) = \sin(F^{-1}(x)),$$
$$cn(x) = \cos(F^{-1}(x)),$$
$$am(x) = F^{-1}(x).$$

Find the first two terms of the asymptotic expansions for the Jacobian functions.

2.1.3. Find an asymptotic expansion for the function

$$F(\mu) = \int_0^{\pi/2} \frac{dx}{\sqrt{1 - \mu \sin^2 x}}.$$

2.1.4. Find the first two terms of the asymptotic expansions of the function

$$F(\lambda) = \int_0^{\lambda} \sqrt{1 + x^2}\, dx$$

and the order of the correction,

(a) by calculating the given integral by means of quadratures,
(b) by expanding the integrand in a series.

2.2 Integration by Parts

Integration by parts is the next method to find asymptotic estimates for integrals. Let
the integrand in (2.1) be a product of two functions:

$$f(z, \mu) = g(z, \mu)h(z, \mu) \tag{2.2.1}$$

under the assumption that $h(z, \mu)$ is analytic in z in some domain containing the path
of integration and $g(z, \mu)$ is integrable in z.
  Denote

$$h_0(z, \mu) \equiv h(z, \mu), \quad h_n(z, \mu) \equiv \frac{\partial^n}{\partial z^n} h(z, \mu),$$

$$g_0(z, \mu) \equiv g(z, \mu), \quad g_{-n+1}(z, \mu) \equiv \frac{\partial}{\partial z} g_{-n}(z, \mu). \tag{2.2.2}$$

Integrating (2.1) n times by parts with respect to z over $[a, b]$, we get the following
expression

$$F(\mu) = \sum_{k=0}^{n-1} (-1)^k \big[g_{-k-1}(b, \mu)h_k(b, \mu) - g_{-k-1}(a, \mu)h_k(a, \mu) \big]$$

$$+ (-1)^n \int_{a(\mu)}^{b(\mu)} g_{-n}(z, \mu)h_n(z, \mu) \, dz \tag{2.2.3}$$

If

(1) relation (2.2.3) holds for all $n \in \mathbb{N}$,
(2) the terms under the summation sign in (2.2.3) form an asymptotic sequence,
(3) for all $n \in \mathbb{N}$, the remainder of the series satisfies the condition in the definition
   of asymptotic expansion,

then (2.2.3) is the required asymptotic expansion.
  The main problem in the application of the method is a successful choice of the
functions g and h.

Example 1
  Find an asymptotic expansion of the function

$$F(\lambda) = \int_1^\lambda \frac{e^x}{x} \, dx.$$

We write the integrand as $f(x) = g(x)h(x)$, where $g(x) = e^x$ and $h(x) = 1/x$. Then, by formula (2.2.3)

$$F(\lambda) = \sum_{k=1}^{n} \left(e^\lambda \frac{k!}{\lambda^k} - e\,k! \right) + n! \int_1^\lambda \frac{e^x}{x^{n+1}}\,dx.$$

The second term under the summation sign has order $O(1)$ and for the integral the following estimates are valid:

$$n! \int_1^\lambda \frac{e^x}{x^{n+1}}\,dx = n! \int_1^{\lambda/2} \frac{e^x}{x^{n+1}}\,dx + n! \int_{\lambda/2}^\lambda \frac{e^x}{x^{n+1}}$$

$$< n! \int_1^{\lambda/2} e^x\,dx + n! \int_{\lambda/2}^\lambda e^x \left(\frac{2}{\lambda} \right)^{n+1} dx = O\left(\frac{e^\lambda}{\lambda^{n+1}} \right).$$

Hence,

$$F(\lambda) = \sum_{k=1}^{n} e^\lambda \frac{k!}{\lambda^k} + O\left(\frac{e^\lambda}{\lambda^{n+1}} \right).$$

The numerical (solid line) and the asymptotic (dashed line) values of the integral versus μ are plotted in Fig. 2.1 for $n = 8$.

Example 2

Find an asymptotic expansion of the function

$$F(\lambda) = \int_\lambda^\infty \frac{e^{-x}}{x^2}\,dx.$$

We assume that $f(x) = g(x)h(x)$, where $g(x) = e^{-x}$ and $h(x) = 1/x^2$. Then an integration by parts gives

Fig. 2.1 Numerical values (*solid line*) and asymptotic values (*dashed line*) for Example 1

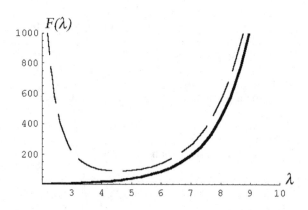

$$F(\lambda) = - \left. e^{-x} \frac{1}{x^2} \right|_\lambda^\infty - 2 \int_\lambda^\infty \frac{e^{-x}}{x^3} \, dx.$$

Integrating by parts again we obtain

$$F(\lambda) = \sum_{k=2}^{n} (-1)^k e^{-\lambda} \frac{(k-1)!}{\lambda^k} + n!(-1)^{n-1} \int_\lambda^\infty \frac{e^{-x}}{x^{n+1}} \, dx.$$

For $\lambda \leq x < \infty$,

$$x^{n+1} \geqslant \lambda^{n+1} \quad \text{and} \quad \frac{1}{x^{n+1}} \leq \frac{1}{\lambda^{n+1}},$$

and, therefore,

$$\int_\lambda^\infty \frac{e^{-x}}{x^{n+1}} \, dx \leq \frac{1}{\lambda^{n+1}} \int_\lambda^\infty e^{-x} \, dx = \frac{e^{-\lambda}}{\lambda^{n+1}}.$$

Hence, we obtain the asymptotic expansion

$$F(\lambda) = e^{-\lambda} \sum_{k=2}^{n} (-1)^k \frac{(k-1)!}{\lambda^k} + O\left(\frac{e^{-\lambda}}{\lambda^{n+1}}\right).$$

This series diverges since the ratio of the mth and the $(m-1)$th terms goes to $-\infty$ as $m \to \infty$:

$$\lim_{m \to \infty} \frac{(-1)^{m-1} m! \lambda^m}{\lambda^{m+1}(-1)^{m-2}(m-1)!} = - \lim_{m \to \infty} \frac{m}{\lambda} = -\infty.$$

The numerical (solid line) and asymptotic (dashed line) values of the integral versus μ are plotted in Fig. 2.2 for $n = 8$.

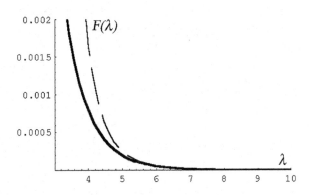

Fig. 2.2 Numerical values (*solid line*) and asymptotic values (*dashed line*) for Example 1

2.2.1 Exercises

Find the asymptotic expansions of the following functions as $\lambda \to \infty$.

2.2.1. $F(\lambda) = \displaystyle\int_0^\infty \frac{e^{-x}}{\lambda + x}\, dx.$

2.2.2. $F(\lambda) = \displaystyle\int_\lambda^\infty \frac{\cos(x - \lambda)x\, dx}{}.$

2.2.3. $F(\lambda) = \displaystyle\int_\lambda^\infty \frac{e^{-x}x\, dx}{}.$

2.3 Laplace Method

Consider the integral

$$F(\lambda) = \int_{a(\lambda)}^{b(\lambda)} f(x, \lambda)\, dx, \quad x \in \mathbb{R}.$$

If the function $f(x, \lambda)$ has a sharp maximum at the point $x_0 \in I$, $I = [a, b]$, which is sharper when λ is larger, then it is good to calculate the integral by means of Laplace method. This method, which is most conveniently applied to integrals of the form

$$F(\lambda) = \int_{a(\lambda)}^{b(\lambda)} \varphi(x)\, e^{\lambda S(x)}\, dx, \quad x \in \mathbb{R}, \tag{2.3.1}$$

is called *a Laplace integral* [12].

We look for the asymptotic expansion of $F(\lambda)$ as $\lambda \to \infty$. A vast literature is devoted to this question, for example the books [14, 20, 22, 23, 66] and handbooks [1, 35]. We mention, in particular, the three volumes of the monograph by Riekstynsh [54], in which integrals of type (1.3.1) and integrals of more general types are studied.

We start with several particular cases to illustrate the application of the method.

Example 1

Let $S(x) = -x$ in (2.3.1). In a neighborhood of the point $x = 0$, we expand the function $\varphi(x)$ in a Taylor series

$$\varphi(x) = \sum_{n=0}^\infty \frac{\varphi^{(n)}(0)}{n!} x^n,$$

and consider the integral

$$F(\lambda) = \int_0^\infty \varphi(x) e^{-\lambda x}\, dx. \tag{2.3.2}$$

Integrating (2.3.2) n times by parts, one gets the asymptotic expansion

$$F(\lambda) \simeq \sum_{n=0}^{N} \frac{\varphi^{(n)}(0)}{\lambda^{n+1}} + O\left(\frac{1}{\lambda^{N+2}}\right), \quad \text{as} \quad \lambda \to \infty.$$

Example 2

Let $S(x) = -x$ in (2.3.1). If in a neighborhood of the point $x = 0$, the function $\varphi(x)$ admits the convergent asymptotic expansion

$$\varphi(x) = \sum_{n=0}^{\infty} a_n x^{n+\alpha}, \quad \alpha > -1,$$

then the integral

$$F(\lambda) = \int_0^b \varphi(x) e^{-\lambda x}\, dx, \quad b > 0, \tag{2.3.3}$$

has the asymptotic expansion

$$F(\lambda) \simeq \sum_{n=0}^{\infty} \frac{\Gamma(n + \alpha + 1)}{\lambda^{n+\alpha+1}}\, a_n, \quad \text{as} \quad \lambda \to \infty, \tag{2.3.4}$$

where $\Gamma(z)$ is the Gamma-function.

Indeed, an integration by parts in (2.3.3) gives

$$F(\lambda) = \sum_{n=0}^{\infty} a_n F_n(\lambda), \quad F_n(\lambda) = \int_0^b x^{n+\alpha} e^{-\lambda x}\, dx.$$

Integrating by parts, one can show that

$$\int_b^{\infty} x^{n+\alpha} e^{-\lambda x}\, dx = o\left(\lambda^N\right),$$

where N is any natural number. Therefore,

$$\begin{aligned}
F_n(\lambda) &= \int_0^{\infty} x^{n+\alpha} e^{-\lambda x}\, dx - \int_b^{\infty} x^{n+\alpha} e^{-\lambda x}\, dx \\
&= \int_0^{\infty} x^{n+\alpha} e^{-\lambda x}\, dx + o\left(\lambda^N\right).
\end{aligned}$$

Taking

$$\int_0^\infty x^{n+\alpha} e^{-\lambda x} dx = \frac{1}{\lambda^{n+\alpha+1}} \int_0^\infty z^{n+\alpha} e^{-z} dz = \frac{\Gamma(n+\alpha+1)}{\lambda^{n+\alpha+1}},$$

into account, we obtain formula (2.3.4).

Note, that depending on a_n, the series (2.3.4) may diverge or converge. However, in the last case, it may not converge to the function $F(\lambda)$, but gives only its asymptotic expansion. For example, applying formula (2.3.4) with $\alpha = 0$ to the integral

$$F(\lambda) = \int_0^{\pi/2} e^{-\lambda x} \sin x \, dx,$$

we get

$$F(\lambda) \simeq \sum_{k=0}^\infty \frac{(-1)^k}{\lambda^{2k+2}} = \frac{1}{1+\lambda^2}. \tag{2.3.5}$$

The exact solution,

$$F(\lambda) = \frac{1}{1+\lambda^2}\left(1 - \lambda e^{-\pi\lambda/2}\right),$$

differs from the function to which series (2.3.5) converges. This is, as in similar further cases, because the series (2.3.4), or (2.3.5), contributes to the integral only in a neighborhood of the point of maximum of the function $e^{\lambda h(x)}$ while exponentially small terms are neglected.

Example 3

Let

$$\varphi(x) = \sum_{n=0}^\infty a_n x^n, \quad S(x) = -x^2, \quad a < 0, \quad b > 0. \tag{2.3.6}$$

Then integral (2.3.1) has the expansion

$$\int_a^b \varphi(x) e^{-\lambda x^2} dx \simeq \sum_{n=0}^\infty \frac{\Gamma(n+1/2)}{\lambda^{n+1/2}} a_{2n}, \quad \text{as} \quad \lambda \to \infty. \tag{2.3.7}$$

Series (2.3.7) is a contribution to integral of the neighborhood of the point $x = 0$, where the function $e^{-\lambda x^2}$ attains its maximum.

Now we go back to the general integral (2.3.1). For that integral one cannot assume $u = \varphi(x)$ and $dv = e^{\lambda S(x)} dx$ to integrate by parts since the last expression cannot be integrated in the general case. We suppose that $S'(x) \neq 0$ in the interval (a, b), and let

$$u = \frac{\varphi(x)}{S'(x)}, \quad dv = e^{\lambda S(x)} S'(x) \, dx.$$

Then

$$F(\lambda) = \frac{e^{\lambda S(x)}\varphi(x)}{\lambda S'(x)}\Bigg|_a^b - \frac{1}{\lambda}\int_a^b e^{\lambda S(x)}\left[\frac{\varphi(x)}{S'(x)}\right]' dx$$

or, as $\lambda \to \infty$,

$$F(\lambda) \simeq \frac{e^{\lambda S(b)}\varphi(b)}{\lambda S'(b)} - \frac{e^{\lambda S(a)}\varphi(a)}{\lambda S'(a)} + O\left(\frac{1}{\lambda^2}\right).$$

If $S(a) \neq S(b)$, then the asymptotic behavior is determined by the end point where $S(x)$ is larger. Suppose, for definiteness, that $S(a) > S(b)$. Continuing the process one gets

$$F(\lambda) = -\sum_{k=0}^{N} \frac{M^k\left(P(a)e^{\lambda S(a)}\right)}{\lambda^{k+1}} + O\left(\frac{e^{\lambda S(a)}}{\lambda^{N+2}}\right), \qquad (2.3.8)$$

where

$$M^k = \left(\frac{-1}{S'(x)}\frac{d}{dx}\right)^k, \quad P(x) = \frac{\varphi(x)}{S'(x)}.$$

Thus far, everything that was said allows us to conclude that the point $x = c \in (a, b)$ corresponding to the largest value of $S(x)$ makes the main contribution in $F(\lambda)$, whether it is an end point or not.

Assume that $c = a$ and in the neighborhood on this point

$$\varphi(x) \simeq \varphi_0(x - a)^\alpha, \quad S'(a) = 0, \quad S''(a) < 0,$$

i.e.

$$S(x) \simeq S(a) + \frac{S''(a)(x - a)^2}{2!}.$$

For the integral to exist, it is necessary that $\alpha > -1$. Then

$$F(\lambda) \simeq \varphi_0 e^{\lambda S(a)} \int_a^{a+\delta} (x - a)^\alpha e^{\lambda S''(a)(x-a)^2/2!} dx.$$

Since $S''(a) < 0$, the error is exponentially small if $a + \delta$ is replaced by ∞. After the change of variables

$$\tau = -\frac{\lambda S''(a)(x - a)^2}{2!},$$

one gets $\tau = 0$ for $x = a$ and

$$F(\lambda) \simeq \frac{\varphi_0}{2} \left[\frac{2}{-\lambda S''(a)} \right]^{(\alpha+1)/2} e^{\lambda S(a)} \int_0^\infty \tau^{(\alpha-1)/2} e^{-\tau} d\tau,$$

or

$$F(\lambda) \simeq \frac{\varphi_0}{2} \left[\frac{2}{-\lambda S''(a)} \right]^{(\alpha+1)/2} e^{\lambda S(a)} \Gamma \left(\frac{\alpha+1}{2} \right).$$

Let now $a < c < b$ and $S''(c) < 0$. As above, one obtains

$$F(\lambda) \simeq \varphi_0 \left[\frac{2}{-\lambda S''(c)} \right]^{(\alpha+1)/2} e^{\lambda S(c)} \Gamma \left(\frac{\alpha+1}{2} \right).$$

In a similar manner we consider cases where a larger number of derivatives are equal to zero at $x = c$ (or $x = a$). Here we consider the case when $\alpha = 0$.
Let

(1) $\varphi(x), S(x) \in C([a, b])$,
(2) $\max_{x \in I} S(x)$ is attained only at the point x_0,
(3) $\varphi(x), S(x) \in C^\infty$ for x close to x_0,

then the following statement is valid [54]:

(A) If the maximum is attained for $x \in (a, b)$ and

$$S^{(j)}(x_0) = 0, \quad 1 \le j \le 2m - 1, \quad S^{(2m)}(x_0) \ne 0, \quad m \ge 1,$$

then the main term of the asymptotic expansion has the form

$$F(\lambda) = \frac{\Gamma(1/2m) \, e^{\lambda S(x_0)}}{m \lambda^{1/2m}} \left[-\frac{(2m)!}{S^{(2m)}(x_0)} \right]^{1/2m} \left[\varphi(x_0) + O \left(\frac{1}{\lambda^{1/2m}} \right) \right]. \quad (2.3.9)$$

(B) If the maximum is attained at one end of the interval ($x_0 = a$) and

$$S^{(j)}(x_0) = 0, \quad 1 \le j \le m - 1, \quad S^{(m)}(x_0) \ne 0, \quad m \ge 2,$$

then the main term of the asymptotic expansion has the form

$$F(\lambda) = \frac{\Gamma(1/m) \, e^{\lambda S(x_0)}}{m \lambda^{1/m}} \left[-\frac{(m)!}{S^{(m)}(x_0)} \right]^{1/m} \left[\varphi(x_0) + O \left(\frac{1}{\lambda^{1/m}} \right) \right]. \quad (2.3.10)$$

Example 4
Find an asymptotic expansion of the function

$$F(\lambda) = \int_a^b e^{-\lambda x^4} dx, \quad a \le 0, \, b > 0.$$

Fig. 2.3 Exact values (*solid line*) and asymptotic values (*dashed line*) of the integral $F(\lambda)$ of Example 4

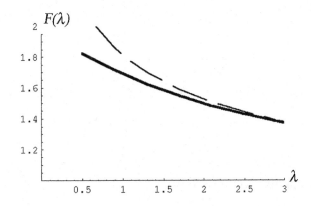

For $a < 0$ and $b > 0$ we should use the asymptotic formula (2.3.9), which defines the main term of the asymptotic expansion

$$\frac{1}{2}\Gamma\left(\frac{1}{4}\right)\lambda^{-1/4}.$$

For $a = -\infty$ and $b = \infty$, the main term of the integral coincides with the exact value of the integral. In Fig. 2.3, we compare the exact (solid line) and asymptotic (dashed line) values of the integral for $a = -1, b = 1$.

For $a = 0, b > 0$ one should use formula (2.3.10), which defines the main term of the asymptotic expansion

$$\frac{1}{4}\Gamma\left(\frac{1}{4}\right)\lambda^{-1/4}.$$

As expected, the value of the integral is twice smaller than in the first case.

2.3.1 Exercises

2.3.1. Find the first two terms of the asymptotic expansion of the function
$$F(\lambda) = \int_0^{\pi^2/4} \exp(\lambda \cos\sqrt{x})\, dx.$$
Find the first term of the asymptotic expansion of the following functions.

2.3.2. $F(\lambda) = \int_0^1 \exp(-1/x - \lambda x)\, dx.$

2.3.3. $F(\lambda) = \int_0^\infty \exp(-\alpha x^{-\alpha} - \lambda x)\, dx, \quad \alpha > 0.$

2.3.4. $F(\lambda) = \int_{-1}^1 (1 - x^2)^\lambda x^2\, dx.$

2.3.5. $F(\lambda) = \int_{-1}^{1} (1 - x^2)^{\lambda} \, dx.$

2.3.6. $F(\lambda) = \int_{0}^{1} (1 - x^2)^{\lambda} x \, dx.$

2.3.7. $F(\lambda) = \int_{0}^{\pi} x^{\lambda} \sin x \, dx, \quad \lambda \in \mathbb{N}.$

2.3.8. $F_{\nu}(\lambda) = \int_{0}^{\infty} \exp(-\lambda \cosh x) \cosh(\nu x) \, dx.$

2.4 Stationary Phase Method

Consider integrals of the form

$$F(\lambda) = \int_{a(\lambda)}^{b(\lambda)} \varphi(x) \exp(i \lambda S(x)) \, dx, \quad x \in \mathbb{R}, \tag{2.4.1}$$

where $\varphi(x)$ is a complex valued function and $S(x)$ a real valued function called *phase function*, and, as before, $\lambda > 0$ is a large parameters.

2.4.1 Integrals Without Stationary Points

Consider a finite interval $I = [a, b]$, in which

$$\varphi(x) \in C^{N+1}(I), \quad S(x) \in C^{N+2}(I), \quad S'(x) \neq 0, \tag{2.4.2}$$

i.e. $\varphi(x)$ and $S(x)$ have $N + 1$ and $N + 2$ continuous derivatives. Then

$$F(\lambda) = \sum_{k=0}^{N} \frac{1}{(i\lambda)^{k+1}} \left[M^k(P(x)) \exp(i\lambda S(x)) \right] \Big|_{a}^{b} + O\left(\frac{1}{\lambda^{N+2}} \right), \tag{2.4.3}$$

where

$$M = \frac{-1}{S'(x)} \frac{d}{dx}, \quad P(x) = \frac{\varphi(x)}{S'(x)}.$$

In this case, the main term of the asymptotic expansion is

$$F(\lambda) = \frac{1}{i\lambda} \left[\frac{\varphi(x)}{S'(x)} \exp(i\lambda S(x)) \right] \Big|_{a}^{b} + O\left(\frac{1}{\lambda^2} \right).$$

Formula (2.4.3) can be proved by integration by parts and the Riemann–Lebesgue lemma [14, 53]

$$\int_a^b \varphi(x)e^{i\lambda x}\,dx = o(1), \quad \text{as} \ \lambda \to \infty$$

for a piecewise continuous function $\varphi(x)$.

Example 1

Find the first term of the asymptotic expansion and an estimate of the error for the function

$$F(\lambda) = \int_0^{\pi/4} \sin(\lambda \sin x)x\,dx.$$

Write the given integral in the form

$$F(\lambda) = \Im \int_0^{\pi/4} \exp(i\lambda \sin x)x\,dx.$$

Applying formula (2.4.3) we get

$$F(\lambda) = \Im\left(\frac{\pi}{4i\lambda}\frac{2}{\sqrt{2}}\exp\left(i\lambda\frac{\sqrt{2}}{2}\right) + O\left(\frac{1}{\lambda^2}\right)\right)$$

$$= -\frac{\pi}{2\sqrt{2}\lambda}\cos\left(\lambda\frac{\sqrt{2}}{2}\right) + O\left(\frac{1}{\lambda^2}\right).$$

In Fig. 2.4, the numerical (solid line) and asymptotic (dashed line) values for the integral are compared.

From relation (2.4.3) and under assumption (2.4.2), it follows that the asymptotic expansion $F(\lambda)$ depends only on the values of the functions $\varphi(x)$ and $S(x)$ and their derivatives at the end of the interval of integration. In neighborhoods of other points, oscillation interference occurs. In Fig. 2.5, the graph of the function $\Re(\exp(S(x)))$ is plotted where $S(x) = i(x-6)^2$ and $S'(x) \neq 0$ in the interval $[-3, 5]$.

If $\varphi(x)$ and $S(x)$ or their derivatives have a finite number of points of discontinuity, the interval of integration $[a, b]$ may be split in parts in such a way that the points of

Fig. 2.4 Numerical values (*solid line*) and asymptotic values (*dashed line*) of the function $F(\lambda)$ of Example 1

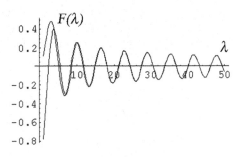

Fig. 2.5 Graph of the
function $\Re(\exp(S(x)))$ of
Example 1

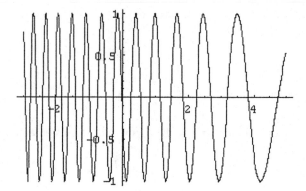

discontinuity coincide with the ends of these parts and then apply formula (2.4.3) to each of the parts.

If $b = \infty$, conditions (2.4.2) are still satisfied and

$$M^k(P(x)) = o(1) \quad \text{as} \quad x \to \infty, \qquad \frac{d}{dx} M^k(P(x)) \in L_1[0, \infty],$$

then

$$F(\lambda) = \sum_{k=0}^{N} \frac{1}{(i\lambda)^{k+1}} \left[M^k(P(x)) \exp(i\lambda S(x)) \right]\Big|_a + O\left(\frac{1}{\lambda^{N+2}}\right). \qquad (2.4.4)$$

Example 2

Find the first two terms of the expansion and the estimate of the error for the integral

$$F(\lambda) = \int_0^\infty \frac{e^{i\lambda x}\, dx}{(1+x)^\alpha}, \qquad \alpha > 0.$$

In this case, $S(x) = x$, $S'(x) = 1$, $M^k(x) = (-1)^k \dfrac{d^k}{dx^k}$,

$$P(x) = (1+x)^{-\alpha}, \quad M^0(P(x)) = (1+x)^{-\alpha},$$

$$M^k(P(x)) = (\alpha + k - 1)M^{k-1}(P(x))(1+x)^{-1}.$$

Therefore formula (2.4.4) holds and

$$F(\lambda) = -\left[\frac{1}{i\lambda} + \alpha \frac{1}{(i\lambda)^2}\right] + O\left(\frac{1}{\lambda^3}\right).$$

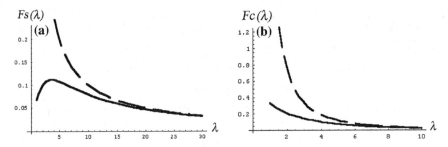

Fig. 2.6 Numerical values (*solid line*) and asymptotic values (*dashed line*) of $F_s(\lambda)$ and $F_c(\lambda)$ of Example 2

Separating the real and imaginary parts, we obtain

$$F_s(\lambda) = \Im(F(\lambda)) = \int_0^\infty \frac{\sin \lambda x \, dx}{(1+x)^\alpha} = \frac{1}{\lambda} + O\left(\frac{1}{\lambda^2}\right),$$

$$F_c(\lambda) = \Re(F(\lambda)) = \int_0^\infty \frac{\cos \lambda x \, dx}{(1+x)^\alpha} = \frac{\alpha}{\lambda^2} + O\left(\frac{1}{\lambda^3}\right).$$

In Fig. 2.6, the numerical (solid line) and asymptotic (dashed line) values of (a) the function $F_c(\lambda) = \Re(F(\lambda))$ and (b) $F_s(\lambda) = \Im(F(\lambda))$ are compared for $\alpha = 3$.

2.4.2 Erdélyi's Lemma [20]

Consider the integral

$$F(\lambda) = \int_0^a x^{\beta-1} \varphi(x) \exp(i\lambda x^\alpha) \, dx. \tag{2.4.5}$$

If

$$\varphi(x) \in C^\infty((0, a]), \quad \alpha \geq 1, \quad \beta > 0, \quad \forall k \quad \varphi^{(k)}(a) = 0,$$

then the following expansion holds

$$F(\lambda) = \frac{1}{\alpha} \sum_{k=0}^\infty \frac{1}{\lambda^{(k+\beta)/\alpha}} \Gamma\left(\frac{k+\beta}{\alpha}\right) \frac{\varphi^{(k)}(0)}{k!} \exp\left(\frac{i\pi(k+\beta)}{2\alpha}\right). \tag{2.4.6}$$

Erdélyi's lemma defines the contribution of the point $x_0 = 0$ in the asymptotic expansion of the integral.

Fig. 2.7 Graph of
$\Re(\exp(S(x)))$ with
stationary point $x = 1$

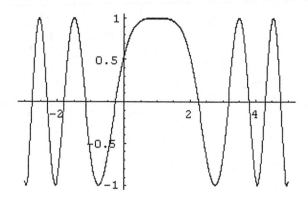

2.4.3 Integrals with Stationary Points

The points $x = c$, where $S'(x) = 0$, are called *stationary points*. With stationary
points, expansion (2.4.3) does not hold. In Fig. 2.7 we plot the graph of the function
$\Re(\exp(S(x)))$, where $S(x) = i(x - 1)^2$, which has the stationary point $x = 1$ in the
interval $[-3, 5]$.

To illustrate the input of the stationary point in the value of the integral we start
with an example with only one stationary point, $x = 0$, where $S''(0) \neq 0$,

$$F(\lambda) = \int_{-1}^{3} \exp(i\lambda x^2)\, dx.$$

The function $S(x) = x^2$ attains its minimum at the point $x_0 = 0$ and $x_0 \in I$.
Represent the initial integral in the form

$$\int_{-1}^{3} \exp(i\lambda x^2)\, dx = \int_{-\infty}^{+\infty} \exp(i\lambda x^2)\, dx - J_1 - J_2,$$

where

$$J_1 = \int_{-\infty}^{-1} \exp(i\lambda x^2)\, dx, \quad J_2 = \int_{3}^{+\infty} \exp(i\lambda x^2)\, dx.$$

It may be shown by integration by parts that that the integrals J_1 and J_2 have order
$O(\lambda^{-1})$, i.e.

$$F(\lambda) = \int_{-\infty}^{+\infty} \exp(i\lambda x^2)\, dx + O(\lambda^{-1}) = 2\int_{0}^{+\infty} \exp(i\lambda x^2)\, dx + O(\lambda^{-1}).$$

We evaluate this integral by Cauchy's theorem, which says that the circulation integral
of an analytic function of a complex variable, in a domain limited by a closed contour,
is equal to zero. The main idea for calculating the integral $F(\lambda)$ is in selecting the

Fig. 2.8 Integral along the closed path $C_1C_2C_3$

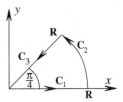

contour in such a way as to transform the initial Fourier integral into a Laplace integral. The integral along the closed path $C_1C_2C_3$ (see Fig. 2.8) is equal to 0.

Besides that it can be shown that for $R \to \infty$ the integral over C_2 converges to 0. Therefore, assuming that $x^2 = z^2 \exp(i\pi/2)$ and using Laplace's method we get

$$\int_0^{+\infty} \exp(i\lambda x^2)\, dx = \exp(i\pi/4) \int_0^{+\infty} \exp(-\lambda z^2)\, dz = \frac{\exp(i\pi/4)}{2}\sqrt{\frac{\pi}{\lambda}}$$

or

$$F(\lambda) = \exp(i\pi/4)\sqrt{\frac{\pi}{\lambda}} + O(\lambda^{-1}).$$

In Fig. 2.9a, b, the numerical (solid line) and asymptotic (dashed line) values of functions $F_c(\lambda) = \Re(F(\lambda))$ and $F_s(\lambda) = \Im(F(\lambda))$ are compared.

If the function $S(x)$ has a finite number of isolated stationary points x_i in the interval $[a, b]$, then the asymptotic expansion of integral (2.4.1) consists in a sum of contributions of the stationary points of $F(\lambda, x_i)$.

As a rule, using the stationary phase method, one limits oneself to the calculation of the first term of an asymptotic expansion.

We define the contribution of a stationary point.

Let $I = [a, b]$ be a finite segment where $S(x) \in C(I)$. Assume that $S(x)$ attains its maximum, $\max_{x \in I} S(x)$, at only the point x_0 and $\varphi(x) \approx C(x - x_0)^{\alpha}$, $\alpha > -1$, in a neighborhood of the point x_0. Then

(A) If $a < x_0 < b$ and $S^{(j)}(x_0) = 0$, $1 \le j \le 2m - 1$, $S^{(2m)}(x_0) \neq 0$, $m \ge 1$, then the main term of the asymptotic expansion has the form

$$F(\lambda, x_0) = m^{-1}\Gamma\left(\frac{1+\alpha}{2m}\right)\left[\frac{(2m)!}{|S^{2m}(x_0)|}\right]^{(1+\alpha)/2m} \lambda^{-(1+\alpha)/2m}$$

$$\times \exp\left(i\lambda S(x_0) + \frac{i\pi(1+\alpha)}{4m}\, \text{sgn}(S^{(2m)}(x_0))\right)\left[\varphi(x_0) + O(\lambda^{-1/2m})\right].$$

$$(2.4.7)$$

(B) If $x_0 = a$ and $S^{(j)}(x_0) = 0$, $1 \le j \le m - 1$, $S^{(m)}(x_0) \neq 0$, $m \ge 2$, then the main term of the asymptotic expansion has the form

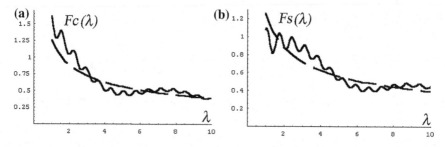

Fig. 2.9 Numerical values (*solid line*) and asymptotic values (*dashed line*) of functions F_c and F_s

$$F(\lambda, x_0) = m^{-1}\Gamma\left(\frac{1+\alpha}{m}\right)\left[\frac{(m)!}{|S^m(x_0)|}\right]^{(1+\alpha)/m}\lambda^{-(1+\alpha)/m}$$

$$\times \exp\left[i\lambda S(x_0) + \frac{i\pi(1+\alpha)}{2m}\operatorname{sgn}(S^m(x_0))\right]\left[\varphi(x_0) + O(\lambda^{-1/m})\right]. \quad (2.4.8)$$

Example 3
Find the first term of the asymptotic expansion of the integral

$$F(\lambda) = \int_0^1 \frac{\exp\left(i\lambda x^3\right)}{\sqrt{x}}\, dx$$

The integrand has only the critical point $x_0 = 0$, the contribution of which is evaluated by Erdélyi's lemma. For $\alpha = 3$, $\beta = 1/2$, $\varphi(x) \equiv 1$, one gets

$$F(\lambda) = \frac{1}{3}\lambda^{-1/6}\Gamma\left(\frac{1}{6}\right)\exp(i\pi/12).$$

In Fig. 2.10 the numerical (solid line) and asymptotic (dashed line) values of the functions (a) $F_c(\lambda) = \Re(F(\lambda))$ and (b) $F_s(\lambda) = \Im(F(\lambda))$ are compared.

Example 4
Find the first term of the asymptotic expansion and estimate the error of Bessel's function of positive integer index n, defined by the integral

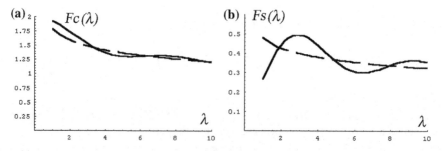

Fig. 2.10 Numerical values (*solid line*) and asymptotic values (*dashed line*) of functions F_c and F_s

Fig. 2.11 Numerical values (*solid line*) and asymptotic values (*dashed line*) of the integral $J_n(\lambda)$ of Example 4

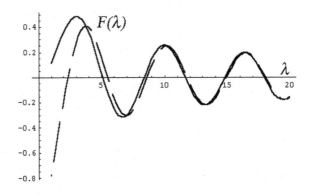

$$J_n(\lambda) = \pi^{-1} \int_0^\pi \cos(\lambda \sin x - nx)\, dx.$$

Represent the given integral as

$$J_n(\lambda) = \pi^{-1} \Re \int_0^\pi \exp(i\lambda \sin x) \exp(-inx)\, dx,$$

For this we have $S(x) = \sin x$, $x_0 = \pi/2$, $S(x_0) = 1$, $S''(x_0) = -1$, $m = 1$, $\varphi(x_0) = \exp(-in\pi/2)$. As a result, we obtain

$$J_n(\lambda) = \Re \left[\pi^{-1} \Gamma\left(\frac{1}{2}\right) \left[\frac{2!}{1}\right]^{1/2} \lambda^{-1/2} \exp\left(i\lambda - \frac{i\pi}{4}\right) \left[\exp(-in\pi/2) + O(\lambda^{-1})\right]\right]$$

$$= \frac{1}{\sqrt{\lambda}} \sqrt{\frac{2}{\pi}} \cos\left(\lambda - \frac{n\pi}{2} - \frac{\pi}{4}\right) + O\left(\lambda^{-1}\right). \quad (2.4.9)$$

In Fig. 2.11 the numerical values (solid line) and asymptotic values (dashed line) of the integral are compared.

2.4.4 Complete Asymptotic Expansions

By van der Korput's lemma [20], the interval of integration may be split into parts to provide only one point in the interval to contribute to the value of the integral. To calculate the contribution of a point, the following complete asymptotic expansions are used:

For case (A), i.e. if the stationary point x_0 is within the interval of integration and the first $2m - 1$ derivatives of the function $S(x)$ at x_0 are equal to zero and $S^{(2m)}(x_0) \neq 0$, then

$$F(\lambda, x_0) = \frac{1}{\lambda^{(1+\alpha)/2m}} \exp\left(i\lambda S(x_0) + \frac{i\pi(1+\alpha)}{4m} \mathrm{sgn}(S^{2m}(x_0))\right) \sum_{k=0}^{\infty} a_k \frac{1}{\lambda^{k/m}}.$$

where

$$a_k = \frac{2^{2k+1/2}}{(2k)!} \Gamma\left(\frac{2k+1+\alpha}{2m}\right) \exp\left[\frac{i\pi(2k+1+\alpha)}{4m} \mathrm{sgn}(S^m(x_0))\right]$$

$$\times \left[\left(h^{-1}(x, x_0)\frac{d}{dx}\right)^{2k} (\varphi(x)h(x, x_0))\right]\Bigg|_{x=x_0},$$

and

$$h(x, x_0) = \left[2\,\mathrm{sgn}(S^{2m}(x_0))[S(x) - S(x_0)]\right]^{1-1/2m}/S'(x).$$

For case (B), if a stationary point x_0 coincides with an end of the interval, the first $m - 1$ derivatives of the function $S(x)$ at the stationary point are equal to zero and $S^{(m)}(x_0) \neq 0$, then

$$F(\lambda, x_0) = \frac{1}{\lambda^{(1+\alpha)/m}} \exp(i\lambda S(x_0)) \sum_{k=0}^{\infty} a_k \lambda^{-k/m},$$

where

$$a_k = \frac{1}{k!m} \Gamma\left(\frac{k+1+\alpha}{m}\right) \exp\left(\frac{i\pi(1+k+\alpha)}{2m} \mathrm{sgn}(S^m(x_0))\right)$$

$$\left[\left(\frac{d}{dx}\right)^k [\varphi(x)h(x, x_0)]\right]\Bigg|_{x=x_0},$$

and

$$h(x, x_0) = \left[-\mathrm{sgn}(S^m(x_0)(S(x) - S(x_0))\right]^{-(k+1)/m} (x - x_0)^{k+1}.$$

So, contributions to integral (2.4.1) come only from neighborhoods of

(i) the ends of the integration interval,
(ii) points of discontinuity for the functions $\varphi(x)$ and $S(x)$ or their derivatives,
(iii) the stationary points.

For example, the integral

$$F(\lambda) = \int_{-\infty}^{+\infty} \frac{e^{i\lambda x}}{1 + x^2} dx = \pi e^{-\lambda}$$

admits the asymptotic expansion (2.4.3) with zero terms, since it has no points of discontinuity nor stationary points and the contributions of the ends are equal to zero since $\varphi^{(n)} \to 0$ as $x \to \pm\infty$.

2.4.5 Exercises

2.4.1. Find the first term of the asymptotic expansion and estimate the error for Bessel's function of the large real index ν for a fixed argument $t > 0$,

$$
J_\nu(\nu t) = \pi^{-1} \int_0^\pi \cos(\nu(x - t \sin x)) \, dx
$$
$$
- \pi^{-1} \sin \nu\pi \int_0^\infty \exp(-\nu(x + t \sinh x)) \, dx
$$

2.4.2. Find the first term of the asymptotic expansion and estimate the error of the integral

$$
F(\lambda) = \int_0^1 \exp(i\lambda x^3) \, dx.
$$

2.4.3. Find the first term of the asymptotic expansion and estimate the error of the integral

$$
F_n(\lambda) = \int_0^1 \exp(i\lambda x^n) \, dx.
$$

2.5 Saddle Point Method

2.5.1 Description of the Method

We seek an asymptotic expansion of the integral

$$
F(\lambda) = \int_\gamma \varphi(z) \, e^{\lambda h(z)} \, dz, \quad \text{as} \quad \lambda \to \infty, \tag{2.5.1}
$$

where the curve γ lies in the complex plane and the functions $\varphi(z)$ and $h(z)$ are analytic in a domain S which contains γ.

The saddle point method includes two stages [20, 24, 53]:

(1) deforming the contour γ into the contour γ_0 is most convenient to find the asymptotic estimates.
(2) calculating asymptotics for the integral over the contour γ_0.

As $\lambda \to \infty$, the absolute value of the integrand attains its maximum at points z where the function $\Re(h(z))$ is maximal. Assume that, among the contours with the same ends as γ, there exists a contour γ_0 where

$$
\min_\gamma \max_{z \in \gamma} \Re(h(z)) = \max_{z \in \gamma_0} \Re(h(z)).
$$

We also assume that one can deform the contour γ to contour γ_0 inside the domain S. Then, according to Cauchy's theorem,

$$F(\lambda) = \int_{\gamma_0} \varphi(z) e^{\lambda h(z)} \, dz. \tag{2.5.2}$$

For simplicity, let the maximum,

$$\max_{z \in \gamma_0} \Re(h(z))$$

be attained only at the point $z = c$. If $z = c$ is an inner point of γ_0, then, from the minimax property of the contour γ_0, it follows that c is a stationary point of the function $f(x, y) = \Re(h(z))$, where $z = x + iy$. At c, $f_x = f_y = 0$, and from the Cauchy–Riemann conditions it follows that $h'(c) = 0$.

The point $z = c$ is called *a saddle point*, and the value of $\Re(h(c))$ is the height of the saddle point. A sable point is called *simple* if $h''(c) \neq 0$. In this case, two straight lines, $\Im(h(z) - h(c)) = 0$, intersect at right angle at point c. On these lines, the function $f(x, y)$ changes in the fastest way. One of the lines, on which $f(x, y)$ decreases away from point $z = c$, is called the *line of fastest descent*.

The contour γ_0 may be deformed such that in the neighborhood of the point $z = c$ it coincides with the line of fastest descent. Then, in this neighborhood, $\Im(h(z)) = \text{const}$ and to estimate integral (2.5.2) it is convenient to use the Laplace method.

If the contour γ_0 lies entirely in the domain of regularity of the functions $\varphi(z)$ and $h(z)$ and the maximum value, $\Re(h(z))$, on γ_0 is attained at the saddle points (or at the ends of the interval) where the contour passes through, then the asymptotic behavior of integral (2.5.1) as $\lambda \to \infty$ is given as the sum of the contributions of the saddle points (or the ends of the contour).

There exists no general algorithm to construct the contour γ_0. For the specific integral (2.5.1) we firstly should find the saddle points and draw the lines of fastest descent through them. After that, one should try to deform the contour γ in such a way that it consists of the parts of the lines of fastest descent and maybe some other curves over which the integrals may be neglected for being asymptotically small.

2.5.2 *Asymptotics of Airy's Functions*

Find the first terms of the expansions for Airy's functions. Consider the integrals [17]

$$w_k(\eta) = C_k \int_{\gamma_k} e^{z^3/3 - \eta z} \, dz, \tag{2.5.3}$$

where γ_k are the contours with ends going to infinity over rays $O A_n$ where $\arg z = \frac{\pi}{3}(-1 + 2n)$, $n = 1, 2, 3$ (Fig. 2.12). Suppose the integrand in (2.5.3) converges to zero as $|z| \to \infty$ along these rays.

Fig. 2.12 Rays OA_n,
$n = 1, 2, 3$

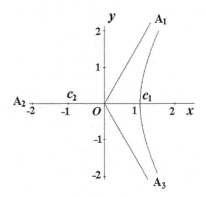

The standard Airy's functions $\mathrm{Ai}(\eta)$ and $\mathrm{Bi}(\eta)$ are real for real η and are defined by the formulas [1, 53]:

$$\mathrm{Ai}(\eta) = w_1(\eta), \quad C_1 = i/2\pi, \quad \gamma_1 = A_1 O A_3,$$
$$\mathrm{Bi}(\eta) = w_2(\eta), \quad C_2 = 1/2\pi, \quad \gamma_2 = A_2 O A_1 + A_2 O A_3. \tag{2.5.4}$$

Airy's functions find applications in the asymptotic integration of differential equations with parameter which contain turning points (see Chap. 5).

Consider the case $\eta > 0$. After the change of variables $z = \sqrt{\eta}\, z_1$, formula (2.5.3) becomes

$$w_k(\eta) = C_k \sqrt{\eta} \int_{\gamma_k} e^{\lambda h(z_1)} \, dz_1, \quad h(z) = \frac{z^3}{3} - z, \quad \lambda = \eta^{3/2}. \tag{2.5.5}$$

Evaluate the integral in (2.5.5) by the saddle point method. The roots of the equation $h'(c) = 0$ are the saddle points $c_{1,2} = \pm 1$. Since $h''(c_k) = 2c_k \neq 0$, the saddle points are the saddle foci for the function $f(x, y) = \Re(h(z))$, where $z = x + iy$. In Fig. 2.13a we plot the graph of the function $\Re f(x, y)$ near the point $c = 0$.

The lines of fastest descent passing through the saddle points c_k, are defined by the equation

$$\Im h(z) = \Im h(c_k). \tag{2.5.6}$$

In the case under consideration, $\Im h(c_k) = 0$ for $k = 1, 2$. Equation (2.5.6) are equivalent to the equations $y = 0$ and $x^2 - y^2/3 = 1$. The lines of fastest descent through c_1 and c_2 are shown in Fig. 2.13b with thick lines and the contour lines for function $\Re(h(z))$ with thin lines.

The paths of integration γ_1 and γ_2 are deformed into the paths $\gamma_{10} = A_1 c_1 A_3$ and $\gamma_{20} = A_2 c_2 c_1 A_1 + A_2 c_2 c_1 A_3$, consisting of the parts of the lines of fastest descent.

The paths γ_{k0} have the rays OA_n as their asymptotes on which the integrand in (2.5.5) goes to zero as $z \to \infty$. Therefore the asymptotic expansions of integrals (2.5.5) consist only of the contributions of the neighborhoods of the saddle points.

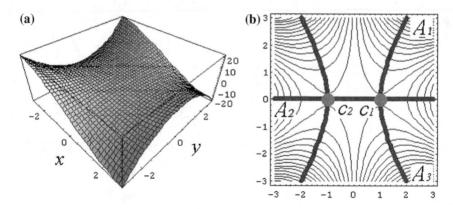

Fig. 2.13 **a** Graph of the function $\Re f(x, y)$. **b** Thick lines of fastest descent and thin contour lines

For Ai(η), we get

$$\mathrm{Ai}(\eta) = w_1(\eta) = C_1\sqrt{\eta} \int_{\gamma_{10}} e^{\lambda h(z)}\, dz \simeq C_1\sqrt{\eta}\, J_1,$$

where

$$J_1 = \int_{\gamma_{10}^*} e^{\lambda h(z)}\, dz$$

is the contribution of the saddle point c_1 and γ_{10}^* is a small arc of the contour γ_{10}, containing the saddle point c_1. On the arc γ_{10}^* we have

$$\tau^2 = h(c_1) - h(z) \simeq -h''(c_1)(z - c_1)^2/2. \tag{2.5.7}$$

From (2.5.7) it follows that on γ_{10}^* the approximate equality $dz \simeq i\, d\tau$ holds. Hence,

$$J_1 \simeq e^{\lambda h(c_1)} \int_{\delta}^{-\delta} e^{-\lambda\tau^2} i\, d\tau = -i\, e^{\lambda h(c_1)} \int_{-\infty}^{\infty} e^{-\lambda\tau^2}\, d\tau = -i\sqrt{\pi/\lambda}\, e^{-2\lambda/3},$$

$$\mathrm{Ai}(\eta) \simeq \frac{i\eta^{1/2}}{2\pi} J_1 = \frac{1}{2\sqrt{\pi}}\eta^{1/4} e^{-\zeta}, \quad \zeta = \frac{2}{3}\eta^{3/2}.$$

In a similar manner, the main term of the asymptotic expansion can be found for the function Bi(η) as $\eta \to \infty$,

$$\mathrm{Bi}(\eta) \simeq \frac{1}{\sqrt{\pi}}\eta^{1/4} e^{\zeta}.$$

A calculation shows that the contribution of the point c_2 is doubled since the path γ_{20} goes through this point twice and the contribution of the point c_1 may be neglected since the height of the saddle point c_1 is smaller than the height of the point c_2.

To find asymptotics for Airy's function as $\eta \to -\infty$ we change the variable $z = \sqrt{-\eta} z_1$ in (2.5.3). Then, formula (2.5.3) becomes

$$w_k(-\eta) = C_k \sqrt{\eta} \int_{\gamma_k} e^{\lambda h(z_1)} \, dz_1, \quad h(z_1) = \frac{z_1^3}{3} + z_1, \quad \lambda = \eta^{3/2}. \qquad (2.5.8)$$

The approximate expressions for integrals (2.5.8) may be obtained by means of the saddle point method (see the solution for Exercise 2.5.1).

2.5.3 Exercises

2.5.1. Find the first terms of the asymptotic expansions for Airy's functions as $\eta \to -\infty$.

Find the first term of the asymptotic expansions of the following integrals as $\lambda \to \infty$.

2.5.2. $F(\lambda) = \int_0^\infty \exp(\lambda(x + ix - x^3)) \, dx.$

2.5.3. $F(\lambda) = \int_{-\infty}^\infty \exp(i\lambda x)(1 + x^2)^{-\lambda} \, dx.$

2.5.4. $F(\lambda) = \int_{-\infty}^{+\infty} \exp(i\lambda(3x - x^3)) \, dx.$

2.6 Answers and Solutions

2.1.1. $F(\mu) = \frac{1}{2} \pi \left[1 - \frac{1}{4} \mu - \frac{3}{64} \mu^2 + O\left(\mu^3\right) \right].$

2.1.2. After a change of variables, the given function is represented in the form

$$F(\mu) = \mu \int_0^1 \frac{dt}{\sqrt{1 - m \sin^2 \mu t}}.$$

The integrand is expanded into the truncated series

$$f(t) = 1 + \frac{m}{2} \sin^2 \mu t + \frac{3m^2}{8} \sin^4 \mu t + O\left(\mu^6\right).$$

In its turn

$$\sin \mu t = \mu t - \frac{(\mu t)^3}{6} + O\left(\mu^5\right), \quad \sin^2 \mu t = (\mu t)^2 - \frac{(\mu t)^4}{3} + O(\mu^6),$$

$$\sin^4 \mu t = (\mu t)^4 + O\left(\mu^6\right).$$

Hence,

$$f(t) = 1 + \frac{m}{2}(\mu t)^2 + \left(-\frac{m}{6} + \frac{3m^2}{8}\right)(\mu t)^4 + O\left(\mu^6\right)$$

and after an integration by parts we obtain

$$F(\mu) = \mu + \frac{m}{6}\mu^3 + \left(-\frac{m}{30} + \frac{3m^2}{40}\right)\mu^5 + O\left(\mu^7\right).$$

To find the inverse function F^{-1} we use the method of undetermined coefficients and represent μ as a truncated series,

$$\mu = F + aF^3 + bF^5 + O\left(F^7\right), \quad \text{as} \quad F \to 0.$$

Equating the coefficients of F of equal powers, we get

$$a = -\frac{m}{6}, \quad b = \frac{m}{30} - \frac{m}{6}3a - \frac{3m^2}{40}.$$

Whence $b = \dfrac{m^2 + 4}{120}$. Therefore

$$\text{am}(\mu) = \mu - \frac{m}{6}\mu^3 + \frac{m^2 + 4}{120}\mu^5 + O\left(\mu^7\right),$$

$$\text{sn}(\mu) = \sin\left(F^{-1}(\mu)\right) = \mu - \frac{m+1}{6}\mu^3 + \frac{m^2 + 14m + 1}{120}\mu^5 + O\left(\mu^7\right),$$

$$\text{cn}(\mu) = \cos\left(F^{-1}(\mu)\right) = 1 - \frac{1}{2}\mu^2 + \frac{4m+1}{24}\mu^4 + O\left(\mu^6\right).$$

2.1.3. $F(\mu) = \dfrac{\pi}{2}\left[1 + \dfrac{1}{4}\mu + \dfrac{9}{64}\mu^2\right] + O\left(\mu^3\right).$

2.1.4.

(a) Using the substitution $t = \sinh x$ one gets

$$\int \sqrt{1+x^2}\,dx = \int \sinh^2 t\,dt = \frac{t}{2} + \frac{\sinh 2t}{4}$$

$$= \frac{1}{2}\ln\left(x + \sqrt{x^2+1}\right) + \frac{x\sqrt{x^2+1}}{2}.$$

Then

$$F(\lambda) = \frac{1}{2}\ln\left(\lambda + \sqrt{\lambda^2+1}\,\right) + \frac{\lambda\sqrt{\lambda^2+1}}{2} = \frac{1}{2}\ln(2\lambda) + \frac{\lambda^2}{2} + \frac{1}{4} + O\left(\lambda^{-2}\right).$$

(b) The integrand can be expanded in a series for $x > 0$

$$\sqrt{1+x^2} = x\left[1 + \frac{1}{2x^2} + O\left(x^{-2}\right)\right] = x + \frac{1}{2x} + O\left(x^{-1}\right).$$

Write the given integral in the form

$$\int_0^\lambda \sqrt{1+x^2}\,dx = \int_0^1 \sqrt{1+x^2}\,dx + \int_1^\lambda \sqrt{1+x^2}\,dx.$$

For the first integral in the right side the estimate $O(1)$ holds. Substitute in the second integral the expansion for the integrand and integrate term by term,

$$F(\lambda) = \frac{\lambda^2}{2} + \frac{1}{2}\ln(\lambda) + O(1).$$

2.2.1. $F(\lambda) = \displaystyle\sum_{n=1}^{N} \frac{(-1)^{n-1}(n-1)!}{\lambda^n} + O\left(\frac{1}{\lambda^{N+1}}\right).$

2.2.2. $F(\lambda) = \displaystyle\sum_{n=1}^{N} \frac{(-1)^{n-1}(2n-1)!}{\lambda^{2n}} + O\left(\frac{1}{\lambda^{2N+2}}\right).$

2.2.3. $F(\lambda) = e^{-\lambda}\left[\displaystyle\sum_{n=1}^{N} \frac{(-1)^{n-1}(n-1)!}{\lambda^n} + O\left(\frac{N!}{\lambda^{N+1}}\right)\right].$

2.3.1. In this case $S(x) = \cos(\sqrt{x})$, $S'(x) = -\dfrac{\sin(\sqrt{x})}{2\sqrt{x}}$, $S'(x) \neq 0$ for $x \in \left(0, \dfrac{\pi^2}{4}\right]$, $S'(+0) = -\dfrac{1}{2}$. Therefore we may use formula (2.3.6). Calculate

$$\frac{d}{dx}\left(\frac{1}{S'(x)}\right) = \frac{\frac{\sin\sqrt{x}}{\sqrt{x}} - \cos\sqrt{x}}{\sin^2\sqrt{x}}.$$

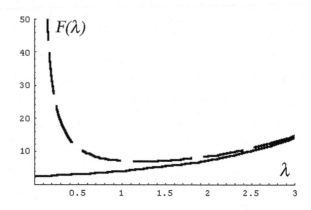

Fig. 2.14 Numerical solution (*solid line*) and one-term asymptotic approximation (*dashed line*)

Hence,

$$F(\lambda) = -\frac{1}{\lambda S'(0)} e^{\lambda S(0)} + \frac{1}{\lambda^2 S'(0)} \frac{d}{dx} \left(\frac{1}{S'(x)}\right)\bigg|_{x=0} e^{(\lambda S(0))} + O\left(\frac{1}{\lambda^3}\right).$$

Calculating $\frac{d}{dx}\left(\frac{1}{S'(x)}\right)_{x=0}$ and substituting $\sin\sqrt{x}$ with $\sqrt{x} - (\sqrt{x},)^3/6$ and $\cos\sqrt{x}$ with $1 - x/2$, we obtain

$$F(\lambda) = e^\lambda \left[\frac{2}{\lambda} + \frac{2}{3\lambda^2} + O\left(\frac{1}{\lambda^3}\right)\right].$$

In Fig. 2.14 the solid line corresponds to the numerical solution and the dashed line to the one-term asymptotic approximation.

2.3.2. The substitution $x = z\lambda^{-1/2}$ transforms the integral into

$$F(\lambda) = \frac{1}{\sqrt{\lambda}} \int_0^{\sqrt{\lambda}} \exp\left(-\sqrt{\lambda}\left(z + \frac{1}{z}\right)\right) dz.$$

So,

$$F(\lambda) = \lambda^{-3/4} \exp\left(-2\sqrt{\lambda}\right) \left[\sqrt{\pi} + O\left(\lambda^{-1}\right)\right].$$

In Fig. 2.15 the solid line corresponds to numerical solution and the dashed line to the one-term asymptotic approximation.

2.3.3. Substituting $x = z \left(\alpha/\lambda\right)^{1/(1+\alpha)}$ we get the integral

$$F(\lambda) = \left(\frac{\alpha}{\lambda}\right)^{1/(1+\alpha)} \int_0^\infty \exp\left(-\lambda^{\alpha/(\alpha+1)} \alpha^{1/(\alpha+1)} \left(z + z^{-\alpha}\right)\right) dz.$$

Fig. 2.15 Numerical solution (*solid line*), one-term asymptotic approximation (*dashed line*)

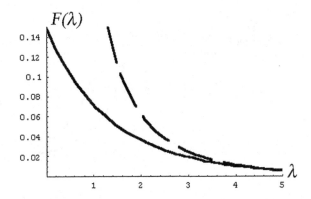

Fig. 2.16 Numerical solution (*solid line*), one-term asymptotic approximation (*dashed line*)

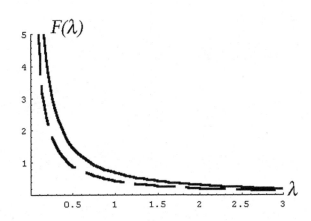

Therefore

$$F(\lambda) \simeq \lambda^{-\frac{\alpha+2}{2(1+\alpha)}} \alpha^{\frac{1}{1+\alpha}} \exp\left(-(\alpha+1)\alpha^{\frac{1-\alpha}{1+\alpha}}\lambda^{\frac{\alpha}{1+\alpha}}\right)\left(\sqrt{\frac{2\pi}{1+\alpha}}\right).$$

In Fig. 2.16 the solid line corresponds to the numerical solution and the dashed line to the one-term asymptotic approximation for $\alpha = 0.3$.

2.3.4. We write the given integral as $F(\lambda) = \int_{-1}^{1} \exp(\lambda S(x))x^2 \, dx$. In this case, $S(x) = \ln(1-x^2)$, $x_0 = 0$, $S(x_0) = 0$, $S''(x_0) = -2$, $m = 1$, $\varphi(x) = x^2$, i.e. $\alpha = 2$. Hence $F(\lambda) = \lambda^{-3/2}(\sqrt{\pi}/2 + O(\lambda^{-3/2}))$. In Fig. 2.17 the solid line corresponds to the numerical solution and the dashed line to the one-term asymptotic approximation.

2.3.5. We write the initial integral in the form

$$F(\lambda) = \int_{-1}^{1} \exp(\lambda S(x)) \, dx.$$

Fig. 2.17 Numerical
solution (*solid line*) and
one-term asymptotic
approximation (*dashed line*)

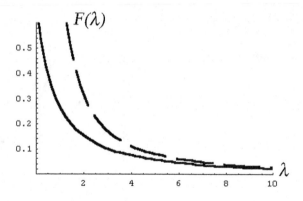

Fig. 2.18 Numerical
solution (*solid line*) and
one-term asymptotic
approximation (*dashed line*)

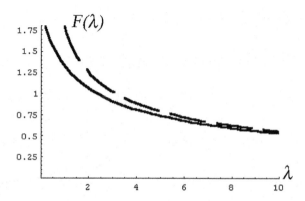

In this case, $S(x) = \ln(1 - x^2)$, $x_0 = 0$, $S(x_0) = 0$, $S''(x_0) = -2$, $m = 1$, $\varphi(x) = 1$, i.e. $\alpha = 0$. Thus, $F(\lambda) = \lambda^{-1/2}(\sqrt{\pi} + O(\lambda^{-1/2}))$. In Fig. 2.18 the solid line corresponds to numerical solution and the dashed line to one-term asymptotic approximation.

2.3.6. Write the given integral as $F(\lambda) = \int_0^1 \exp(\lambda S(x))x\,dx$. In this case, $S(x) = \ln(1 - x^2)$, $x_0 = 0$, $S(x_0) = 0$, $S''(x_0) = -2$, $m = 1$, $\varphi(x) = x$, i.e. $\alpha = 1$. Note that now the point of maximum lies on the boundary (this point coincides with the interval end point). So, we have $F(\lambda) = \lambda^{-1}(1/2 + O(\lambda^{-1}))$. In Fig. 2.19 the solid line corresponds to the numerical solution and the dashed line to the one-term asymptotic approximation.

2.3.7. We represent the initial integral in the form

$$F(\lambda) = \int_0^\pi \exp(nS(x)) \sin x\,dx.$$

In this case, $S(x) = \ln x$, $x_0 = b = \pi$ (note that the function $S(x)$ attains its maximum at the upper limit), $S(x_0) = \ln \pi$, $S'(x_0) = 1/\pi$, $m = 1$, $h(x, x_0) = x$. Hence, $\varphi(x_0) = 0$, i.e. the asymptotic expansion does not include the term for $k = 0$.

Fig. 2.19 Numerical
solution (*solid line*) and
one-term asymptotic
approximation (*dashed line*)

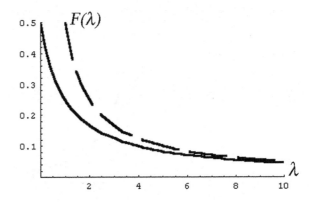

Fig. 2.20 Numerical
solution (*solid line*),
one-term asymptotic
approximation (*dashed line*)

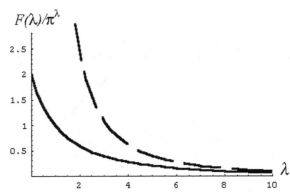

Calculate $x \frac{d}{dx}(x \sin x)|^{x=\pi} = -\pi^2$. Thus,

$$F(\lambda) = \lambda^{-2} \exp(\lambda \ln \pi) \left[\pi^2 + O\left(\lambda^{-1}\right)\right] = \pi^{\lambda+2}\lambda^{-2}\left[1 + O\left(\lambda^{-1}\right)\right].$$

In Fig. 2.20 the solid line corresponds to the numerical solution and the dashed line
to the one-term asymptotic approximation.

2.3.8. The function $S(x) = \cosh(x)$ has a critical point at $x_0 = 0$. $S(x_0) = 1$,
$S''(x_0) \neq 0$.

$$F_\nu(\lambda) = \sqrt{\frac{\pi}{2}} \frac{\exp(-\lambda)}{\sqrt{\lambda}} \left[1 + O\left(\lambda^{-1/2}\right)\right].$$

In Fig. 2.21 the solid line corresponds to the numerical solution and the dashed line
to the one-term asymptotic approximation.

2.4.1. $J_\nu(\nu t) = \sqrt{\frac{2}{\pi\nu\sqrt{t^2 - 1}}} \cos\left(\nu \arccos \frac{1}{t} - \nu\sqrt{t^2 - 1} + \frac{\pi}{4}\right) + O\left(\nu^{-1}\right)$. In
Fig. 2.22 the solid line corresponds to the numerical solution and the dashed line to
the one-term asymptotic approximation for $t = 1.3$.

Fig. 2.21 Numerical solution (*solid line*), one-term asymptotic approximation (*dashed line*)

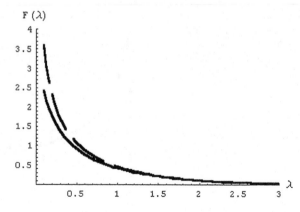

Fig. 2.22 Numerical solution (*solid line*), one-term asymptotic approximation (*dashed line*)

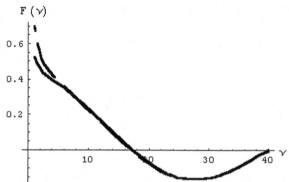

2.4.2. $S(x) = x^3$, $x_0 = 0$, $S''(x_0) = 0$, $S'''(x_0) = 6$, $S(x_0) = 0$. Therefore

$$F(\lambda) = \frac{1}{3}\Gamma\left(\frac{1}{3}\right)\lambda^{-1/3}\exp\left(\frac{i\pi}{6}\right)\left[1 + O\left(\lambda^{-1}\right)\right].$$

The numerical (solid line) and asymptotic (dashed line) values of the functions $F_c(\lambda) = \Re(F(\lambda))$ and $F_s(\lambda) = \Im(F(\lambda))$ are compared in Fig. 2.23.

2.4.3. $S(x) = x^n$, $x_0 = 0$, $S^{(n-1)}(x_0) = 0$, $S^{(n)}(x_0) = n!$, $S(x_0) = 0$. Therefore,

$$F(\lambda) = \frac{1}{n}\Gamma\left(\frac{1}{n}\right)\lambda^{-1/n}\exp\left(\frac{i\pi}{2n}\right)\left[1 + O\left(\lambda^{-1/n}\right)\right].$$

2.5.1. Let $h(z) = z + z^3/3$. The saddle points $c_{1,2} = \pm i$ for integrals (2.5.8) are obtained from the condition $h'(c) = 0$. The lines of fastest descent passing through the saddle points are found from Eq. (2.5.6), which, for this case, are $x^2 y - y^3 + 3y = \pm 2$, where $x = \Re(z)$ and $y = \Im(z)$. In Fig. 2.24a we plot the function $\Re(h(z))$ in a neighborhood of $z = 0$, and in Fig. 2.24b the lines of fastest descent passing through the saddle points c_1 and c_2 and the contour lines of the function $\Re(h(z))$.

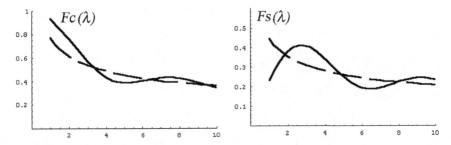

Fig. 2.23 Numerical (*solid line*) and asymptotic (*dashed line*) values of F_c and F_s

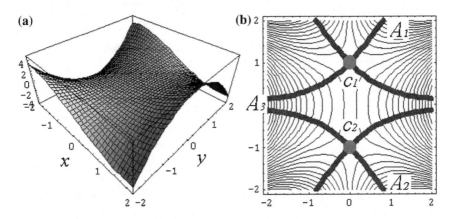

Fig. 2.24 For $\Re(h(z))$, **a** plot in a neighborhood of $z = 0$, **b** lines of fastest descent through c_1 and c_2 and contour lines

The integration paths γ_1 and γ_2 are deformed into the paths $\gamma_{10} = A_1 c_1 A_2 + A_2 c_2 A_3$ (consisting of two lines of fastest descent with the end points going to infinity) and $\gamma_{20} = A_2 c_1 A_1 + A_2 c_2 A_3$, respectively.

Let $J_k = \int_{\gamma_{k0}^*} e^{\lambda h(z)} dz$, $k = 1, 2$, be the contributions of the saddle points c_1 and c_2, corresponding to integrations from A_2 to A_1 and from A_2 to A_3, respectively. Taking into account that for the inverse direction the sign of the contribution J_k is reversed, one gets

$$\mathrm{Ai}(-\eta) \simeq C_1 \sqrt{\eta} \, (J_2 - J_1), \quad \mathrm{Bi}(-\eta) \simeq C_2 \sqrt{\eta} \, (J_2 + J_1), \quad \text{as} \quad \eta \to \infty.$$

To calculate the contribution J_1 of the integral over the arc γ_{10}^* we change the variable (2.5.7). With $dz \simeq \exp(i\pi/4) d\tau$ one finds

$$J_1 \simeq e^{\lambda h(c_1) + i\pi/4} \int_{-\infty}^{\infty} e^{-\lambda \tau^2} d\tau = \sqrt{\pi/\lambda} \, e^{i(2\lambda/3 + \pi/4)}.$$

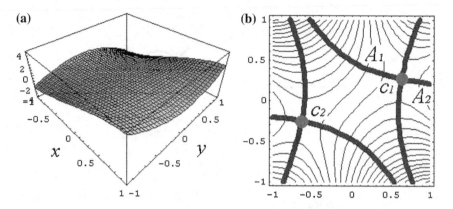

(a)   **(b)**

Fig. 2.25 For $\Re(h(z))$, **a** plot in a neighborhood of $z = 0$, and **b** lines of fastest descent through c_1 and c_2 and contour lines

Fig. 2.26 The path γ_0

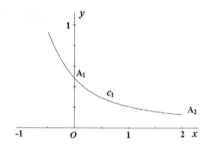

Similarly, we obtain

$$J_2 \simeq \sqrt{\pi/\lambda}\, e^{-i(2\lambda/3 + \pi/4)}.$$

Hence,

$$\mathrm{Ai}(-\eta) \simeq \frac{1}{\sqrt{\pi}}\, \eta^{-1/4} \sin\left(\zeta + \frac{\pi}{4}\right), \quad \mathrm{Bi}(-\eta) \simeq \frac{1}{\sqrt{\pi}}\, \eta^{-1/4} \cos\left(\zeta + \frac{\pi}{4}\right),$$

where $\zeta = \frac{2}{3}(-\eta)^{3/2}$ and $\lambda = (-\eta)^{3/2}$.

2.5.2. Let $h(z) = z + iz - z^3$. The equation $h'(z) = 0$ has two roots: $c_{1,2} = \pm 2^{1/4} 3^{-1/2} e^{i\pi/8}$. In Fig. 2.25a we plot the graph of the function $\Re(h(z))$ in a neighborhood of $z = 0$, and in Fig. 2.25b the lines of fastest descent through the saddle points c_1 and c_2 and the contour lines of the function $\Re(h(z))$. The equation for the line of fastest descent through c_1 has the form $y^3 - 3x^2 y + x + y = 2^{7/4} 3^{-3/2} \sin(3\pi/8)$.

For the path γ_0 we chose the contour consisting of the segment OA_1 and a part of the line of fastest descent $A_1 c_1 A_2$ (see Fig. 2.26).

The main term of the asymptotics of $F(\lambda)$ as $\lambda \to \infty$ is equal to the contribution J_1 of the saddle point c_1, since, compared to it, the contribution of the segment OA_1 is exponentially small.

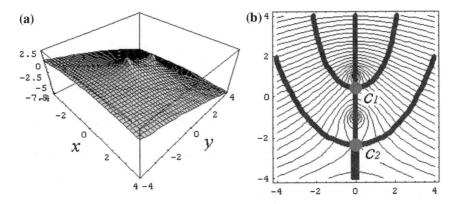

Fig. 2.27 For $\Re(h(z))$, **a** plot in a neighborhood of $z = 0$, **b** lines of fastest descent through c_1 and c_2 and contour lines

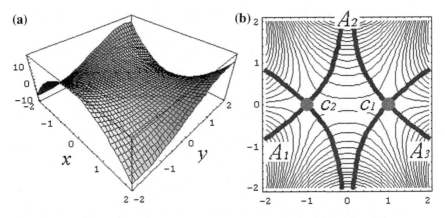

Fig. 2.28 For $\Re(h(z))$, **a** plot in a neighborhood of $z = 0$, **b** lines of fastest descent through c_1 and c_2 and contour lines

After the change of variable (2.5.7), in view of $dz \simeq 2^{-1/8}3^{-1/4}\,e^{i\pi/16}\,d\tau$, one gets

$$F(\lambda) \simeq J_1 = \sqrt{\pi/\lambda}\,2^{-1/8}3^{-1/4}\,\exp\left(2^{7/4}3^{-3/2}\,e^{3\pi i/8}\lambda - i\pi/16\right).$$

2.5.3. Let $h(z) = iz - \ln\left(1 + z^2\right)$. Then $F(\lambda)$ is of the form (2.5.1) and $\varphi(z) \equiv 1$. In this case, there are two simple saddle points: $c_1 = i(\sqrt{2} - 1)$ and $c_2 = -i(\sqrt{2} + 1)$. In Fig. 2.27a we have the graph of the function $\Re(h(z))$ in a neighborhood of $z = 0$, and in Fig. 2.27b we have the lines of fastest descent passing through the saddle points c_1 and c_2 and the contour lines of the function $\Re(h(z))$. The equation for the line of fastest descent through c_1 is $\sin x(1 + x^2 - y^2) = 2xy\cos x$.

The main term of the asymptotic expansion of $F(\lambda)$ is equal to the contribution J_1 of the saddle point c_1. Changing the variable (2.5.7) we get

$$F(\lambda) \simeq J_1 = \lambda^{-1/2} \exp(-\lambda c)(2c)^{-\lambda}[\pi(1 - c)]^{1/2},$$

where $c = \sqrt{2} - 1$.

2.5.4. The saddle points for the function $h(z) = i(3z - z^3)$ are obtained from the equation $h'(z) = 0$ with roots $c_{1,2} = \pm 1$.

In Fig. 2.28a, we plot the graph of the function $\Re(h(z))$ in a neighborhood of $z = 0$, and in Fig. 2.28b the lines of fastest descent through the saddle points c_1 and c_2 and the contour lines of the function $\Re(h(z))$.

The main term of the expansion of $F(\lambda)$ is equal to the sum of the contributions of the saddle points c_1 and c_2,

$$F(\lambda) \simeq J_1 + J_2 = 2\pi^{1/2}(3\lambda)^{-1/2} \cos(2\lambda - \pi/4).$$

Chapter 3
Regular Perturbation of Ordinary Differential Equations

In this chapter we find asymptotic solutions of regularly perturbed equations and systems of equations, to which problems in mechanics are reduced. We consider Cauchy problems, problems for periodic solutions and boundary value problems.

3.1 Introduction

A system of differential equations of first order

$$\frac{d\mathbf{y}}{dt} = \mathbf{f}(t, \mathbf{y}, \varepsilon), \tag{3.1.1}$$

where $\mathbf{y}$ and $\mathbf{f}$ are vector functions, t an independent scalar variable and ε a small parameter, is said to be *regularly perturbed* if

$$\mathbf{f}(t, \mathbf{y}, \varepsilon) \simeq \sum_{k=0}^{\infty} \mathbf{f}_k(t, \mathbf{y})\varepsilon^k. \tag{3.1.2}$$

In this case, the solution of the system can be found in the form of an asymptotic series

$$\mathbf{y}(t, \varepsilon) \simeq \sum_{k=0}^{\infty} \mathbf{y}_k(t)\varepsilon^k. \tag{3.1.3}$$

Asymptotic expansions of type (3.1.3) are called *direct expansions* [50].

Substituting (3.1.2) and (3.1.3) in (3.1.1) and equating coefficients of equal powers of ε in the right and in the left sides of system (3.1.1), we get a sequence of equations

© Springer International Publishing Switzerland 2015
S.M. Bauer et al., *Asymptotic Methods in Mechanics of Solids*,
International Series of Numerical Mathematics 167,
DOI 10.1007/978-3-319-18311-4_3

to find the vector functions $y_k(t)$. The differential system determining $y_0(t)$ has the form:

$$\frac{dy_0}{dt} = f(t, y_0, 0) \qquad (3.1.4)$$

and it is called *the generating system* for system (3.1.1). The systems of equations for $y_1(t)$, $y_2(t)$, ... is linear. If the generating system (3.1.4) is also linear, then (3.1.1) is called a *quasilinear system*.

System (3.1.1) is said to be *autonomous* if the vector function f does not depend explicitly on t. For an autonomous quasilinear system, the generating system is a linear system with constant coefficients.

A differential equation of the form

$$L(x, t, \varepsilon) = 0, \qquad (3.1.5)$$

where L is a differential operator and $x(t, \varepsilon)$ is a scalar function will be called *regularly perturbed* if it is reducible to a regularly perturbed system of equations.

For example, the equation

$$\frac{d^2x}{dt^2} = f(t, x, \varepsilon),$$

where the function f is expanded as

$$f(t, x, \varepsilon) \simeq \sum_{k=0}^{\infty} f_k(t, x)\varepsilon^k,$$

is regularly perturbed since after the change of variables $x = y_1$, $\dot{x} = y_2$ it reduces to the system

$$\frac{dy_1}{dt} = y_2, \quad \frac{dy_2}{dt} = f(t, y_1, \varepsilon).$$

Solution of regularly perturbed differential equation may be represented in the form

$$x(t, \varepsilon) \simeq \sum_{k=0}^{\infty} x_k(t)\varepsilon^k,$$

where

$$L(x, t, 0) = 0$$

is the generating equation for Eq. (3.1.5).

3.2 Cauchy Problems

The Cauchy problem for system (3.1.1) consists in this system with the initial condition $y(t_*) = y_*$. Initial conditions for the nth order equation (3.1.5) have the form

$$x = x_*, \quad \frac{dx}{dt} = x_*^{(1)}, \quad \ldots, \quad \frac{d^{n-1}x}{dt^{n-1}} = x_*^{(n-1)} \quad \text{for } t = t_*.$$

3.2.1 Motion of a Material Point in a Gravity Field

Assume that a material point of mass m with initial velocity v_0 at angle α with respect to the horizon moves freely under the action of gravity $P = mg$ and air resistance $R = -\nu v f(v)$ (Fig. 3.1).

According to Newton's second law of motion,

$$m\frac{dv}{dt} = P - \nu v f(v), \quad v = \frac{dr}{dt}. \tag{3.2.1}$$

Projecting the vector equalities (3.2.1) on the coordinate axes Ox and Oy, we obtain the following system of differential equations

$$m\frac{dv_x}{dt} = -\nu v_x f(v), \quad m\frac{dv_y}{dt} = -mg - \nu v_y f(v),$$
$$\frac{dx}{dt} = v_x, \quad \frac{dy}{dt} = v_y. \tag{3.2.2}$$

The initial conditions for system (3.2.2) are:

$$v_x = v_0 \cos \alpha, \quad v_y = v_0 \sin \alpha, \quad x = y = 0 \quad \text{for } t = 0. \tag{3.2.3}$$

We introduce the dimensionless variables by means of the formulas

$$t = \frac{v_0}{g}\tau, \quad x = \frac{v_0^2}{g}\xi, \quad y = \frac{v_0^2}{g}\eta, \quad u = \frac{v_x}{v_0}, \quad w = \frac{v_y}{v_0}.$$

Fig. 3.1 Motion of point under gravity with initial velocity v_0 at angle α

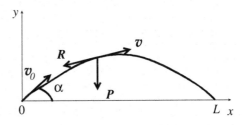

In the non-dimensional variables, system (3.2.2) and boundary conditions (3.2.3) become

$$\frac{du}{d\tau} = -\varepsilon u f, \quad \frac{dw}{d\tau} = -1 - \varepsilon w f, \quad \frac{d\xi}{d\tau} = u, \quad \frac{d\eta}{d\tau} = w, \tag{3.2.4}$$

$$u = \cos \alpha, \quad w = \sin \alpha, \quad \xi = \eta = 0 \quad \text{for } \tau = 0, \tag{3.2.5}$$

where

$$\varepsilon = \frac{\nu v_0}{mg}$$

is a non-dimensional parameter.

Assume, that $\varepsilon \ll 1$ and consider the case $f(v) \equiv 1$ when the resistance force is proportional to velocity. We seek a solution of system (3.2.4) in the form

$$u = u_0 + \varepsilon u_1 + \varepsilon^2 u_2 + \cdots . \tag{3.2.6}$$

Formulas for w, ξ, and η are obtained when u in (3.2.6) is replaced with the corresponding variables.

The generating system

$$\frac{du_0}{d\tau} = 0, \quad \frac{dw_0}{d\tau} = -1, \quad \frac{d\xi_0}{d\tau} = u_0, \quad \frac{d\eta_0}{d\tau} = w_0$$

with initial conditions

$$u_0 = \cos \alpha, \quad w_0 = \sin \alpha, \quad \xi_0 = \eta_0 = 0 \quad \text{for } \tau = 0$$

describes the motion of the material point without resistance ($f = 0$) and it has solution

$$u_0 = \cos \alpha, \quad w_0 = -\tau + \sin \alpha,$$

$$\xi_0 = \tau \cos \alpha, \quad \eta_0 = -\frac{\tau^2}{2} + \tau \sin \alpha. \tag{3.2.7}$$

In the zeroth approximation the trajectory is a parabola.

Next approximations ($k = 1, 2, \ldots$) are found after solving the systems

$$\frac{du_k}{d\tau} = -u_{k-1}, \quad \frac{dw_k}{d\tau} = -w_{k-1}, \quad \frac{d\xi_k}{d\tau} = u_k, \quad \frac{d\eta_k}{d\tau} = w_k, \quad k = 1, 2, \ldots$$

with zero initial conditions. The solution of the system of first approximation ($k = 1$) is

$$u_1 = -\tau \cos \alpha, \quad w_1 = \frac{\tau^2}{2} - \tau \sin \alpha,$$

$$\xi_1 = -\frac{\tau^2}{2} \cos \alpha, \quad \eta_1 = \frac{\tau^3}{6} - \frac{\tau^2}{2} \sin \alpha. \tag{3.2.8}$$

Fig. 3.2 Motion with resistance (*dashed line*) and motion without resistance (*solid line*)

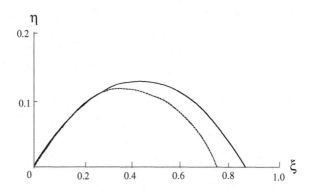

In Fig. 3.2, we plot trajectories for the motion with resistance (dashed line) and without resistance (solid line) for $\varepsilon = 0.2$, $\alpha = \pi/6$.

3.2.2 Duffing Equation

Consider the construction of the asymptotic expansion for the solution of the Cauchy problem for Duffing equation [50]

$$\ddot{x} + \omega^2 x + \varepsilon b x^3 = 0, \quad \ddot{x} = \frac{d^2 x}{dt^2}, \tag{3.2.9}$$

which, in particular, describes vibrations of a mass on a non-linear spring. Here x and t are dimensionless variables and ε is a small parameter.

Assume that for $t = 0$ the initial conditions are

$$x(0) = a, \quad \dot{x}(0) = 0. \tag{3.2.10}$$

We seek the solution for problem (3.2.9)–(3.2.10) in the form of a series

$$x = x_0 + \varepsilon x_1 + \varepsilon^2 x_2 + \cdots . \tag{3.2.11}$$

Substitute (3.2.11) into (3.2.9) and (3.2.10) and equate the coefficients of equal powers of ε in the right and left sides. Equating the terms not containing ε (the zeroth approximation), one gets the generating equation

$$\ddot{x}_0 + \omega^2 x_0 = 0 \tag{3.2.12}$$

with initial conditions

$$x_0(0) = a, \quad \dot{x}_0(0) = 0. \tag{3.2.13}$$

The equations of first and second approximations to determine x_1 and x_2 are

$$\ddot{x}_1 + \omega^2 x_1 + bx_0^3 = 0, \tag{3.2.14}$$
$$\ddot{x}_2 + \omega^2 x_2 + 3bx_1 x_0^2 = 0. \tag{3.2.15}$$

From formulas (3.2.10) and (3.2.13) it follows that

$$x_k(0) = \dot{x}_k(0) = 0 \quad \text{for} \quad k \geq 1. \tag{3.2.16}$$

Substituting the general solution of the generating equation (3.2.12)

$$x_0 = M_0 \cos z + N_0 \sin z, \quad z = \omega t,$$

into the initial conditions (3.2.13), we find $M_0 = a$, $N_0 = 0$ and, therefore,

$$x_0 = a \cos z. \tag{3.2.17}$$

Substituting (3.2.17) into (3.2.14) and using the formula

$$\cos^3 z = \frac{1}{4}(\cos 3z + 3 \cos z),$$

lead to the equation

$$\ddot{x}_1 + \omega^2 x_1 = -\frac{a^3 b}{4}(\cos 3z + 3 \cos z). \tag{3.2.18}$$

Its general solution is the sum of a partial solution and a general solution of the corresponding homogeneous equation. The partial solution for equation (3.2.18) according to the principle of superposition for linear equations may be written as a sum of solutions of the equations

$$\ddot{x} + \omega^2 x = -\frac{a^3 b}{4} \cos 3z \quad \text{and} \quad \ddot{x} + \omega^2 x = -\frac{3a^3 b}{4} \cos z.$$

We seek the solution of the first equation in the form $x = A \cos 3z$, and of the second equation in the form $x = Bt \sin z$. After evaluating of the constants A and B we obtain

$$x_1 = M_1 \cos z + N_1 \sin z - \frac{a^3 b}{32\omega^2}(12z \sin z - \cos 3z).$$

The constants M_1 and N_1 are found from the initial conditions (3.2.16):

$$M_1 = -\frac{a^3 b}{32\omega^2}, \quad N_1 = 0.$$

Fig. 3.3 Numerical solution
(*solid line*) and asymptotic
solution (*dashed line*)

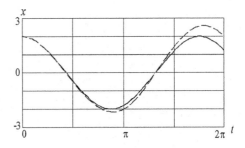

Hence the two-term approximation for the solution of equation (3.2.9) has the form

$$x \simeq a \cos z - \varepsilon \, \frac{a^3 b \sin z \, (\sin 2z + 6z)}{16\omega^2}. \tag{3.2.19}$$

In a similar manner, after substituting the expressions for x_0 and x_1 in Eq. (3.2.15), one gets x_2.

In Fig. 3.3 the solid line is the solution of equation (3.2.9) obtained with a Runge–Kutta method for $\omega = b = 1$, $\varepsilon = 0.1$, and $a = 2$. The dashed line is the solution obtained by formula (3.2.19). The error of the asymptotic formula increases with t since the two-term approximation contains the term with the multiplier εt and for $t \sim \varepsilon^{-1}$ the direct asymptotic expansion (3.2.19) becomes non-uniform.

Remark One may seek a solution for the Cauchy problem for Eq. (3.2.9) in the form

$$x = M(\varepsilon) \cos z + N(\varepsilon) \sin z + \varepsilon x_1 + \varepsilon^2 x_2 + \cdots. \tag{3.2.20}$$

where

$$M = M_0 + \varepsilon M_1 + \cdots, \quad N = N_0 + \varepsilon N_1 + \cdots,$$

and the function x_k is a particular solution of the equation of the kth approximation. This construction method of asymptotic expansion appears to be more convenient in some cases (see Sect. 4.1).

3.2.3 Exercises

3.2.1. Consider the time and range (the length of segment OL) for the trajectory of a material point with accuracy $O(\varepsilon^2)$ for a problem of Sect. 3.2.1 (Fig. 3.2) with $f(v) \equiv 1$. Find the angle α_* for which the trajectory range in maximal.

3.2.2. Find an exact solution of system (3.2.4) for $f(v) \equiv 1$ and compare it with the asymptotic expansion (3.2.6).

3.2.3. Construct the first approximation for the solutions u and v of system (3.2.4) when the resistance force is proportional to the velocity squared $f(v) = \sqrt{u^2 + w^2}$.

Compare the numerical solution of system (3.2.4) with the zeroth and first approximations for different values of ε.

3.2.4. Find a two-term asymptotic representation of the solutions of the system of differential equations

$$t\dot{x} = x + \varepsilon y, \quad t\dot{y} = (2 - x)y$$

with initial conditions

$$x(1) = 1, \quad y(1) = e^{-1}.$$

Compare the obtained solution with the numerical solution for the system.

3.2.5. Find a two-term asymptotic representation of the solution of the differential equation

$$\ddot{x} + \omega^2 x = \varepsilon \dot{x}^2 x$$

with initial conditions

$$x(0) = 0, \quad \dot{x}(0) = v_0.$$

Compare the obtained approximate solution with the numerical solution.

3.2.6. The system of non-dimensional equations

$$\ddot{x} = -\varepsilon \dot{y} \sin \varphi, \quad \ddot{y} = \varepsilon(\dot{x} \sin \varphi - \dot{z} \cos \varphi), \quad \ddot{z} = 2 + \varepsilon \dot{y} \cos \varphi$$

with zero initial conditions describes the drop of a material point on the Earth surface from height h taking into account Coriolis forces. Here x is the vertical deviation southward, y is the vertical deviation westward, φ is the latitude ($\varphi < 0$ in the southern hemisphere). The motions ends when $z = 1$. It is assumed that

$$\varepsilon = 2\omega\sqrt{2h/g}$$

is a small parameter, where ω is the Earth angular velocity, and g is the acceleration due to gravity.

   Find the main terms of the asymptotic expansions for the functions $x(t)$, $y(t)$, and $z(t)$. Construct the exact solution of the system and compare it to the approximate solution.

3.3 Periodic Solutions

The asymptotic expansions (3.1.3) can be used for approximating the periodic solutions of some non-autonomous systems of differential equations when their right sides contain periodic functions of the independent variable. These problems are considered in Sects. 3.3.1 and 3.3.2.

For autonomous systems, where the right sides do not contain explicit functions of the independent variable, direct expansions (3.1.3) appear to be inconvenient for evaluating periodic solutions because of the so-called secular terms in the expansions. Secular terms lead to non-uniform asymptotic expansions for large values of the independent variable [50]. In this case, asymptotic expansions of a more complex form are used, as constructed in Sect. 3.3.3.

3.3.1 Solution of Non-autonomous Quasilinear Equations Without Resonance

Consider the quasilinear equation

$$\ddot{x} + \omega^2 x = h \sin t + \varepsilon \varphi(x, \dot{x}) \tag{3.3.1}$$

The generating equation

$$\ddot{x} + \omega^2 x = h \sin t$$

for $\omega \neq 1$ has the periodic solution

$$x_0 = A \sin t, \quad A = \frac{h}{\omega^2 - 1}. \tag{3.3.2}$$

We seek a periodic solution of equation (3.3.1) with a period 2π of the form

$$x = x_0 + \varepsilon x_1 + \varepsilon^2 x_2 + \cdots . \tag{3.3.3}$$

Substitute (3.3.3) into (3.3.1). Then, from

$$\varphi(x, \dot{x}) = \varphi(x_0 + \varepsilon x_1 + \cdots, \dot{x}_0 + \varepsilon \dot{x}_1 + \cdots) = \varphi_0 + \varepsilon \varphi_1 + \varepsilon^2 \varphi_2 + \cdots ,$$

where

$$\varphi_0 = \varphi(x_0, \dot{x}_0), \quad \varphi_1 = \frac{\partial \varphi_0}{\partial x_0} x_1 + \frac{\partial \varphi_0}{\partial \dot{x}_0} \dot{x}_1, \quad \ldots ,$$

one finds

$$\ddot{x}_n + \omega^2 x_n = \varphi_{n-1}(t), \quad n = 1, 2, \ldots \tag{3.3.4}$$

Expand the function $\varphi_0(t) = \varphi(A \sin t, A \cos t)$ in a Fourier series

$$\varphi_0(t) = A_{00} + \sum_{k=1}^{\infty} (A_{0k} \cos kt + B_{0k} \sin kt). \tag{3.3.5}$$

Remark For the expansion of some functions into a Fourier series it is handier to use the formulas:

$$\sin t = \frac{e^{it} - e^{-it}}{2i}, \quad \cos t = \frac{e^{it} + e^{-it}}{2}.$$

For example,

$$\sin^3 t = \frac{(e^{it} - e^{-it})^3}{(2i)^3} = -\frac{e^{3it} - 3e^{it} + 3e^{-it} - e^{-3it}}{8i} = \frac{1}{4}(3 \sin 3t - \sin 3t).$$

$$(3.3.6)$$

For $\omega \neq k$, the equation of the first approximation (3.3.4) has the periodic solution

$$x_1 = \frac{A_{00}}{\omega^2} + \sum_{k=1}^{\infty} \left(\frac{A_{0k}}{\omega^2 - k^2} \cos kt + \frac{B_{0k}}{\omega^2 - k^2} \sin kt \right).$$

The functions $x_2, x_3, \ldots$ are sequentially evaluated from equation (3.3.4):

$$x_{n+1} = \frac{A_{n0}}{\omega^2} + \sum_{k=1}^{\infty} \left(\frac{A_{nk}}{\omega^2 - k^2} \cos kt + \frac{B_{nk}}{\omega^2 - k^2} \sin kt \right), \quad n = 1, 2, \ldots,$$

where A_{nk} and B_{nk} are the Fourier coefficients of $\varphi_n(t)$.

The asymptotic expansion (3.3.3) is not uniform in the parameter ω. Uniformity is violated due to the appearance of small denominators for $|\omega^2 - k^2| = O(\varepsilon)$. The case $\omega^2 = k^2 - \varepsilon\delta$, with $\delta = O(1)$, is called *resonance* and will be considered further.

Find the first terms of the asymptotic expansion of the 2π-periodic solution (3.3.3) of Duffing equation

$$\ddot{x} + \omega^2 x + \varepsilon b x^3 = h \sin t \qquad (3.3.7)$$

in the non-resonance case.

The zeroth approximation has the form (3.3.2). Substitute (3.3.2) in the equation of the first approximation

$$\ddot{x}_1 + \omega^2 x_1 = -b x_0^3$$

and use formula (3.3.6). Since $\omega \neq k$, the differential equation

$$\ddot{x}_1 + \omega^2 x_1 = -\frac{b}{4} A^3 (3 \sin t - \sin 3t)$$

admits the periodic solution

$$x_1 = -\frac{b}{4} A^3 \left(\frac{3 \sin t}{\omega^2 - 1} - \frac{\sin 3t}{\omega^2 - 9} \right).$$

3.3.2 Solution of Non-autonomous Quasilinear Equations with Resonance

Consider the case of the main resonance $\omega^2 = 1 - \varepsilon\delta$ for Eq. (3.3.1)

$$\ddot{x} + x = h \sin t + \varepsilon\delta x + \varepsilon\varphi(x, \dot{x}). \qquad (3.3.8)$$

The generating equation

$$\ddot{x} + x = h \sin t$$

has the general solution

$$x_0 = M \cos t + N \sin t - \frac{h}{2} t \cos t,$$

which is not periodic for any value of the constants M and N. However, one can construct a 2π-periodic solution of equation (3.3.8) if $h = \varepsilon a$. Indeed, in this case, the generating equation

$$\ddot{x} + x = 0$$

has a two-parameter family of 2π-periodic solutions of the form

$$x_0 = M \cos t + N \sin t. \qquad (3.3.9)$$

For some values of the constants M and N, which are found from the existence conditions for periodic solutions of the equation of first approximation

$$\ddot{x}_1 + x_1 = a \sin t + \delta x_0 + \varphi_0(t), \qquad (3.3.10)$$

equation (3.3.8) may have periodic solutions of the form (3.3.3).

For Eq. (3.3.10) to have a 2π-periodic solution it is necessary and sufficient that the expansion of its right side in Fourier series does not contain terms of the form $A \cos t$ and/or $B \sin t$. Equating to zero the coefficients of functions $\cos t$ and $\sin t$ in the expansion of the right side of (3.3.10) in a Fourier series, we get a system of two equations for evaluating M and N:

$$P(M, N) = 0, \quad Q(M, N) = 0, \qquad (3.3.11)$$

with

$$P = M\delta + A_{01}(M, N), \quad Q = a + N\delta + B_{01}(M, N),$$

where

$$A_{01} = \frac{1}{\pi} \int_0^{2\pi} \varphi(M \cos t + N \sin t, -M \sin t + N \cos t) \cos t \, dt,$$

$$B_{01} = \frac{1}{\pi} \int_0^{2\pi} \varphi(M \cos t + N \sin t, -M \sin t + N \cos t) \sin t \, dt,$$

are the Fourier coefficients of $\varphi_0(t)$ in expansion (3.3.5).

Let the system of equations (3.3.11) have the solution $M = M_0$ and $N = N_0$. Then

$$x_0 = M_0 \cos t + N_0 \sin t,$$

and the equation for evaluating x_1,

$$\ddot{x}_1 + x_1 = A_{00} + \sum_{k=2}^{\infty} (A_{0k} \cos kt + B_{0k} \sin kt),$$

has a periodic solution

$$x_1 = M_1 \cos t + N_1 \sin t + \Phi_1(t),$$

where

$$\Phi_1 = A_{00} + \sum_{k=2}^{\infty} \frac{A_{0k} \cos kt + B_{0k} \sin kt}{1 - k^2}.$$

From the existence condition for periodic solution for the equation of second approximation,

$$\ddot{x}_2 + x_2 = \delta x_1 + \frac{\partial \varphi_0}{\partial x_0} x_1 + \frac{\partial \varphi_0}{\partial \dot{x}_0} \dot{x}_1,$$

we obtain a system of two linear algebraic equations for the constants M_1 and N_1:

$$\frac{\partial P}{\partial M} M_1 + \frac{\partial P}{\partial N} N_1 = D_{11}, \quad \frac{\partial Q}{\partial M} M_1 + \frac{\partial Q}{\partial N} N_1 = D_{12}, \qquad (3.3.12)$$

where

$$D_{11} = -\frac{1}{\pi} \int_0^{2\pi} \left(\frac{\partial \varphi_0}{\partial x_0} \Phi_1 + \frac{\partial \varphi_0}{\partial \dot{x}_0} \dot{\Phi}_1 \right) \cos t \, dt,$$

$$D_{12} = -\frac{1}{\pi} \int_0^{2\pi} \left(\frac{\partial \varphi_0}{\partial x_0} \Phi_1 + \frac{\partial \varphi_0}{\partial \dot{x}_0} \dot{\Phi}_1 \right) \sin t \, dt.$$

The partial derivatives of M and N in formulas (3.3.12) are calculated at $M = M_0$ and $N = N_0$. The system of equation (3.3.12) has a unique solution if its determinant,

$$D = \frac{\partial P}{\partial M} \frac{\partial Q}{\partial N} - \frac{\partial P}{\partial N} \frac{\partial Q}{\partial M},$$

is non-vanishing. In this case, the construction of asymptotic solutions may be continued. The functions x_k are recursively calculated by the formulas

$$x_k = M_k \cos t + N_k \sin t + \Phi_k(t), \quad k = 1, 2, \ldots,$$

and the constants M_k and N_k are found after solving the system of linear equations

$$\frac{\partial P}{\partial M} M_k + \frac{\partial P}{\partial N} N_k = D_{k1}, \quad \frac{\partial Q}{\partial M} M_k + \frac{\partial Q}{\partial N} N_k = D_{k2}, \quad k = 1, 2, \ldots$$

The peculiarity of the resonance case compared to the non-resonance case is in the excitation of the oscillations with amplitude of order 1 under the action of a periodic force with small amplitude of order ε.

Example 1

Consider the Duffing equation for the case $\omega^2 = 1 - \varepsilon\delta$, $h = \varepsilon a$, $a \neq 0$:

$$\ddot{x} + x = \varepsilon(a \sin t + \delta x - bx^3). \tag{3.3.13}$$

The zeroth approximation for the solution of equation (3.3.13) has the form (3.3.9). Substitute the expression for x_0 into the equation of first approximation

$$\ddot{x}_1 + x_1 = a \sin t + \delta x_0 - bx_0^3$$

and transform it into

$$\ddot{x}_1 + x_1 = P \cos t + Q \sin t - \frac{b}{4} \left(M^3 - 3MN^2 \right) \cos 3t + \frac{b}{4} \left(N^3 - 3M^2N \right) \sin 3t, \tag{3.3.14}$$

where

$$P = M \left[\delta - \frac{3b}{4} \left(M^2 + N^2 \right) \right], \quad Q = a + N \left[\delta - \frac{3b}{4} \left(M^2 + N^2 \right) \right].$$

Equation (3.3.14) has a periodic solution if $P = Q = 0$. Assume that $M \neq 0$. Then from the equations $P = 0$ and $Q = 0$ it follows that

$$\delta - \frac{3b}{4} \left(M^2 + N^2 \right) = 0, \quad a = 0,$$

which contradicts the condition $a \neq 0$. Therefore $M = 0$. To obtain N one gets a cubic equation

$$\frac{3b}{4} N^3 - \delta N - a = 0. \tag{3.3.15}$$

Let N_0 be a real root of equation (3.3.15). Then

$$x_0 = N_0 \sin t, \tag{3.3.16}$$

and Eq. (3.3.14) becomes

$$\ddot{x}_1 + x_1 = \frac{b}{4} N_0^3 \sin 3t.$$

The constants M_1 and N_1 in the 2π-periodic solution

$$x_1 = M_1 \cos t + N_1 \sin t - \frac{b}{32} N_0^3 \sin 3t \tag{3.3.17}$$

are found from the existence condition for periodic solutions of the second approximation equation

$$\ddot{x}_2 + x_2 = \delta x_1 - 3bx_0^2 x_1.$$

Substitute (3.3.16) and (3.3.17) in this equation, transform its right side to a linear combination of the functions $\cos kt$ and $\sin kt$, and equate to zero the coefficients of $\sin t$ and $\cos t$. Thus, we get

$$M_1 \left(\delta - \frac{3b}{4} N_0^2 \right) = 0, \quad N_1 \left(\delta - \frac{9b}{4} N_0^2 \right) = \frac{3}{128} b^2 N_0^5. \tag{3.3.18}$$

From first equation in (3.3.18) it follows that $M_1 = 0$. Indeed, if not, then from the equality

$$\delta - \frac{3b}{4} N_0^2 = 0,$$

we get $a = 0$ by (3.3.15).

If $\delta \neq 9bN_0^2/4$, then

$$N_1 = \frac{3b^2 N_0^5}{32(4\delta - 9bN_0^2)},$$

and a periodic solution of equation (3.3.13) can be represented in the form (3.3.3).

Similarly, one constructs asymptotic expansions of 2π-periodic solutions of the type $\omega^2 = k^2 - \varepsilon\delta$ for $k \neq 1$ in the resonance case (see Exercise 3.3.4).

3.3.3 Poincaré's Method

The quasilinear equation

$$\ddot{x} + \omega^2 x = \varepsilon\varphi(x, \dot{x}) \tag{3.3.19}$$

is said to be *autonomous* since its right side does not depend explicitly on time.

An application of the direct asymptotic expansion (3.3.3) for solving Eq. (3.3.19) usually leads to the appearance in this expansion of *secular terms* of type $A\varepsilon t \sin t$ and $A\varepsilon t \cos t$ (see formula (3.2.19) and the solution of Exercise 3.2.5). As a result the direct asymptotic expansion is non-uniform for large values of $t \sim \varepsilon^{-\alpha}$, $\alpha \geq 1$. The term "secular term" was introduced for the solution of celestial problems, where the value of εt often plays an important role only for values of t equal to centuries.

The appearance of secular terms comes from the fact that, as a rule, the period of the solution of equation (3.3.19) depends on the parameter ε. For example, the direct asymptotic expansion of the function

$$x = \sin[(\omega_0 + \varepsilon\omega_1)t] = \sin\omega_0 t + \varepsilon\omega_1 t \cos\omega_0 t + \cdots$$

contains the secular term $\varepsilon\omega_1 t \cos\omega_0 t$.

To get a uniform asymptotic expansion one may use Poincaré's method. Assume that the desired solution of equation (3.3.19) has period

$$T(\varepsilon) = \frac{2\pi}{\omega} g(\varepsilon), \quad g(\varepsilon) = 1 + \varepsilon g_1 + \varepsilon^2 g_2 + \cdots$$

where g_k are unknown constants. Under the change of variable:

$$t = \frac{\tau}{\omega} g(\varepsilon), \tag{3.3.20}$$

Eq. (3.3.19) becomes

$$\frac{d^2 x}{d\tau^2} + x g^2 = \varepsilon\varphi\left(x, \frac{\omega}{g}\frac{dx}{d\tau}\right)\frac{g^2}{\omega^2}. \tag{3.3.21}$$

Let $x(t)$ be the periodic solution of equation (3.3.19) with period T. Then

$$x\left[\frac{\tau + 2\pi}{\omega} g(\varepsilon)\right] = x(t + T) = x(t) = x\left[\frac{\tau}{\omega} g(\varepsilon)\right],$$

and, therefore, the function $x[\tau g(\varepsilon)/\omega]$, which is a solution of equation (3.3.21), has period 2π.

For the periodic solution of equation (3.3.21) there exists a value $\tau = \tau_0$ such that

$$\frac{dx}{d\tau} = 0, \quad \text{for } \tau = \tau_0.$$

If we change the variable $\tau' = \tau - \tau_0$ in Eq. (3.3.21), then the equality

$$\frac{dx}{d\tau'} = 0, \quad \text{for } \tau' = 0,$$

holds and the form of the equation does not change. Therefore, without loss of generality, we may assume that $\tau_0 = 0$.

We seek a solution of equation (3.3.21) in the form

$$x(\tau, \varepsilon) = x_0(\tau) + \varepsilon x_1(\tau) + \varepsilon^2 x_2(\tau) + \cdots, \tag{3.3.22}$$

where

$$x_i(\tau + 2\pi) = x_i(\tau), \quad \frac{dx_i}{d\tau} = 0, \quad \text{for } \tau = 0, \quad i = 0, 1, 2, \ldots \tag{3.3.23}$$

We substitute (3.3.22) in Eq. (3.3.21) and equate to zero the coefficients of equal powers of ε. Then, the generating equation

$$\frac{d^2 x_0}{d\tau^2} + x_0 = 0$$

has the family of solutions

$$x_0 = M \cos \tau, \tag{3.3.24}$$

satisfying conditions (3.3.23).

The first approximation equation

$$\frac{d^2 x_1}{d\tau^2} + x_1 = \frac{\varphi_0}{\omega^2} - 2g_1 x_0 \tag{3.3.25}$$

includes the 2π-periodic function

$$\varphi_0(\tau) = \varphi(M \cos \tau, -\omega M \sin \tau).$$

We expand the function φ_0/ω^2 in the Fourier series:

$$\frac{\varphi_0}{\omega^2} = A_0 + \sum_{k=1}^{\infty} (A_k \cos k\tau + B_k \sin k\tau).$$

Equation (3.3.25) has a periodic solution if

$$B_1(M) = \frac{1}{\pi\omega^2} \int_0^{2\pi} \varphi(M\cos\tau, -\omega M\sin\tau)\sin\tau\, d\tau = 0 \qquad (3.3.26)$$

and

$$A_1(M) - 2g_1 M = \frac{1}{\pi\omega^2} \int_0^{2\pi} \varphi(M\cos\tau, -\omega M\sin\tau)\cos\tau\, d\tau - 2g_1 M = 0.$$

Let $M = M_0 \neq 0$ be a solution of equation (3.3.26). Then

$$g_1 = \frac{A_1(M_0)}{2M_0},$$

and the zeroth approximation periodic solution of equation (3.3.19) has the form

$$x_0 = M_0 \cos\frac{\omega t}{1 + \varepsilon g_1}.$$

The equation of first approximation

$$\frac{d^2 x_1}{d\tau^2} + x_1 = A_0 + \sum_{k=2}^{\infty}(A_k\cos k\tau + B_k\sin k\tau)$$

admits the solution

$$x_1 = M_1\cos\tau + N_1\sin\tau + \Phi_1, \qquad (3.3.27)$$

where

$$\Phi_1 = A_0 + \sum_{k=2}^{\infty}\frac{A_k\cos k\tau + B_k\sin k\tau}{1 - k^2}.$$

From the condition

$$\frac{dx_1}{d\tau} = 0, \quad \text{for } \tau = 0$$

we obtain

$$N_1 = -\sum_{k=2}^{\infty}\frac{k A_k}{1 - k^2}.$$

To find M_1 one should consider the equation of second approximation

$$\frac{d^2 x_2}{d\tau^2} + x_2 = -2g_2 x_0 - 2g_1 x_1 + \frac{1}{\omega^2}\frac{\partial\varphi_0}{\partial x_0}x_1 + \frac{1}{\omega}\frac{\partial\varphi_0}{\partial x_{0\tau}}\frac{dx_1}{d\tau} + \psi_2, \qquad (3.3.28)$$

where

$$\psi_2 = -g_1 x_0^2 - \frac{g_1}{\omega} \frac{\partial \varphi_0}{\partial x_{0\tau}} \frac{dx_0}{d\tau} + 2g_1 \varphi_0, \quad x_{0\tau} = \frac{dx_0}{d\tau}.$$

Substitute (3.3.24) and (3.3.27) in (3.3.28), expand the right side in Fourier series and equate to zero the coefficients of $\cos \tau$ and $\sin \tau$. After these transformations, we find two equations for M_1 and g_2:

$$\frac{dB_1}{dM} M_1 = -\frac{1}{\pi} \int_0^{2\pi} \varphi_2 \sin \tau d\tau = 0,$$

$$-2g_2 M_0 + 2\frac{dg_1}{dM} M_0 M_1 = -\frac{1}{\pi} \int_0^{2\pi} \varphi_2 \cos \tau \, d\tau, \qquad (3.3.29)$$

where

$$\varphi_2 = \frac{1}{\omega^2} \frac{\partial \varphi_0}{\partial x_0} \Phi_1 + \frac{1}{\omega} \frac{\partial \varphi_0}{\partial x_{0\tau}} \frac{d\Phi_1}{d\tau} - 2g_1 \Phi_1 + \psi_2.$$

The derivatives with respect to M in (3.3.29) are calculated at $M = M_0$. If

$$\frac{dB_1}{dM} \neq 0 \quad \text{at} \quad M = M_0,$$

then, from the first equation in (3.3.29), one gets M_1, and, from the second, g_2. In this case, one can continue the construction of the periodic solution of equation (3.3.19) and find x_k recursively for $k \geq 2$.

Let φ be an analytic function. Then the obtained series is convergent for small enough ε. Hence, the quasilinear equation (3.3.19) has a periodic solution for sufficiently small ε if Eq. (3.3.26) has a simple root $M = M_0$. It follows that the number of roots of equation (3.3.26) depends on the properties of the function φ.

If the function φ does not depend on $\dot{x}$, then Eq. (3.3.26) is satisfied identically for any M, and, therefore, for sufficiently small ε, Eq. (3.3.19) has an infinite number of periodic solutions. Assume that Eq. (3.3.19) describes the motion of a material point of unit mass. Then, in this case, the law of conservation of energy is valid, i.e. the mechanical system is *conservative*.

If $\varphi \dot{x} < 0$, then $B_1 < 0$, and Eq. (3.3.26) has no solution. The corresponding mechanical system is dissipative, since its mechanical energy decreases.

If Eq. (3.3.26) has a finite number of solutions, the mechanical system if called a *self-oscillating system*.

One of the simple equations describing the motion of a self-oscillating system is the Van der Pol equation [50]:

$$\ddot{x} - \varepsilon(1 - x^2)\dot{x} + x = 0. \qquad (3.3.30)$$

Assume that $\varepsilon \ll 1$ and find an asymptotic expansion of the periodic solution for equation (3.3.30) by the Poincaré method.

After the substitution $t = \tau g(\varepsilon)$, Eq. (3.3.30) becomes:

$$\frac{d^2x}{d\tau^2} + xg^2 = \varepsilon \left(1 - x^2\right) \frac{dx}{d\tau} g. \qquad (3.3.31)$$

We seek a solution of (3.3.31) in the form (3.3.22). We substitute the solution of the generating equation (3.3.24) in the equation of first approximation

$$\frac{d^2x_1}{d\tau^2} + x_1 = \left(1 - x_0^2\right) \frac{dx_0}{d\tau} - 2g_1 x_0.$$

After these transformations one gets

$$\frac{d^2x_1}{d\tau^2} + x_1 = \left(\frac{M^2}{4} - 1\right) M \sin \tau - 2g_1 M \cos \tau + \frac{1}{4} M^3 \sin 3\tau. \qquad (3.3.32)$$

Equation (3.3.32) has a periodic solution if

$$\left(\frac{M^2}{4} - 1\right) M = 0, \quad 2g_1 M = 0.$$

The value $M = 0$ corresponds to the trivial solution $x = 0$. Assuming that $M \neq 0$ one obtains

$$M = M_0 = \pm 2, \quad g_1 = 0.$$

The values $M_0 = 2$ and $M_0 = -2$ correspond to the same periodic solution, so we analyze only the case $M_0 = 2$. In this case,

$$x_0 = 2 \cos \tau, \qquad (3.3.33)$$

and Eq. (3.3.32) becomes

$$\frac{d^2x_1}{d\tau^2} + x_1 = 2 \sin 3\tau.$$

Its solution

$$x_1 = M_1 \cos \tau + N_1 \sin \tau - \frac{1}{4} \sin 3\tau \qquad (3.3.34)$$

contains the unknown constants M_1 and N_1. From the condition

$$\frac{dx_1}{d\tau} = 0, \quad \text{for } \tau = 0$$

we find $N_1 = 3/4$.

The constants M_1 and g_2 are determined from the existence condition for a periodic solution of the equation of second approximation

$$\frac{d^2x_2}{d\tau^2} + x_2 = -2g_2x_0 + \left(1 - x_0^2\right)\frac{dx_1}{d\tau} - 2x_0\frac{dx_0}{d\tau}x_1. \qquad (3.3.35)$$

Substitute (3.3.33) and (3.3.34) in (3.3.35), represent the right side in the form of a linear combination of the functions $\sin k\tau$ and $\cos k\tau$ and equate to zero the coefficients of $\sin \tau$ and $\cos \tau$. As a result we get

$$M_1 = 0, \quad g_2 = \frac{1}{16}.$$

Therefore the two-term asymptotic expansion of the periodic solution of the Van der Pol equation (3.3.30) has the form

$$x = 2\cos\tau + \frac{\varepsilon}{4}(3\sin\tau - \sin 3\tau), \quad \tau = \frac{16t}{16 + \varepsilon^2}. \qquad (3.3.36)$$

The construction of periodic solutions by means of numerical methods appears to be essentially a more difficult problem than the solution of the Cauchy problem. That is why we limit ourselves with finding the phase curves. *The phase curve* for Eq. (3.3.1) is a trajectory of a representative point with coordinates $x(t)$ and $\dot{x}(t)$ in the phase plane $(x, \dot{x})$ when the variable t changes. Plotting a sufficiently large number of phase curves permits us to judge the qualitative behavior of the solutions of the equation under consideration for different initial conditions. The closed phase curves correspond to periodic solutions. The phase curves corresponding to periodic solutions of self-oscillating systems are called *limit cycles*.

In Fig. 3.4, we plot with dashed line the limit cycle for equation (3.3.30) found by the asymptotic formula (3.3.36) for $\varepsilon = 0.3$. The solid line is a phase curve obtained by means of the numerical integration of the Cauchy problem with the initial conditions $x = 3$, $\dot{x} = 0$. For sufficiently large values of t, the representative point traces out a curve which, within the accuracy of the plot, seems closed. This

Fig. 3.4 Limit cycle (*dashed line*), phase curve (*solid line*)

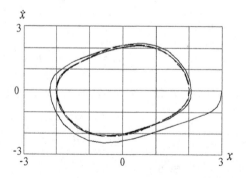

closed line may be treated as a limit cycle obtained by means of the numerical method. In practice, the limit cycles found by asymptotic and numerical methods coincide.

It is easy to verify that for other initial conditions the phase curves for Eq. (3.3.30) approach the limit cycle; this implies its asymptotic stability (see Sect. 3.4.2).

3.3.4 Exercises

3.3.1. Find the function x_2 in the expansion (3.3.3) for Eq. (3.3.7).

3.3.2. Find the asymptotic expansion of the 2π-periodic solution of the equation

$$\ddot{x} + \omega^2 x = h \sin t + \varepsilon \dot{x}^2 x$$

in the non-resonance case.

3.3.3. Find the two-term asymptotic expansion of the 2π-periodic solution of the equation

$$\ddot{x} + x = \varepsilon(a \sin t + \delta x + \dot{x}^2 x), \quad a \neq 0.$$

3.3.4. Find the main term of the asymptotic expansion of the 2π-periodic solution of the equation

$$\ddot{x} + 9x = \varepsilon(\delta x - bx^3) + h \sin t, \quad h \neq 0.$$

3.3.5. Use Poincaré's method to find a uniform asymptotic expansion of the periodic solutions of the equations

$$(1) \quad \ddot{x} + x = \varepsilon(1 - x^4)\dot{x},$$
$$(2) \quad \ddot{x} + x = \varepsilon(1 - x^2)\dot{x} - \varepsilon x^3,$$

for $\varepsilon \ll 1$. For the first equation, find the limit cycle with the obtained asymptotic formulas. Study the behavior of the phase curves near the limit cycle with a numerical method.

3.4 Transient Regimes

Periodic solutions of systems of differential equations (3.1.1) and solutions of the form $y = y_0$, where y_0 is a constant vector, are called *stationary*. Solutions which approache some stationary regime as $t \to \infty$ are called *transient regimes*. Such solutions cannot be found with Poincare's method since they are not periodic. Due to its non-uniformity, the method of direct asymptotic expansions can obtain the transient regimes only for a relatively small t interval of length $O(1)$, which, as a rule, is not so interesting for the applications.

In this section, we consider the averaging method [13] and the multiscale method [42] which can obtain uniform solutions for Cauchy problems with arbitrary initial conditions in intervals of order $O(\varepsilon^{-1})$.

3.4.1 Van der Pol Method

Approximate solutions of the equation

$$\ddot{x} + \omega^2 x = \varepsilon\varphi(x, \dot{x}). \tag{3.4.1}$$

can be obtained by a method proposed by the Dutch engineer Van der Pol. We seek a solution of equation (3.4.1) in the form

$$x = a(t)\cos z, \quad z = \omega t + \beta(t). \tag{3.4.2}$$

Then

$$\dot{x} = -a\omega \sin z + \dot{a}\cos z - a\dot{\beta}\sin z.$$

Solution (3.4.2) contains two unknown functions $a(t)$ and $z(t)$. One of the conditions to determine them is that solution (3.4.2) must satisfy Eq.(3.4.1). The second condition is our choice. We require that

$$\dot{a}\cos z - a\dot{\beta}\sin z = 0. \tag{3.4.3}$$

In this case,

$$\dot{x} = -a\omega \sin z, \quad \ddot{x} = -a\omega^2 \cos z - \dot{a}\omega \sin z - a\omega\dot{\beta}\cos z. \tag{3.4.4}$$

Substituting (3.4.2) and (3.4.4) into (3.4.1) one gets

$$-\dot{a}\omega \sin z - a\omega\dot{\beta}\cos z = \varepsilon\varphi(a\cos z, -a\omega \sin z). \tag{3.4.5}$$

Formulas (3.4.3) and (3.4.5) imply that

$$\dot{a} = -\frac{\varepsilon}{\omega}\varphi(a\cos z, -a\omega \sin z)\sin z,$$

$$\dot{\beta} = -\frac{\varepsilon}{\omega a}\varphi(a\cos z, -a\omega \sin z)\cos z. \tag{3.4.6}$$

The change of variables (3.4.2) and (3.4.4) allows us to reduce Eq.(3.4.1) to the equivalent system of two equation (3.4.6). The functions $a(t)$ and $\beta(t)$ are called *the slow variables* ($\dot{a} \sim \dot{\beta} \sim \varepsilon$), the function $z(t)$ *the fast variable* ($\dot{z} \sim 1$), and the above transformation *separation of slow and fast variables*.

Replace the right sides of system (3.4.6) with their average values over the z interval $[0, 2\pi]$. We thus obtain *the shortened system*

$$\dot{a} = -\frac{\varepsilon}{\omega} B(a), \quad \dot{\beta} = -\frac{\varepsilon}{\omega a} A(a), \tag{3.4.7}$$

where

$$A(a) = \frac{1}{2\pi} \int_0^{2\pi} \varphi(a \cos z, -\omega a \sin z) \cos z \, dz,$$

$$B(a) = \frac{1}{2\pi} \int_0^{2\pi} \varphi(a \cos z, -\omega a \sin z) \sin z \, dz,$$

for which the solution can be found by quadratures. The transition from (3.4.6) to (3.4.7) is called *the averaging*.

The approximate solution (3.4.2), where a and β are determined from system (3.4.7), is the main term of the asymptotic expansion of the solution of equation (3.4.1) in the t interval of length $O(1/\varepsilon)$.

The Van der Pol method is also convenient for determining the zeroth approximation of the solution. However, it does not permit to find the next approximations. To construct them one may use the multiscale method (see Sect. 3.4.3).

Apply the Van der Pol method to approximate the solution of the equation

$$\ddot{x} + \Lambda^2 x = \varepsilon[\varphi(x, \dot{x}) + h \sin \omega t], \tag{3.4.8}$$

where the right side contains a periodic function of t.

We seek a solution of the form (3.4.2). By (3.4.3) we get the system

$$\dot{a} \cos z - a\dot{\beta} \sin z = 0,$$
$$-\dot{a}\omega \sin z - a\omega\dot{\beta} \cos z = a(\omega^2 - \Lambda^2) \cos z + \varepsilon\varphi + \varepsilon h \sin \omega t \tag{3.4.9}$$

to find $a(t)$ and $\beta(t)$.

The transition to the shortened system is justified if $\omega^2 - \Lambda^2 = O(\varepsilon)$ which corresponds to the case of the main resonance (see Sect. 3.3.2). Taking

$$\sin \omega t = \sin(z - \beta) = \sin z \cos \beta - \sin \beta \cos z$$

into account, after averaging (3.4.9) with respect to the variable z, one finds

$$\dot{a}\omega = -\varepsilon \left[B(a) + \frac{h}{2} \cos \beta \right],$$

$$a\omega\dot{\beta} = -\varepsilon \left[A(a) - \frac{h}{2} \sin \beta \right] - \frac{a}{2} \left(\omega^2 - \Lambda^2 \right). \tag{3.4.10}$$

In the case $\dot{a} = \dot{\beta} = 0$, the solution (3.4.2) is periodic

$$x(t) = a_0 \cos(\omega t + \beta_0), \qquad (3.4.11)$$

where the constants a_0 and β_0 are evaluated from the system of equations

$$B(a_0) + \frac{h}{2} \cos \beta_0 = 0,$$
$$A(a_0) - \frac{h}{2} \sin \beta_0 + \frac{a_0}{2\varepsilon} \left(\omega^2 - \Lambda^2 \right) = 0. \qquad (3.4.12)$$

Solution (3.4.11) with accuracy of order $O(\varepsilon)$ coincides with the approximate solution (3.3.9).

If we exclude β_0 from system (3.4.12) then we have the equation

$$F(a_0, \omega) = B^2 + \left[A + \frac{a_0}{2\varepsilon} (\omega^2 - \Lambda^2) \right]^2 - \frac{h^2}{4} = 0, \qquad (3.4.13)$$

which connects the amplitude of the periodic solution, a_0, with its frequency ω. The dependence of a_0 on ω is called the *amplitude frequency response*.

Consider the equation

$$\ddot{x} + x = \varepsilon \left(h \sin \omega t - n\dot{x} - bx^3 \right), \qquad (3.4.14)$$

which, when compared to Eq. (3.3.13), contains in the right side the additional term $-\varepsilon n\dot{x}$ emerging, in particular, in problems where the resistance forces are proportional to velocity. The functions A and B for Eq. (3.4.14) are obtained by the formulas

$$A(a) = \frac{1}{2\pi} \int_0^{2\pi} \left(na\omega \sin z - ba^3 \cos^3 z \right) \cos z \, dz = -\frac{3}{8} ba^3,$$

$$B(a) = \frac{1}{2\pi} \int_0^{2\pi} \left(na\omega \sin z - ba^3 \cos^3 z \right) \sin z \, dz = \frac{na\omega}{2}.$$

The equation for the amplitude frequency response has the form

$$n^2 a_0^2 \omega^2 + \left[\frac{a_0}{\varepsilon} (\omega^2 - 1) - \frac{3}{4} ba_0^3 \right]^2 - h^2 = 0, \quad a_0 > 0.$$

In Fig. 3.5, the amplitude frequency responses for Eq. (3.4.14) are plotted for different values of b for $h = n = 1$ and $\varepsilon = 0.02$.

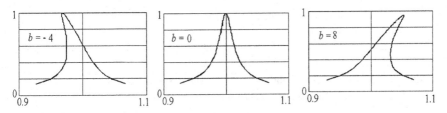

Fig. 3.5 Amplitude frequency responses for Eq. (3.4.14)

The case $b = 0$ corresponds to a linear equation. If $b \neq 0$ then, for some values of ω, Eq. (3.4.14) can have more than one periodic solutions.

3.4.2 Stability of Stationary Solutions

In real systems which are modeled with differential equations, there usually exist small random perturbations. In many cases, these perturbations may be taken into account when making small changes in the initial conditions. If small perturbations in the initial conditions lead to significant perturbations of solutions in the considered time interval, such unstable solutions are a bad description of the real process. In this case, one should study the behavior of the solutions with initial conditions close to the initial conditions of the solution under consideration, i.e. to study the solution stability.

Let system (3.1.1) have a solution $y(t)$ satisfying the initial conditions $y(0) = y_0$. Denote $\eta = \tilde{y} - y$, $\eta_0 = \tilde{y}_0 - y_0$, where $\tilde{y}$ is any other solution satisfying the initial conditions $\tilde{y}_0(0) = \tilde{y}_0$. The solution $y(t)$ is said to be *stable in the sense of Lyapunov*, if, for any $\delta > 0$, one can find $\delta_0 > 0$ such that, if $|\eta_0| < \delta_0$, then, for any values of t, the inequality $|\eta| < \delta$ also holds. A stable solution is called *asymptotically stable* if $|\eta| \to 0$ as $t \to \infty$.

The study of stability of solutions of equations (3.4.1) and (3.4.8) is a rather difficult problem. In this section we limit ourselves to study the stability of solutions of shortened systems (3.4.7) and (3.4.10). Generally, from the stability of a solution for a shortened equation it does not follow that the corresponding solution of the given equation is stable in the sense of Lyapunov. Nevertheless, information on the stability of solutions of shortened systems can be useful when solving many applied problems.

Consider the first of equations of system (3.4.7)

$$\omega \dot{a} = -\varepsilon B(a). \tag{3.4.15}$$

To study the stability of the stationary solution $a = a_0$ satisfying equation

$$B(a_0) = 0, \tag{3.4.16}$$

one considers the solution of equation (3.4.15)

$$a = a_0 + \xi. \tag{3.4.17}$$

Substitute (3.4.17) in (3.4.15) and expand the right side of the obtained identity in a series of powers of ξ. Taking (3.4.16) into account, we find

$$\omega\dot{\xi} = -\varepsilon B_0\xi + \cdots, \quad B_0 = \left.\frac{dB}{da}\right|_{a=a_0}. \tag{3.4.18}$$

By the theorem on the stability of a linear approximation, the trivial solution of equation (3.4.18) is asymptotically stable if the trivial solution of the linear equation

$$\omega\dot{\xi} = -\varepsilon B_0\xi \tag{3.4.19}$$

is asymptotically stable. The solution of equation (3.4.19) has the form

$$\xi = C\exp\left(-\frac{\varepsilon}{\omega}B_0 t\right).$$

Obviously, $\xi \to 0$ as $t \to \infty$ if $B_0 > 0$. Therefore, the stationary solution $a = a_0$ is asymptotically stable if

$$\left.\frac{dB}{da}\right|_{a=a_0} > 0. \tag{3.4.20}$$

For Eq. (3.3.30),

$$B(a) = \frac{a}{2}\left(\frac{a^2}{4} - 1\right).$$

The equation $B(a) = 0$ has the roots $a_0 = 0$ and $a_0 = 2$. From the inequalities

$$\left.\frac{dB}{da}\right|_{a=0} = -\frac{1}{2} < 0, \quad \left.\frac{dB}{da}\right|_{a=2} = 1 > 0,$$

it follows that the trivial solution $a_0 = 0$ is unstable and the limit cycle $a_0 = 2$ is stable.

Consider now the stability of the stationary solutions of the shortened system (3.4.10). Write this system in the form

$$\omega\dot{a} = X(a, \beta), \quad a\omega\dot{\beta} = Y(a, \beta), \tag{3.4.21}$$

where

$$X = -\varepsilon\left[B(a) + \frac{h}{2}\cos\beta\right], \quad Y = -\varepsilon\left[A(a) - \frac{h}{2}\sin\beta\right] - \frac{a}{2}(\omega^2 - \Lambda^2).$$

The stationary solution $a = a_0$, $\beta = \beta_0$ of system (3.4.21) is a solution of the system of equations

$$X(a_0, \beta_0) = 0, \quad Y(a_0, \beta_0) = 0. \tag{3.4.22}$$

Consider the solution of system (3.4.21)

$$a = a_0 + \xi, \quad \beta = \beta_0 + \eta. \tag{3.4.23}$$

Substitute (3.4.23) in (3.4.21) and expand the right sides of the obtained system in series of powers of ξ and η. Taking (3.4.22) into account, one gets

$$\omega\dot{\xi} = \frac{\partial X}{\partial a}\xi + \frac{\partial X}{\partial \beta}\eta + \cdots, \quad a_0\omega\dot{\eta} = \frac{\partial Y}{\partial a}\xi + \frac{\partial Y}{\partial \beta}\eta + \cdots. \tag{3.4.24}$$

The partial derivatives in (3.4.24) are calculated at $a = a_0$, $\beta = \beta_0$. From the theorem on the stability of linear approximation, it follows that the trivial solution of system (3.4.24) is asymptotically stable if the trivial solution of the linear system is also stable, which is obtained from (3.4.24) when omitting the non-linear terms. Hence, the trivial solution of the linear system is asymptotically stable if the roots Λ_1 and Λ_2 of the characteristic equation

$$\left(\frac{1}{\omega}\frac{\partial X}{\partial a} - \Lambda\right)\left(\frac{1}{a_0\omega}\frac{\partial Y}{\partial \beta} - \Lambda\right) - \frac{1}{a_0\omega^2}\frac{\partial X}{\partial \beta}\frac{\partial Y}{\partial a} = 0 \tag{3.4.25}$$

have negative real parts. The last condition is satisfies if the coefficients of the quadratic equation (3.4.25) are positive, i.e.

$$-\frac{1}{a_0\omega}\left(a_0\frac{\partial X}{\partial a} + \frac{\partial Y}{\partial \beta}\right) > 0, \quad \frac{1}{a_0\omega^2}\left(\frac{\partial X}{\partial a}\frac{\partial Y}{\partial \beta} - \frac{\partial X}{\partial \beta}\frac{\partial Y}{\partial a}\right) > 0. \tag{3.4.26}$$

From the relations

$$\frac{\partial X}{\partial a} = -\varepsilon\frac{dB}{da}\bigg|_{a=a_0}, \quad \frac{\partial Y}{\partial \beta} = \frac{\varepsilon h}{2}\cos\beta_0 = -\varepsilon B(a_0),$$

$$\frac{\partial Y}{\partial a} = -\frac{d}{da}\left[\varepsilon A(a) + \frac{a}{2}(\omega^2 - \Lambda^2)\right], \quad \frac{\partial X}{\partial \beta} = \frac{\varepsilon h}{2}\sin\beta_0 = -\varepsilon A(a_0) + \frac{a}{2}(\omega^2 - \Lambda^2)$$

it follows that inequalities (3.4.26) are equivalent to the inequalities

$$\frac{d(aB)}{da}\bigg|_{a=a_0} > 0, \quad \frac{dF}{da}\bigg|_{a=a_0} > 0, \tag{3.4.27}$$

where the function $F(\omega, a)$ is evaluated by formula (3.4.13).

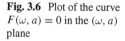

Fig. 3.6 Plot of the curve $F(\omega, a) = 0$ in the (ω, a) plane

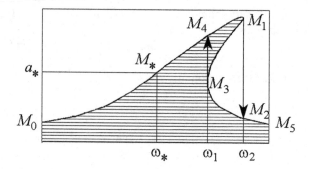

If inequalities (3.4.27) are satisfied, then the stationary solution $a = a_0$, $\beta = \beta_0$ of the shortened system (3.4.21) is asymptotically stable.

For Eq. (3.4.14) the function $B(a) = na\omega/2$, and the first inequality in (3.4.27) imply that $na_0\omega > 0$. This inequality is satisfied for all stationary solutions if $n > 0$. To check the second inequality in (3.4.27) it is convenient to plot the curve $F(\omega, a) = 0$ in the (ω, a) plane. Consider the case $b > 0$ (Fig. 3.6).

The inequality $F(\omega, a) < 0$ is satisfied in the shaded area under the curve and $F(\omega, a) > 0$ above the curve. Consider the point M_* with coordinates (a_*, ω_*) that lies on the curve above the shaded area. Obviously, $F < 0$ for $a_* < 0$ and $F > 0$ for $a_* > 0$. This is why

$$\frac{dF}{da} > 0 \quad \text{for } a = a_*, \quad \omega = \omega_*,$$

and the solution corresponding to this point is asymptotically stable. The unstable solutions correspond to the points of the curve $F = 0$ which lie under the shaded area. Thus there is one stable periodic solution in both intervals $(0, \omega_1)$ and (ω_2, ∞) and, in the interval (ω_1, ω_2), there are three periodic solutions two of which are stable.

The character of vibrations depends on the way ω changes. If ω increases, then the amplitude changes according to the rule given by the broken line $M_0M_1M_2M_5$. For $\omega = \omega_2$, the amplitude sharply goes down. This phenomenon is called *the oscillation drop*. When the frequency decreases, the law of variation of amplitude is given by the broken line $M_5M_3M_4M_0$. In this case an oscillation jump occurs when $\omega = \omega_1$.

3.4.3 Multiscale Method

Introduce the variables

$$\tau = t, \quad \tau_1 = \varepsilon t, \quad \tau_2 = \varepsilon^2 t, \dots$$

and seek a solution of equation (3.4.1) in the form of an asymptotic series

$$x = x_0(\tau, \tau_1, \tau_2, \ldots) + \varepsilon x_1(\tau, \tau_1, \tau_2, \ldots) + \cdots . \qquad (3.4.28)$$

Applying the chain rule one obtains

$$\frac{dx}{dt} = \frac{\partial x}{\partial \tau} + \varepsilon \frac{\partial x}{\partial \tau_1} + \varepsilon^2 \frac{\partial x}{\partial \tau_2} + \cdots ,$$

$$\frac{d^2x}{dt^2} = \frac{\partial^2 x}{\partial \tau^2} + 2\varepsilon \frac{\partial^2 x}{\partial \tau \partial \tau_1} + \varepsilon^2 \frac{\partial^2 x}{\partial \tau_1^2} + 2\varepsilon^2 \frac{\partial^2 x}{\partial \tau \partial \tau_2} + \cdots . \qquad (3.4.29)$$

After substituting (3.4.28) into (3.4.1), taking (3.4.29) into account, and equating the coefficients of equal powers of ε on the right and left sides of (3.4.1), we find the equation for the k-approximation

The equation for the zeroth approximation

$$\frac{\partial^2 x_0}{\partial \tau^2} + \omega^2 x_0 = 0$$

has the solution

$$x_0 = a \cos z, \qquad (3.4.30)$$

where

$$a = a(\tau_1, \tau_2, \ldots), \quad z = \omega \tau + \beta(\tau_1, \tau_2, \ldots).$$

Substitution of (3.4.30) into the equation for the first approximation

$$\frac{\partial^2 x_1}{\partial \tau^2} + \omega^2 x_1 = -2 \frac{\partial^2 x_0}{\partial \tau \partial \tau_1} + \varphi \left(x_0, \frac{\partial x_0}{\partial \tau} \right)$$

leads to the equality

$$\frac{\partial^2 x_1}{\partial \tau^2} + \omega^2 x_1 = 2\omega \frac{\partial a}{\partial \tau_1} \sin z + 2a\omega \frac{\partial \beta}{\partial \tau_1} \cos z + \varphi(a \cos z, -a\omega \sin z). \quad (3.4.31)$$

The periodic function $\varphi(a \cos z, -a\omega \sin z)$ is expanded in a Fourier series,

$$\varphi = A_0 + \sum_{k=1}^{\infty} (A_k \cos kz + B_k \sin kz).$$

For the asymptotic expansion (3.4.28) to have no secular terms it is necessary that the coefficients of $\sin z$ and $\cos z$ in the right side of equation (3.4.31) be equal to zero. This gives

$$2\omega \frac{\partial a}{\partial \tau_1} + B_1(a) = 0, \quad 2a\omega \frac{\partial \beta}{\partial \tau_1} + A_1(a) = 0, \tag{3.4.32}$$

where

$$A_1(a) = \frac{1}{\pi} \int_0^{2\pi} \varphi(a\cos z, -\omega a \sin z) \cos z \, dz,$$

$$B_1(a) = \frac{1}{\pi} \int_0^{2\pi} \varphi(a\cos z, -\omega a \sin z) \sin z \, dz.$$

In the zeroth approximation one should assume that $a = a(\tau_1)$ and $\beta = \beta(\tau_1)$. Indeed,

$$a(\tau_1, \tau_2, \ldots) = a(\tau_1, 0, \ldots) + \frac{\partial a}{\partial \tau_2} \tau_2 + \cdots = a(\tau_1) + \frac{\partial a}{\partial \tau_2} \varepsilon^2 t + \cdots$$

and for $t \sim 1/\varepsilon$ the error in replacing $a(\tau_1, \tau_2, \ldots)$ with $a(\tau_1)$ is of order $O(\varepsilon)$. Hence, the system of zeroth approximation (3.4.32) coincides with the shortened system (3.4.7) obtained by means of the averaging method. The approximate formula for the solution of equation (3.4.1) has the form

$$x = a(\varepsilon t) \cos[\omega t + \beta(\varepsilon t)] + O(\varepsilon).$$

Now we come to the construction of the first approximation. To find solutions with accuracy of order $O(\varepsilon^2)$ one should assume that $a = a(\tau_1, \tau_2)$ and $\beta = \beta(\tau_1, \tau_2)$. Considering τ_2 as a parameter, rewrite (3.4.32) as

$$d\tau_1 = -2\omega \frac{da}{B_1(a)}, \quad \frac{d\beta}{d\tau_1} = -\frac{A_1(a)}{2\omega a}.$$

After integration we get

$$\tau_1 = -2\omega \int \frac{da}{B_1(a)} + c_1(\tau_2), \quad \beta = -\frac{1}{2\omega} \int \frac{A_1(a)}{a} d\tau_1 + c_2(\tau_2). \tag{3.4.33}$$

Equating to zero the coefficients of $\sin z$ and $\cos z$ in the right side of the equation of second approximation we find two equations for the evaluation of the functions $c_1(\tau_2)$ and $c_2(\tau_2)$. In more detail, we consider the problem of constructing the first approximation for the Van der Pol equation

$$\ddot{x} - \varepsilon(1 - x^2)\dot{x} + x = 0.$$

Substituting the zeroth approximation for the solution of this equation

$$x_0 = a(\tau_1, \tau_2) \cos[\tau + \beta(\tau_1, \tau_2)] \tag{3.4.34}$$

into the equation of the first approximation

$$\frac{\partial^2 x_1}{\partial \tau^2} + x_1 = -2 \frac{\partial^2 x_0}{\partial \tau \partial \tau_1} + \left(1 - x_0^2\right) \frac{\partial x_0}{\partial \tau},$$

we obtain, after transformations,

$$\frac{\partial^2 x_1}{\partial \tau^2} + \omega^2 x_1 = 2 \frac{\partial a}{\partial \tau_1} \sin z + 2a \frac{\partial \beta}{\partial \tau_1} \cos z + a \left(\frac{a^2}{4} - 1\right) \sin z + \frac{a^3}{4} \sin 3z.$$

The equations for a and β are

$$\frac{\partial a}{\partial \tau_1} = \frac{a}{2} \left(1 - \frac{a^2}{4}\right), \quad \frac{\partial \beta}{\partial \tau_1} = 0. \tag{3.4.35}$$

The solutions of equations (3.4.35),

$$a = \frac{2}{\sqrt{1 + c(\tau_2)e^{-\tau_1}}}, \quad \beta = \beta(\tau_2),$$

are represented in terms of the unknown functions $c(\tau_2)$ and $\beta(\tau_2)$. To construct a solution with accuracy of order $O(\varepsilon)$ we assume that $c(\tau_2) = c_0$, $\beta(\tau_2) = \beta_0$,

$$x = \frac{2 \cos(t + \beta_0)}{\sqrt{1 + c_0 e^{-\varepsilon t}}} + O(\varepsilon). \tag{3.4.36}$$

For $c_0 \neq 0$, the solution (3.4.36) is a transient regime approaching a periodic solution with amplitude $a = 2$ as $t \to \infty$.

From equalities (3.4.35) it follows that

$$\frac{\partial^2 x_1}{\partial \tau^2} + x_1 = \frac{a^3}{4} \sin 3z. \tag{3.4.37}$$

Assume that the arbitrary constants in solution (3.4.28) are functions of ε (see Sect. 3.3.2). Then x_1 is a particular solution of equation (3.4.37) and, therefore,

$$x_1 = -\frac{a^3}{32} \sin 3z. \tag{3.4.38}$$

Substitute (3.4.34) and (3.4.38) in the equation of second approximation

$$\frac{\partial^2 x_2}{\partial \tau^2}+x_2 = -2\frac{\partial^2 x_1}{\partial\tau\partial\tau_1}-2\frac{\partial^2 x_0}{\partial\tau\partial\tau_2}-\frac{\partial^2 x_0}{\partial\tau_1^2}+\left(1-x_0^2\right)\left(\frac{\partial x_1}{\partial\tau}+\frac{\partial x_0}{\partial\tau_1}\right)-2x_0\frac{\partial x_0}{\partial\tau}x_1,$$

represent the right side in the form of a linear combination of the functions $\sin kz$ and $\cos kz$ and equate to zero the coefficients of $\sin z$ and $\cos z$. Thus, we obtain

$$\frac{\partial a}{\partial \tau_2}=0,\quad 2a\frac{d\beta}{d\tau_2}+\frac{a^5}{128}-\frac{\partial^2 a}{\partial\tau_1^2}+\frac{\partial a}{\partial\tau_1}\left(1-\frac{3a^2}{4}\right)=0. \qquad (3.4.39)$$

From the first equality in (3.4.39) it follows that $c(\tau_2) = c_0$, and from the second equality in (3.4.39) with the help of formula (3.4.35) we get the following equation for $\beta(\tau_2)$:

$$\frac{d\beta}{d\tau_2}=\frac{1}{8}\left(-1+a^2-\frac{7a^4}{32}\right).$$

The multiscale method is a generalization of Poincaré's method. For periodic solutions the methods give similar results. In the case $c_0 = 0$ the multiscale methods gives the asymptotic expansion

$$x\simeq 2\cos z-\frac{\varepsilon}{4}\sin 3z,\quad z=t-\frac{\varepsilon^2 t}{16}+\beta_0$$

for a periodic solution, which for $\beta_0 = -3\varepsilon/8$ coincides with expansion (3.3.36) obtained with the Poincaré method with accuracy of order $O(\varepsilon^2)$.

3.4.4 Exercises

3.4.1. Find by the Van der Pol method the approximate solution of equation

$$\ddot{x}+x=\varepsilon\left(1-x^2\right)\dot{x}-\varepsilon bx^3,\quad \varepsilon\ll 1.$$

3.4.2. Study the stability of the stationary solutions of the equation

$$\ddot{x}+x=\varepsilon(1-x^4)\dot{x},\quad \varepsilon\ll 1.$$

3.4.3. Write the shortened system (3.4.10) for the equation

$$\ddot{x}+x=\varepsilon\left(h\sin\omega t+\dot{x}-\dot{x}^3/3\right).$$

3.4.4. Obtain the amplitude frequency response for the equation

$$\ddot{x} + x = \varepsilon \left(h \sin \omega t + 8x\dot{x}^2 - n\dot{x} \right).$$

3.4.5. Find the zeroth and first approximations for the solution of the equations of Exercise 3.4.1 for $b = 1$ and Exercise 3.4.2 by means of the multiscale method.

3.5 Boundary Value Problems

We limit ourselves to boundary value problems for linear ordinary differential equations. Non-homogeneous regularly perturbed boundary value problems are considered in Sect. 3.5.1. Solutions of boundary value problems are discussed in Sect. 3.5.2. In both cases, we use direct asymptotic expansions. In Sect. 3.5.3, boundary value problems describing media with periodic structure are solved by the multiple scale method.

3.5.1 Non-homogeneous Boundary Value Problems

Consider the linear differential equation of the nth order

$$Ly = f(x), \tag{3.5.1}$$

with the boundary conditions

$$G_i y\big|_{x=0} = g_i, \quad i = 1, 2, \ldots n_1, \quad H_j y\big|_{x=l} = h_j, \quad j = 1, 2, \ldots n_2, \tag{3.5.2}$$

where $n_1 + n_2 = n$ and

$$L = \sum_{k=0}^{n} a_k(x, \varepsilon) \frac{d^k}{dx^k}, \quad a_n = 1,$$

$$G_i = \sum_{k=0}^{n-1} b_{ik} \frac{d^k}{dx^k}, \quad H_j = \sum_{k=0}^{n-1} c_{jk} \frac{d^k}{dx^k}. \tag{3.5.3}$$

If

$$a_k(x, \varepsilon) \simeq \sum_{m=0}^{\infty} a_{km}(x)\varepsilon^m, \quad k = 0, 1, \ldots, n - 1, \tag{3.5.4}$$

then

$$L = \sum_{m=0}^{\infty} L_m \varepsilon^m, \quad L_m = \sum_{k=0}^{n} a_{km} \frac{d^k}{dx^k},$$

and the solution of the boundary value problem (3.5.1)–(3.5.2) can be searched in the form of a direct asymptotic expansion

$$y(x, \varepsilon) \simeq \sum_{m=0}^{\infty} y_m(x) \varepsilon^m. \tag{3.5.5}$$

The function $y_0(x)$ is a solution of the generating equation

$$L_0 y = f(x) \tag{3.5.6}$$

satisfying the boundary conditions (3.5.2). For $m \geq 1$, the function y_m satisfies the mth approximation equation

$$L_0 y_m = -\sum_{k=1}^{m} L_k y_{m-k}$$

and the homogeneous boundary conditions (3.5.2)

$$G_i y\Big|_{x=0} = H_j y\Big|_{x=l} = 0.$$

Example 1

Use the asymptotic expansions (3.5.5) to find the longitudinal displacements, $u(x)$, of a bar of length l under force P applied to the bar end (see Fig. 3.7). Assume that the cross-sectional area of the bar is given by the formula

$$S = S_0(1 - \varepsilon x/l), \quad 0 \leq x \leq l, \quad \varepsilon \ll 1. \tag{3.5.7}$$

The function $u(x)$ satisfies the equation

$$\frac{d}{dx}\left(ES\frac{du}{dx}\right) = 0 \tag{3.5.8}$$

Fig. 3.7 Bar $u(x)$ of length l
under force P

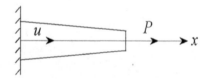

and the boundary conditions are

$$u = 0 \quad \text{for} \quad x = 0, \quad ES\frac{du}{dx} = P \quad \text{for} \quad x = l. \qquad (3.5.9)$$

where E is Young's modulus.

From Eq. (3.5.8) and the second boundary condition in (3.5.9) it follows that

$$ES\frac{du}{dx} = P, \quad 0 \le x \le l. \qquad (3.5.10)$$

We seek a solution of equation (3.5.10) in the form

$$u = u_0 + \varepsilon u_1 + \cdots$$

From (3.5.7) and (3.5.9) we get

$$ES_0\frac{du_0}{dx} = P, \quad u_0 = \frac{Px}{ES_0}.$$

The function u_0 describes the longitudinal displacements of the bar with cross-section of constant area S_0.

The equation of first approximation

$$\frac{du_1}{dx} = \frac{x}{l}\frac{du_0}{dx}$$

with boundary condition $u_1 = 0$ at $x = 0$ has the solution

$$u_1 = \frac{Px^2}{2lES_0}.$$

Therefore

$$u(x) \simeq \frac{Px}{ES_0}\left(1 + \frac{\varepsilon x}{2l}\right).$$

The same result can be obtained if we expand the exact solution for this problem in a power series of the small parameter ε,

$$u = -\frac{Pl}{ES_0\varepsilon}\ln(1 - \varepsilon x/l).$$

Example 2

Consider the bending of a beam of length l under the action of a uniformly distributed load q (Fig. 3.8).

Fig. 3.8 Bending of beam of
length l under a uniformly
distributed load q

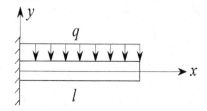

The equation for the deflection of a beam, $w(x)$, has the form

$$\frac{d^2}{dx^2}\left(EJ\frac{d^2w}{dx^2}\right) = -q, \tag{3.5.11}$$

where E is Young's modulus and J is the moment of inertia of the cross-section.

Assume that the left end of the beam, $x = 0$, is clamped and the right end, $x = l$, is free. Then

$$w = \frac{dw}{dx} = 0 \quad \text{for} \quad x = 0, \qquad \frac{d^2w}{dx^2} = \frac{d^3w}{dx^3} = 0 \quad \text{for} \quad x = l. \tag{3.5.12}$$

Let the beam have a rectangular cross-section of constant height b_0 but with variable width a according to the linear law: $a = a_0(1 - \varepsilon x/l)$. In this case, the moment of inertia is

$$J = \frac{ab_0^3}{12} = J_0\left(1 - \frac{\varepsilon x}{l}\right), \qquad J_0 = \frac{a_0b_0^3}{12}. \tag{3.5.13}$$

We seek a solution of equation (3.5.11) in the form

$$w = w_0 + \varepsilon w_1 + \cdots. \tag{3.5.14}$$

The function w_0 satisfies the equation

$$EJ_0\frac{d^4w}{dx^4} = -q,$$

Its general solution,

$$w_0 = -\frac{q}{EJ_0}\frac{x^4}{4!} + A\frac{x^3}{3!} + B\frac{x^2}{2!} + Cx + D, \tag{3.5.15}$$

contains four arbitrary constants A, B, C, and D. After evaluation of these constants with the help of the boundary conditions (3.5.12), we find

$$w_0 = -\frac{qx^2}{24EJ_0}\left(x^2 - 4xl + 6l^2\right).$$

The function

$$w_1 = -\frac{qx^3}{120 E J_0 l}\left(3x^2 - 10xl + 10l^2\right)$$

is obtained from the solution of the equation of first approximation

$$\frac{d^4 w_1}{dx^4} = \frac{d^2}{dx^2}\left(\frac{x}{l}\frac{d^2 w_0}{dx^2}\right) \tag{3.5.16}$$

with the boundary conditions (3.5.12).

3.5.2 Eigenvalue Problems

Consider the homogeneous linear differential equation

$$Ly = \Lambda M_0 y \tag{3.5.17}$$

with the homogeneous boundary conditions

$$G_i y\big|_{x=0} = 0, \quad i = 1, 2, \ldots n_1, \quad H_j y\big|_{x=l} = 0, \quad j = 1, 2, \ldots n_2. \tag{3.5.18}$$

Here Λ is the required parameter, the operators L, G and H are defined by formulas (3.5.3), and

$$M_0 = \sum_{k=0}^{m} b_k(x)\frac{d^k}{dx^k}, \quad m < n.$$

Assume that expansions (3.5.4) hold. Then the solution of the eigenvalue problem (3.5.17)–(3.5.18) can be represented in the form

$$y(x, \varepsilon) \simeq \sum_{m=0}^{\infty} y_m(x)\varepsilon^m, \quad \Lambda \simeq \sum_{m=0}^{\infty} \Lambda_m \varepsilon^m. \tag{3.5.19}$$

For some Λ_0, let the eigenvalue problem of the zeroth approximation,

$$L_0 y = \Lambda_0 M_0 y, \quad G_i y\big|_{x=0} = H_j y\big|_{x=l} = 0, \tag{3.5.20}$$

have nontrivial solution y_0.

In the first approximation we get

$$L_0 y_1 - \Lambda_0 M_0 y_1 = -L_1 y_0 + \Lambda_1 M_0 y_0. \tag{3.5.21}$$

Denote (y, z) the scalar product of the functions y and z and assume that for any y and z satisfying the boundary conditions (3.5.18) the operators L_0 and M_0 are self-adjoint, i.e.

$$(L_0 y, z) = (y, L_0 z), \quad (M_0 y, z) = (y, M_0 z).$$

Scalar multiply Eq. (3.5.21) by y_0. Due to the self-adjointness of L_0 and M_0 we get

$$(y_1, L_0 y_0 - \Lambda_0 M_0 y_0) = -(L_1 y_0, y_0) + \Lambda_1 (M_0 y_0, y_0).$$

From (3.5.20) it follows that

$$\Lambda_1 = \frac{(L_1 y_0, y_0)}{(M_0 y_0, y_0)}.$$

This last equality is a solvability condition for the eigenvalue problem of first approximation.

In a similar manner, Λ_k and y_k may be found for $k > 1$.

Example 3

Use the asymptotic method to obtain the approximate values of the frequencies and modes of the longitudinal vibrations of a bar of length l and cross-sectional area which changes according to (3.5.7).

The equation of the longitudinal vibrations of a bar has the form

$$\frac{\partial}{\partial x}\left(ES \frac{\partial u}{\partial x}\right) - \rho S \frac{\partial^2 u}{\partial t^2} = 0, \tag{3.5.22}$$

where ρ is the material density. Assume that the left end of the bar is fixed and the right end is free. Then

$$u = 0 \quad \text{for} \quad x = 0, \quad \frac{\partial u}{\partial x} = 0 \quad \text{for} \quad x = l.$$

We seek a solution of equation (3.5.22) in the form

$$u(x, t) = u(x) \sin(\omega t + \alpha).$$

We have the eigenvalue problem for $u(x)$ and ω:

$$(Su')' + \Lambda Su = 0, \quad u(0) = u'(l) = 0,$$

where $\Lambda = \rho \omega^2 / E$. Here the prime denotes derivative with respect to x.

The equation of zeroth approximation

$$u_0'' + \Lambda_0 u_0 = 0 \tag{3.5.23}$$

has the solution

$$u_0 = A \sin \alpha x + B \cos \alpha x, \quad \alpha^2 = \Lambda_0.$$

Taking the boundary conditions into account, we obtain

$$u_{0n} = A \sin \alpha_n x, \quad \alpha_n = (\pi/2 + \pi n)/l, \quad n = 0, 1, 2, \ldots$$

The found modes u_{0n} and frequencies $w_{0n} = \alpha_n \sqrt{E/\rho}$ describe the vibrations of a bar of constant cross-section of area S_0.

We multiply the equations of first approximation,

$$u_1'' + \Lambda_0 u_1 = u_0'/l - \Lambda_1 u_0,$$

by u_0 and integrate by parts over the interval $[0, l]$. Taking (3.5.23) into account, we find

$$\Lambda_1 = \frac{I_1}{lI}, \quad I_1 = \int_0^l u_0' u_0 \, dx, \quad I = \int_0^l u_0^2 \, dx.$$

Evaluating the integrals we obtain the correction of first approximation $\Lambda_1 = 1/l^2$. Note that, in this problem, Λ_1 does not depend on n.

If the boundary conditions (3.5.18) contain a small parameter, an integration by parts of the equation of first approximation may result in the appearance of nonzero terms outside the integral.

Example 4

Consider the longitudinal vibrations of a bar with fixed left end and right end tighten with a spring of stiffness c. Let the cross-sectional area of the bar, S_0, be independent of the coordinate x. Then the eigenvalue problem for the bar has the form

$$u'' + \Lambda u = 0, \quad u(0) = 0, \quad u'(l) = -\frac{\varepsilon u(l)}{l}, \quad \varepsilon = \frac{cl}{E S_0}.$$

Assume that $\varepsilon \ll 1$ and seek a solution of the eigenvalue problem in the form (3.5.19). The solution of the first-approximation problem coinsides with the solution of Example 3. In the first approximation we get

$$u_1'' + \Lambda_0 u_1 = -\Lambda_1 u_0, \quad u_1(0) = 0, \quad u_1'(l) = -u_0(l)/l.$$

Multiplying the first-approximation equation by u_0 and integrating by parts over the interval $[0, l]$, then, by equality (3.5.23) we obtain

$$\Lambda_1 \int_0^l u_0^2 \, dx = -\left[u_1' u_0 - u_1 u_0' \right]_0^l.$$

From this formula and the boundary conditions for u_0 and u_1 it follows that

$$\Lambda_1 = 2\frac{u_0^2(l)}{A^2 l^2} = \frac{2}{l^2}.$$

Regularly perturbed eigenvalue problems arise, in particular, in the buckling analysis.

Example 5

Find the effect of gravity on the critical axial load T for a stand-up free-supported beam.

The buckling equation

$$EJ\frac{d^4 w}{dx^4} + \rho g S\frac{d}{dx}\left(x\frac{dw}{dx}\right) + T\frac{d^2 w}{dx^2} = 0$$

and the boundary conditions

$$w = \frac{d^2 w}{dx^2} = 0 \quad \text{for} \quad x = 0, \quad x = l, \tag{3.5.24}$$

are rewritten in the non-dimensional form

$$\frac{d^4 w}{d\xi^4} + \varepsilon\frac{d}{d\xi}\left(\xi\frac{dw}{d\xi}\right) + \Lambda\frac{d^2 w}{d\xi^2} = 0$$

$$w = \frac{d^2 w}{d\xi^2} = 0 \quad \text{for} \quad \xi = 0, \quad \xi = \pi.$$

Assume that $\varepsilon \ll 1$ and seek a solution in the form (3.5.19). In the zeroth approximation we obtain the classical Euler problem

$$\frac{d^4 w_0}{d\xi^4} + \Lambda_0\frac{d^2 w_0}{d\xi^2} = 0, \quad w = \frac{d^2 w}{d\xi^2} = 0 \quad \text{for} \quad \xi = 0, \quad \xi = \pi.$$

The first eigenvalue, $\Lambda_0 = 1$, provides the critical load, $T = EJ\pi^2/l^2$, and the buckling mode, $w_0 = A\sin\xi$. From the first-approximation equation

$$\frac{d^4 w_1}{d\xi^4} + \Lambda_0\frac{d^2 w_1}{d\xi^2} = -\frac{d}{d\xi}\left(\xi\frac{dw_0}{d\xi}\right) + \Lambda_1\frac{d^2 w_0}{d\xi^2},$$

we find $\Lambda_1 = -\pi/2$.

3.5.3 Boundary Value Problems for Equations with Highly Oscillating Coefficients

The multiple scale method described in Sect. 3.4.3 can be used for solving boundary value problems for differential equations with highly-oscillating periodic coefficients [4, 7, 55].

As an example, we consider the problem of the longitudinal deformation of a bar with highly oscillating cross-sectional area. The equation for the longitudinal displacement of a bar, $u(x)$, under a uniformly distributed load q has the form

$$\frac{d}{dx}\left(a\,\frac{du}{dx}\right) = q, \quad a = ES > 0. \tag{3.5.25}$$

Let the bar ends be fixed. Then

$$u(0) = u(l) = 0. \tag{3.5.26}$$

Assume that a is a periodic function of the variable $\xi = x/\varepsilon$ with period 1 and $\varepsilon \ll 1$. Replace Eq. (3.5.25) with the equivalent system of two equations of first order

$$\frac{du}{dx} = \frac{w}{a}, \quad \frac{dw}{dx} = q. \tag{3.5.27}$$

We seek the solution of the boundary value problem (3.5.26) and (3.5.27) in the form

$$u = u_0 + \varepsilon u_1 + \cdots, \quad w = w_0 + \varepsilon w_1 + \cdots, \tag{3.5.28}$$

where

$$u_i(x, \xi + 1) = u_i(x, \xi), \quad w_i(x, \xi + 1) = w_i(x, \xi), \quad i = 0, 1, 2, \ldots$$

Substituting expressions (3.5.28) in Eq. (3.5.27) and taking

$$\frac{du_i}{dx} = \frac{\partial u_i}{\partial x} + \frac{1}{\varepsilon}\frac{\partial u_i}{\partial \xi}, \quad \frac{dw_i}{dx} = \frac{\partial w_i}{\partial x} + \frac{1}{\varepsilon}\frac{\partial w_i}{\partial \xi},$$

into account, we get, in the zeroth approximation,

$$\frac{\partial u_0}{\partial \xi} = 0, \quad \frac{\partial w_0}{\partial \xi} = 0.$$

Therefore $u_0(x, \xi) = u_0(x)$ and $w_0(x, \xi) = w_0(x)$.

The system of equations of first approximation has the form

$$\frac{du_0}{dx} + \frac{\partial u_1}{\partial \xi} = \frac{w_0}{a}, \quad \frac{dw_0}{dx} + \frac{\partial w_1}{\partial \xi} = q. \tag{3.5.29}$$

Introduce the following notation for the *average value* of the function $f(x, \xi)$ for the period

$$\langle f(x, \xi) \rangle = \int_0^1 f(x, \xi) \, d\xi$$

and apply the averaging operator $\langle \cdot \rangle$ to the equations of system (3.5.29). By the equalities

$$\left\langle \frac{\partial u_1}{\partial \xi} \right\rangle = \left\langle \frac{\partial w_1}{\partial \xi} \right\rangle = 0,$$

following from the periodicity of u_1 and w_1 we obtain the system

$$\frac{du_0}{dx} = \langle 1/a \rangle w_0, \quad \frac{dw_0}{dx} = q,$$

which can be reduced to the equation

$$\hat{a} \frac{d^2 v_0}{dx^2} = q, \quad \hat{a} = \langle 1/a \rangle^{-1}. \tag{3.5.30}$$

Thus, in the zeroth approximation, we come to the problem of evaluating the displacements of a bar with some averaged cross-sectional area.

The solution of equation (3.5.30) with the boundary conditions (3.5.26) has the form

$$u_0 = \frac{q}{2\hat{a}} x(x - l).$$

The construction of the next approximations is considered in [4].

The above described method can also be used for solving eigenvalue problems. Consider the equation

$$-\frac{d^2 u}{dx^2} + cu \sum_{i=1}^{n-1} \delta \left(x - \frac{i}{n} \right) = \Lambda u \tag{3.5.31}$$

with boundary conditions

$$u(0) = u(1) = 0. \tag{3.5.32}$$

Here $\delta(x)$ is the Dirac delta function, c is a constant coefficient and Λ is the required eigenvalue. The problem of evaluating the frequencies and modes of a vibrating string supported by $n - 1$ springs uniformly distributed along the string can be reduced to the eigenvalue problem (3.5.31) and (3.5.32).

Assume that $n \gg 1$ and introduce the variable $\xi = nx$. Noting that $\delta(ax) = \delta(x)/a$, replace Eq. (3.5.31) with the equivalent equation

$$-\frac{d^2u}{dx^2} + A(\xi)u = \Lambda u, \qquad (3.5.33)$$

where

$$A(\xi) = cn \sum_{i=1}^{n-1} \delta(\xi - i).$$

Solutions of (3.5.33) are searched as asymptotic expansions in negative powers of the large parameter n:

$$u(x, \xi) = u_0(x, \xi) + n^{-2}u_1(x, \xi) + \cdots, \quad \Lambda = \Lambda_0 + n^{-2}\Lambda_1 + \cdots, \quad (3.5.34)$$

where $u_i(x, \xi + 1) = u_i(x, \xi)$. Further we assume that the coefficient c is small and $cn \sim 1$.

Substitute (3.5.34) in (3.5.33). In the zeroth approximation one gets

$$\frac{\partial^2 u_0}{\partial \xi^2} = 0, \quad \frac{\partial u_0}{\partial \xi} = C_0(x).$$

Apply the averaging operator $\langle \cdot \rangle$ to the last equality. The periodicity of u_0 implies that $C_0(x) = 0$ and $u_0(x, \xi) = v_0(x)$. Taking the equality $\langle A(\xi) \rangle = cn$ into account, we find that the averaged equation of first approximation

$$-\frac{\partial^2 u_1}{\partial \xi^2} - \frac{d^2 v_0}{dx^2} + A(\xi)v_0 = \Lambda_0 v_0 \qquad (3.5.35)$$

results in the equation

$$\frac{d^2 v_0}{dx^2} + (\Lambda_0 - cn)v_0 = 0. \qquad (3.5.36)$$

From the physical point of view, the transition from Eq. (3.5.33) to equation (3.5.36) means that the springs supporting the string are replaced with an elastic foundation.

The non-trivial solution of equation (3.5.36) satisfying the boundary conditions $v_0(0) = v_0(1) = 0$ has the form $v_{0k} = \sin k\pi x$ and the corresponding eigenvalue is $\Lambda_{0k} = (\pi k)^2 + cn$.

To construct the next approximation, represent u_1 as the sum

$$u_1(x, \xi) = v_1(x) + w_1(x, \xi),$$

where $\langle w_1(x, \xi) \rangle = 0$. It follows from (3.5.35) that

$$\frac{\partial w_1}{\partial \xi} = cn(i - \xi)v_0 + C_{1i}, \quad i \leq \xi \leq i + 1. \tag{3.5.37}$$

Averaging (3.5.37) we get $C_{1i} = 1/2$. Integrating (3.5.37) with respect to ξ permits to find

$$w_1 = cnv_0(2i\xi - \xi^2 + \xi)/2 + D_{1i}, \quad i \leq \xi \leq i + 1.$$

Averaging the last equality gives the value

$$D_{1i} = -\frac{cnv_0}{2}\left[i(i + 1) + \frac{1}{6}\right].$$

Therefore,

$$w_1 = -\frac{cnv_0}{2}\left[(\xi - i)(\xi - i - 1) + \frac{1}{6}\right], \quad i \leq \xi \leq i + 1. \tag{3.5.38}$$

After applying the operator $\langle \cdot \rangle$ to the equation of second approximation

$$-\frac{\partial^2 u_2}{\partial \xi^2} - \frac{\partial^2 u_1}{\partial x^2} + A(\xi)u_1 = \Lambda_0 v_1 + \Lambda_1 v_0,$$

we obtain the equation for evaluating $v_1(x)$:

$$-\frac{d^2 v_1}{dx^2} + (cn - \Lambda_0)v_1 = \frac{c^2 n^2}{12}v_0 + \Lambda_1 v_0.$$

Multiply the last equation by v_0 and integrate by parts over the interval $(0, 1)$. Taking (3.5.36) and the boundary conditions $v_1(0) = v_1(1) = 0$ into account, we get the correction of first approximation

$$\Lambda_1 = -\frac{c^2 n^2}{12}.$$

3.5.4 Exercises

3.5.1. Find the two-term asymptotic approximation for the displacements of the beam considered in Sect. 3.5.1 in the case $q = 0$ and the vertically directed force P is applied on the beam right end $x = l$. Neglect the weight of the beam.

3.5.2. Consider the beam described in Sect. 3.5.1 with freely supported ends. Find the first two terms of the asymptotic expansion of the function $w(x)$.

Fig. 3.9 Two joint bars
made of the same material

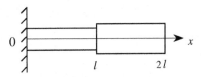

3.5.3. For Example 4 of Sect. 3.5.2 consider the case where the value of $E S_0/(cl)$ is a small parameter. Find Λ_0 and Λ_1 for the lowest eigenvalue for the corresponding boundary value problem.

3.5.4. Consider the longitudinal vibrations of two joint bars made of the same material (Fig. 3.9).

The left end, $x = 0$, of the first bar is fixed and the right end of the second bar, $x = 2l$, is free. The areas of the cross-sections for the first and the second bar, S_1 and S_2, respectively, are connected by the relation $S_2 = (1 + \varepsilon)S_1$, where $\varepsilon \ll 1$. For the eigenvalue problem describing the vibrations of the bars find the coefficients Λ_0 and Λ_1 for the series expansion of the first eigenvalue.

3.5.5. The transverse vibrations of a beam (see Sect. 3.5.1) are described by the equation

$$\frac{\partial^2}{\partial x^2}\left(EJ\,\frac{\partial^2 w}{\partial x^2}\right) + \rho S\,\frac{\partial^2 w}{\partial t^2} = 0.$$

Representing its solution in the form

$$w(x, t) = w(x)\sin(\omega t + \alpha),$$

we get an ordinary differential equation for evaluating the frequencies and vibrations modes

$$\frac{d^2}{dx^2}\left(\beta\,\frac{d^2 w}{dx^2}\right) = \Lambda\beta w,$$

where

$$\beta = 1 - \varepsilon\,\frac{x}{l}, \quad \Lambda = \frac{\rho S_0 \omega^2}{E J_0}, \quad S_0 = a_0 b_0.$$

Assume that the ends of the beam are freely supported (3.5.24). Representing the solution in the form (3.5.19), find the coefficients Λ_0 and Λ_1 of the series expansion of the lowest eigenvalue.

3.5.6. The free vibrations of a rectangular membrane are given by the equation

$$T\left(\frac{\partial^2 w}{\partial x^2} + \frac{\partial^2 w}{\partial y^2}\right) + \rho\omega^2 w = 0, \quad 0 \le x \le a, \quad 0 \le y \le b,$$

where $w(x, y)$ is the deflection of the membrane, a and b are the lengths of the membrane edges, ω is the vibrations frequency, T is the tensile stress, and ρ is the material density. The boundary conditions have the form

$$w = 0 \quad \text{for} \quad x = 0, \quad x = a, \quad y = 0, \quad y = b.$$

Assuming that $c = \rho/T = c_0[1 + \varepsilon g(x)]$, where ε is a small parameter find the first two terms of the asymptotic expansion for the vibrations frequencies of the membrane.

3.5.7. The non-dimensional system of equations

$$u'' + \left(\frac{B'}{B} u\right)' - \frac{\nu}{R_2} w' - \left(\frac{1}{R_2}\right)' w = -\Lambda u,$$

$$\frac{\nu}{R_2} u' + \frac{B'}{R_2 B} u - \frac{w}{R_2^2} = -\Lambda w \tag{3.5.39}$$

describes the axisymmetric membrane vibrations of a truncated conic shell. The prime denotes derivatives with respect to the non-dimensional length of the cone, u and w are the projections of the displacements of the points on the shell neutral surface, $B(s) = 1 - s \sin \beta$ is the distance between the axis of symmetry of the shell and its neutral surface, $R_2(s) = B(s) \cos^{-1} \beta$ is the radius of curvature, 2β is the angle at the vertex of the cone, ν is the Poisson ratio, and Λ is the required spectral parameter proportional to the squared frequency.

The section of the shell with the plane passing through the axis of symmetry is shown in Fig. 3.10.

For clamped shell edges, the boundary conditions for the membrane vibrations system have the form $u(0) = u(l) = 0$. Consider the case $\beta \ll 1$ and find the first two terms of the asymptotic expansion in powers of β for the lowest eigenvalue Λ.

3.5.8. Find the two-term asymptotic expansion in powers of the parameter $n^{-2} \ll 1$ for the function $u(x)$ satisfying the equation

$$-\frac{d^2 u}{dx^2} + cu \sum_{i=1}^{n-1} \delta \left(x - \frac{i}{n}\right) = \sin \pi x,$$

where $c \sim 1/n$, and the boundary conditions are $u(0) = u(1) = 0$.

Fig. 3.10 Section of shell
with plane passing through
the axis of symmetry

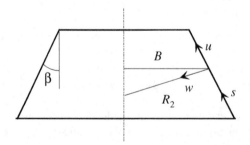

The function $u(x)$ describes the deflection of a string supported by springs under a distributed load.

3.5.9. Find the first two terms of the asymptotic expansion in powers of the parameter $n^{-4} \ll 1$ for the eigenvalue Λ of the eigenvalue problem

$$\frac{d^4 u}{dx^4} + cu \sum_{i=1}^{n-1} \delta\left(x - \frac{i}{n}\right) = \Lambda u,$$

$$u = \frac{d^2 u}{dx^2} = 0 \quad \text{for} \quad x = 0, \quad x = 1,$$

which describes the free vibrations of a freely supported beam stiffened with springs. Assume $cn \sim 1$ in the construction of the asymptotic expansion.

3.6 Answers and Solutions

3.2.1. We seek the flight time in the form $\tau^* \simeq \tau_0^* + \varepsilon \tau_1^*$. Substitute the expression for τ^* in the equation

$$\eta \simeq \eta_0 + \varepsilon \eta_1 = 0,$$

where η_0 and η_1 are defined by formulas (3.2.7) and (3.2.8). Equating the coefficients of equal powers of ε we obtain

$$\tau_0^* = 2 \sin \alpha, \quad \tau_1^* = -\frac{(\tau_0^*)^2}{6}, \quad \tau^* \simeq 2 \sin \alpha \left(1 - \frac{\varepsilon \sin \alpha}{3}\right).$$

In dimensional variables we have

$$t^* = \frac{v_0}{g} \tau^* \simeq \frac{2 v_0 \sin \alpha}{g} \left(1 - \frac{\nu v_0 \sin \alpha}{3 mg}\right).$$

Substituting $\tau = \tau^*$ in the formula for ξ we find the flight distance

$$\xi^* = \xi(\tau^*) \simeq \xi_0(\tau^*) + \varepsilon \xi_1(\tau^*) \simeq \cos \alpha (\tau_0^* + \varepsilon \tau_1^* - \varepsilon (\tau_0^*)^2 / 2).$$

Therefore

$$\xi^* \simeq \tau_0^* \cos \alpha \left(1 - \frac{2 \varepsilon \tau_0^*}{3}\right) = \frac{\sin 2\alpha (3 - 4\varepsilon \sin \alpha)}{3}.$$

The angle α_* is evaluated by means of the equation

$$\frac{d\xi^*}{d\alpha} \simeq \frac{2(3 \cos 2\alpha + \varepsilon(\sin \alpha - 3 \sin 3\alpha))}{3} = 0,$$

The approximate solution of this equation is

$$\alpha_* \simeq \frac{\pi}{4} - \varepsilon \frac{\sqrt{2}}{6}.$$

3.2.2. For $f(v) \equiv 1$ the exact solution of system (3.2.4) with initial conditions (3.2.5) is defined by the formulas

$$u = e^{-\varepsilon\tau} \cos \alpha, \quad w = e^{-\varepsilon\tau} \sin \alpha + \frac{1}{\varepsilon}(e^{-\varepsilon\tau} - 1),$$

$$\xi = \frac{\cos \alpha}{\varepsilon}(1 - e^{-\varepsilon\tau}), \quad \eta = \frac{1 + \varepsilon \sin \alpha}{\varepsilon^2}(1 - e^{-\varepsilon\tau}) + \frac{\tau}{\varepsilon}.$$

3.2.3. The system of first approximation for evaluating u_1 and w_1 is

$$\frac{du_1}{d\tau} = -u_0 \sqrt{u_0^2 + w_0^2}, \quad \frac{dw_1}{d\tau} = -w_0 \sqrt{u_0^2 + w_0^2},$$

where u_0 and w_0 are determined by formulas (3.2.7).

Taking the initial conditions $u_1(0) = w_1(0) = 0$ into account, we get

$$u_1 = -I \cos \alpha, \quad w_1 = (s^3 - 1)/3,$$

where

$$I = \int_0^\tau s(\tau) \, d\tau = \frac{1}{2} \left[(\tau - \sin \alpha)s + \sin \alpha + \cos^2 \alpha \ln \frac{s + \tau - \sin \alpha}{1 - \sin \alpha} \right],$$

$$s(\tau) = \sqrt{\tau^2 - 2 \sin \alpha\tau + 1}.$$

In Fig. 3.11 the solid line is the graph of the numerical solution u of system (3.2.4) and the dashed line is the graph of the corresponding approximate asymptotic solution for $\varepsilon = 0.05$, $\alpha = \pi/4$.

3.2.4. The solution of the generating system

$$t\dot{x}_0 = x_0, \quad t\dot{y}_0 = (2 - x)y_0$$

with the initial conditions

$$x_0(1) = 1, \quad y_0(1) = e^{-1}$$

has the form

$$x_0 = t, \quad y_0 = t^2 e^{-t}.$$

Fig. 3.11 Numerical solution u (*solid line*) and approximate asymptotic solution (*dashed line*)

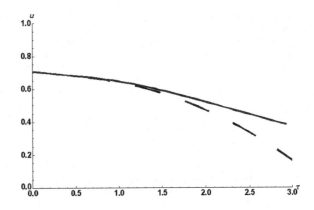

The solution of the system of first approximation,

$$t\dot{x}_1 = x_1 + y_0, \quad t\dot{y}_1 = (2 - x_0)y_1 - x_1 y_0,$$

is represented as the sum of the general solution of the homogeneous system coinciding with the generating system and the particular solution of the non-homogeneous system of first approximation that can be found by the variation of constants:

$$x_1 = Ct + C_1(t)t, \quad y_1 = Dt^2 e^{-t} + D_1(t)t^2 e^{-t}.$$

After substituting the expressions for x_1 and y_1 in the system of first approximation, we obtain

$$\dot{C}_1 = e^{-t}, \qquad C_1 = -e^{-t},$$
$$\dot{D}_1 = e^{-t} - e^{-1}, \quad D_1 = -e^{-t} - te^{-1}.$$

From the initial conditions $x_1(1) = y_1(1) = 0$, we find

$$C = e^{-1}, \quad D = 2e^{-1}.$$

Therefore,

$$x \simeq x_0 + \varepsilon x_1 = t[1 + \varepsilon(e^{-1} - e^{-t})],$$
$$y \simeq y_0 + \varepsilon y_1 = t^2 e^{-t}\{1 + \varepsilon[(2 - t)e^{-1} - e^{-t}]\}.$$

In Fig. 3.12, the numerical solution y for $\varepsilon = 0.5$ is plotted with the solid line and the approximate asymptotic solution with the dashed line.

3.2.5. The solution of the Cauchy problem for the generating equation

$$\ddot{x}_0 + \omega^2 x_0 = 0,$$

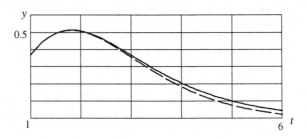

Fig. 3.12 Numerical solution y (*solid line*) and approximate asymptotic solution (*dashed line*)

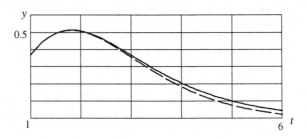

with initial conditions $x_0(0) = 0$ and $\dot{x}_0(0) = v_0$, is

$$x_0 = a \sin z, \quad a = \frac{v_0}{\omega}, \quad z = \omega t.$$

The general solution of the equation of first approximation,

$$\ddot{x}_1 + \omega^2 x_1 = a^3 \omega^2 \cos^2 z \sin z = \frac{a^3}{4} \omega^2 (\sin z + \sin 3z),$$

is

$$x_1 = M_1 \cos z + N_1 \sin z - \frac{a^3}{32} (4z \cos z + \sin 3z).$$

From the initial conditions $x_1(0) = \dot{x}_1(0) = 0$ we find

$$M_1 = 0, \quad N_1 = \frac{7a^3}{32}.$$

Therefore,

$$x \simeq a \left[\sin z + \frac{\varepsilon a^2}{32} (7 \sin z - \sin 3z - 4z \cos z) \right].$$

To find the numerical solution, it is convenient to replace the initial equation with a system of two first-order equations in normal form:

$$\dot{x} = y, \quad \dot{y} = -\omega^2 x + \varepsilon y^2 x.$$

In Fig. 3.13, the numerical solution is plotted with a solid line and the two-term asymptotic approximation with a dashed line for $\omega = 1$, $v_0 = 2$ and $\varepsilon = 0.1$.

3.2.6. We represent the approximate solution in the form

$$x = x_0 + \varepsilon x_1 + \varepsilon^2 x_2 + \cdots, \quad y = y_0 + \varepsilon y_1 + \varepsilon^2 y_2 + \cdots, \quad z = z_0 + \varepsilon z_1 + \varepsilon^2 z_2 + \cdots.$$

Fig. 3.13 Numerical
solution (*solid line*) and
two-term asymptotic
approximation (*dashed line*)

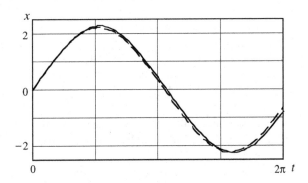

After substituting this solution into the system and taking

$$x_k = y_k = z_k = \dot{x}_k = \dot{y}_k = \ddot{y}_k = 0, \quad k = 0, 1, 2, \ldots \quad \text{for} \quad t = 0,$$

into account, we find

$$x_0 = x_1 = 0, \quad x_2 = \frac{t^4}{12} \sin \varphi \cos \varphi, \quad y_0 = 0, \quad y_1 = -\frac{t^3}{3} \cos \varphi, \quad z_0 = t^2.$$

To obtain the exact solution, we integrate the first and the third equations of the
system. Taking the initial conditions $\dot{x}(0) = \dot{z}(0) = 0$ into account, we get $\dot{x} = -\varepsilon y \sin \varphi$ and $\dot{z} = 2t + \varepsilon y \cos \varphi$. Substitutig these expressions into the second
equation and transforming it, we obtain

$$\ddot{y} + \varepsilon^2 y = -2\varepsilon t \cos \varphi.$$

The function

$$y = \frac{2}{\varepsilon} \cos \varphi \left(\frac{\sin \varepsilon t}{\varepsilon} - t \right) \simeq -\varepsilon \frac{t^3}{3} \cos \varphi$$

is a solution of this equation with zero initial conditions. From the first and the third
equations of the system we find

$$x = 2 \sin \varphi \cos \varphi \left(\frac{\cos \varepsilon t - 1}{\varepsilon^2} + \frac{t^2}{2} \right) \simeq \varepsilon^2 \frac{t^4}{12} \sin \varphi \cos \varphi,$$

$$z = t^2 - 2 \cos^2 \varphi \left(\frac{\cos \varepsilon t - 1}{\varepsilon^2} + \frac{t^2}{2} \right) \simeq t^2.$$

3.3.1. The equation of second approximation

$$\ddot{x}_2 + \omega^2 x_2 = -3bx_0^2 x_1$$

has the solution

$$x_2 = \frac{3}{4} b^2 A^5 \left[\left(\frac{9}{\omega^2 - 1} + \frac{1}{\omega^2 - 9} \right) \frac{\sin t}{4(\omega^2 - 1)} \right.$$
$$\left. - \frac{1}{4} \left(\frac{3}{\omega^2 - 1} + \frac{2}{\omega^2 - 9} \right) \frac{\sin 3t}{\omega^2 - 9} + \frac{\sin 5t}{4(\omega^2 - 9)(\omega^2 - 25)} \right].$$

3.3.2. We seek a solution of the equation in the form (3.3.3). The zeroth approximation x_0 for $\omega \neq 1$ is defined by formula (3.3.2). For $\omega \neq 3$ the equation of first approximation

$$\ddot{x}_1 + \omega^2 x_1 = -bA^3 \sin t \cos^2 t$$

has the 2π-periodic solution

$$x_1 = -\frac{b}{4} A^3 \left(\frac{\sin t}{\omega^2 - 1} + \frac{\sin 3t}{\omega^2 - 9} \right).$$

3.3.3. After substituting the zeroth approximation (3.3.8) into the equation of first approximation

$$\ddot{x}_1 + x_1 = a \sin t + \delta x_0 + x_0 \dot{x}_0^2$$

we get

$$\ddot{x}_1 + x_1 = P \cos t + Q \sin t + \frac{1}{4} \left(3MN^2 - M^3 \right) \cos 3t + \frac{1}{4} \left(N^3 - 3M^2 N \right) \sin 3t,$$

where

$$P = M \left[\delta + \frac{1}{4} \left(M^2 + N^2 \right) \right], \quad Q = a + N \left[\delta + \frac{1}{4} \left(M^2 + N^2 \right) \right].$$

This equation has a periodic solution for x_1 if $P = Q = 0$. Assuming that $M \neq 0$ we arrive at a contradiction with the condition $a \neq 0$. Therefore $M = 0$ and we get the cubic equation

$$\frac{1}{4} N^3 + \delta N + a = 0$$

for N.

Let N_0 be a real root of this equation. The equation of first order

$$\ddot{x}_1 + x_1 = \frac{1}{4} N_0^3 \sin 3t$$

has the solution

$$x_1 = M_1 \cos t + N_1 \sin t - \frac{1}{32} N_0^3 \sin 3t.$$

From the existence condition for a periodic solution of the equation of second approximation,

$$\ddot{x}_2 + x_2 = \delta x_1 + \dot{x}_0^2 x_1 + 2x_0 \dot{x}_0 \dot{x}_1,$$

we obtain the two equations

$$M_1 \left(\delta + \frac{1}{4} N_0^2 \right) = 0, \quad N_1 \left(\delta + \frac{3}{4} N_0^2 \right) = -\frac{5N_0^5}{128}.$$

From the first equation and the condition $a \neq 0$, it follows that $M_1 = 0$. For $\delta \neq -3N_0^2/4$, from the second equation we find

$$N_1 = -\frac{5N_0^5}{32 \left(4\delta + 3N_0^2 \right)}.$$

3.3.4. The generating equation

$$\ddot{x}_0 + 9x_0 = h \sin t$$

has a family of 2π-periodic solutions

$$x_0 = M \cos 3t + N \sin 3t + H \sin t, \quad H = h/8.$$

The constants M and N are found from the existence condition for the periodic solution of the equation of first approximation

$$\ddot{x}_1 + 9x_1 = \delta x_0 - bx_0^3.$$

Substituting the expression for x_0 into this equation, we expand the right side in a Fourier series and equate to zero the coefficients of $\cos 3t$ and $\sin 3t$. Thus, we find

$$\frac{3bM}{4} \left(M^2 + N^2 + 2H^2 \right) - \delta M = 0, \quad \frac{3bN}{4} \left(M^2 + N^2 + 2H^2 \right) - \delta N - \frac{b}{4} H^3 = 0.$$

Assuming that $M \neq 0$ we arrive at a contradiction with the condition $h \neq 0$. Therefore, $M = 0$ and to find N we have the cubic equation

$$3bN^3 + \left(6bH^2 - 4\delta \right) N - bH^3 = 0.$$

Let N_0 be a real root of this equation. Then

$$x_0 = N_0 \sin 3t + H \sin t.$$

3.3.5.

(1) After the change of variables (3.3.20) the equation becomes

$$\ddot{x} + g^2 x = \varepsilon g(1 - x^4)\dot{x}, \quad (\dot{}) \equiv d()/d\tau.$$

We seek its solution in the form (3.3.22). Substituting the solution of the generating equation (3.3.24) into the equation of first approximation

$$\ddot{x}_1 + x_1 = (1 - x_0^4)\dot{x}_0 - 2g_1 x_0$$

and taking the existence condition for 2π-periodic solution $x_1(\tau)$ into account, we get

$$\left(\frac{M^4}{8} - 1\right) M = 0, \quad 2g_1 M = 0.$$

The periodic solution corresponds to $M = M_0 = \sqrt[4]{8}$, $g_1 = 0$. The solution of the equation of first approximation,

$$\ddot{x}_1 + x_1 = \frac{M_0}{2}(3 \sin 3\tau + \sin 5\tau),$$

is

$$x_1 = M_1 \cos \tau + N_1 \sin \tau - \frac{3}{16} M_0 \sin 3\tau - \frac{1}{48} M_0 \sin 5\tau.$$

From the condition $\dot{x}_1 = 0$ for $\tau = 0$ we obtain $N_1 = 2M_0/3$. The constants M_1 and g_2 are evaluated from the existence condition for a periodic solution of the equation of second approximation

$$\ddot{x}_2 + x_2 = -2g_2 x_0 + (1 - x_0^4)\dot{x}_1 - 4x_0^3 \dot{x}_0 x_1.$$

After transformations we arrive at $M_1 = 0$, $g_2 = 7/48$.

In Fig. 3.14 the limit cycle obtained by means of the two-term asymptotic approximation for $\varepsilon = 0.2$ is plotted with dashed line. The solid line is a phase curve found by a numerical integration of the Cauchy problem with initial conditions $x = 2.5$, $\dot{x} = 0$.

(2) The change of variable (3.3.20) transforms the equation into

$$\ddot{x} + g^2 x = \varepsilon g \left(1 - x^2\right) \dot{x} - \varepsilon g^2 x^3.$$

From the existence condition for a 2π-periodic solution of the equation of first approximation,

$$\ddot{x}_1 + x_1 = (1 - x_0^2)\dot{x}_0 - 2g_1 x_0 - x_0^3,$$

Fig. 3.14 Limit cycle (*dashed line*) and phase curve (*solid line*)

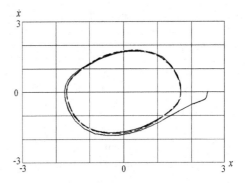

where $x_0 = M \cos \tau$, we obtain $M = 2$ and $g_1 = -3/2$. We substitute the functions x_0 and

$$x_1 = M_1 \cos \tau - \frac{3}{4} \sin \tau + \frac{1}{4} (\cos 3\tau - \sin 3\tau)$$

into the equation of second approximation

$$\frac{d^2 x_2}{d\tau^2} + x_2 = -2g_1 x_1 x_0 + \left(1 - x_0^2\right) \left(\frac{dx_1}{d\tau} + g_1 \frac{dx_0}{d\tau}\right)$$
$$- x_0 \left(2g_2 + g_1^2 + 2x_0 \frac{dx_0}{d\tau} x_1 + 3x_0 x_1 + 2g_1 x_0^2\right).$$

The existence conditions for a periodic solution provide the following system for evaluating M_1 and g_2:

$$6M_1 + 4g_2 = 13, \quad 2M_1 = -1,$$

which has the solution $M_1 = -1/2$ and $g_2 = 4$.

3.4.1. We seek a solution of equation (3.4.2) in the form

$$x = a(t) \cos z, \quad z = t + \beta(t).$$

The shortened system

$$\dot{a} = \frac{\varepsilon a}{2} \left(1 - \frac{a^2}{4}\right), \quad \dot{\beta} = -\frac{3\varepsilon}{8} b a^2,$$

has the solution

$$a = \frac{2}{\sqrt{1 + c_0 e^{-\varepsilon t}}}, \quad \beta = \frac{3b}{2} \ln \left(e^{\varepsilon t} + c_0\right) + \beta_0,$$

where c_0 and β_0 are arbitrary constants. Since $a \to 2$ as $t \to \infty$, the stationary solution $a = 2$ is stable.

3.4.2. In the case at hand, Eq. (3.4.15) has the form

$$\dot{a} = -\varepsilon B(a), \quad B(a) = \frac{a}{2}\left(\frac{a^4}{8} - 1\right).$$

The stationary solutions $a = 0$ and $a = a_0 = \sqrt[4]{8}$ correspond to the state of equilibrium and to a limit cycle. Using condition (3.4.20) we find that the state of equilibrium is unstable and the limit cycle is stable since

$$\left.\frac{dB}{da}\right|_{a=0} = -\frac{1}{2}, \quad \left.\frac{dB}{da}\right|_{a=a_0} = 2.$$

3.4.3.

$$\dot{a}\omega = \varepsilon\left[\frac{a\omega}{2}\left(1 - \frac{a^2\omega^2}{4}\right) - \frac{h}{2}\cos\beta\right],$$

$$a\omega\dot{\beta} = \frac{\varepsilon h}{2}\sin\beta - \frac{a}{2}\left(\omega^2 - 1\right).$$

3.4.4. For the equation under consideration

$$A(a) = a^3, \quad B(a) = na\omega/2.$$

Substituting these expressions into (3.4.13) we get the equation of the amplitude-frequency response

$$n^2 a_0^2\omega^2 + \left[2a_0^3 + \frac{a_0}{\varepsilon}(\omega^2 - 1)\right]^2 - h^2 = 0.$$

To plot the amplitude-frequency response characteristic (AFRC) in the plane (ω, a) one should find all the real values of a_0 satisfying the equation of the amplitude-frequency response for different values of ω. As it was noted above (see Sect. 3.4.1) plotting the AFRC with this method is justified only in a small neighborhood of the straight line $\omega = 1$. Calculating a_0 for a given ω amounts to solve a cubic equation. A more convenient way to construct the AFRC is to find ω for the given value of a_0, since, in this case, one needs to solve a quadratic equation.

Introduce the new variables $y = \omega^2$, $\xi = a^2$. Then the equation for the amplitude-frequency response can be transformed into $y^2 - py + q = 0$, where

$$p = 2\gamma - n^2\varepsilon^2, \quad q = \gamma^2 - \varepsilon^2 h^2/\xi, \quad \gamma = 1 - 2\varepsilon\xi.$$

Fig. 3.15 Amplitude-frequency response characteristic for $h = 1$, $n = 1$ and $\varepsilon = 0.02$

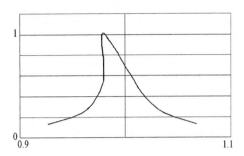

The roots of the last equation,

$$y_{1,2} = \left(p \pm \sqrt{D}\right)/2, \quad \text{where} \quad D = \varepsilon^2 \left(4h^2/\xi - 4\gamma n^2 + \varepsilon^2 n^4\right),$$

are positive real numbers for $0 < \xi \le \xi_1$, where $\xi_1 = \dfrac{c}{4\varepsilon}\left(1 - \sqrt{1 - \dfrac{8\varepsilon h^2}{c^2 n^2}}\right)$ is a

root of the equation $D(\xi) = 0$ with $c = 1 - \varepsilon^2 n^2/4$.

The curves $y_1(\xi)$ and $y_2(\xi)$ representing the amplitude-frequency response characteristic in the plane (ξ, y) intersect at the point $\xi = \xi_1$. To plot the AFRC in the plane (a, ω) one should come back to the initial variables a and ω.

The AFRC for $h = 1$, $n = 1$ and $\varepsilon = 0.02$ is plotted in Fig. 3.15.

3.4.5.

(1) We seek a solution of the equation

$$\ddot{x} + x = \varepsilon \left(1 - x^2\right)\dot{x} - \varepsilon b x^3$$

in the form (3.4.28), where

$$x_0 = a(\tau_1, \tau_2)\cos z, \quad z = \tau + \beta(\tau_1, \tau_2).$$

The equation of first approximation

$$\frac{\partial^2 x_1}{\partial \tau^2} + x_1 = -2\frac{\partial^2 x_0}{\partial \tau \partial \tau_1} + \left(1 - x_0^2\right)\frac{\partial x_0}{\partial \tau} - x_0^3$$

has a periodic solution $x_1 = \dfrac{a^3}{32}\left(\cos 3z - \sin 3z\right)$ if

$$\frac{\partial a}{\partial \tau_1} = \frac{a}{2}\left(1 - \frac{a^2}{4}\right), \quad \frac{\partial \beta}{\partial \tau_1} = \frac{3}{8}a^2,$$

$$a = \frac{2}{\sqrt{1 + c_0(\tau_2)e^{-\tau_1}}}, \quad \beta = \frac{3}{2}\ln\left[e^{\tau_1} + c_0(\tau_2)\right] + \beta_0(\tau_2).$$

From the existence condition for a periodic solution for the equation of second approximation,

$$\frac{\partial^2 x_2}{\partial \tau^2} + x_2 = -2\frac{\partial^2 x_1}{\partial \tau \partial \tau_1} - 2\frac{\partial^2 x_0}{\partial \tau \partial \tau_2} - \frac{\partial^2 x_0}{\partial \tau_1^2} + (1 - x_0^2)\left(\frac{\partial x_1}{\partial \tau} + \frac{\partial x_0}{\partial \tau_1}\right)$$
$$- 2x_0 \frac{\partial x_0}{\partial \tau} x_1 - 3x_0^2 x_1,$$

we get

$$\frac{\partial a}{\partial \tau_2} = \frac{a^3}{32}(a^2 - 6), \quad \frac{\partial \beta}{\partial \tau_2} = \frac{1}{8}\left(-1 + a^2 - \frac{11a^4}{16}\right).$$

With the first equation one finds the function $c_0(\tau_2)$ and with the second $\beta_0(\tau_2)$.

(2) For the equation $\ddot{x} + x = \varepsilon\left(1 - x^4\right)\dot{x}$ in the first approximation we have

$$\frac{\partial^2 x_1}{\partial \tau^2} + x_1 = -2\frac{\partial^2 x_0}{\partial \tau \partial \tau_1} + \left(1 - x_0^4\right)\frac{\partial x_0}{\partial \tau}.$$

The equations for a and β as function in τ_1,

$$\frac{\partial a}{\partial \tau_1} = \frac{a}{2}\left(1 - \frac{a^4}{8}\right), \quad \frac{\partial \beta}{\partial \tau_1} = 0,$$

have the solutions

$$a^4 = \frac{8}{1 + c_0(\tau_2)e^{-2\tau_1}}, \quad \beta = \beta_0(\tau_2).$$

We substitute the periodic solutions of the zeroth and first approximations

$$x_0 = a\cos z, \quad x_1 = -\frac{a^5}{128}\left(\frac{\sin 5z}{3} + 3\sin 3z\right)$$

into the right side of the equation of second approximation,

$$\frac{\partial^2 x_2}{\partial \tau^2} + x_2 = -2\frac{\partial^2 x_1}{\partial \tau \partial \tau_1} - 2\frac{\partial^2 x_0}{\partial \tau \partial \tau_2} - \frac{\partial^2 x_0}{\partial \tau_1^2} + (1 - x_0^4)\left(\frac{\partial x_1}{\partial \tau} + \frac{\partial x_0}{\partial \tau_1}\right)$$
$$- 4x_0^3 \frac{\partial x_0}{\partial \tau} x_1,$$

and equate to zero the coefficients of $\sin z$ and $\cos z$. We then obtain

$$\frac{\partial a}{\partial \tau_2} = 0, \quad \frac{\partial \beta}{\partial \tau_2} = -\frac{1}{8} + \frac{3}{4}\frac{a^4}{8} - \frac{37}{48}\left(\frac{a^4}{8}\right)^2.$$

From the first equation it follows that c_0 does not depend on τ_2, and from the second equation we find $\beta_0(\tau_2)$.

3.5.1. The boundary value problem consists of a fourth order equation $(EJw'')'' = 0$ and 4 boundary conditions:

$$w = w' = 0, \quad \text{for} \quad x = 0, \quad w'' = 0, \quad (EJw'')' = P \quad \text{for} \quad x = l,$$

where the moment of inertia J is given by formula (3.5.13).

An integration of the bending equations with the boundary conditions at $x = l$ gives

$$EJw'' = P(x - l), \quad 0 \le x \le l.$$

Represent w in the form (3.5.14). Then $EJ_0 w_0'' = P(x - l)$, and therefore

$$w_0 = \frac{Px^2}{6EJ_0}(x - 3l).$$

The equation of first approximation

$$\frac{d^4 w_1}{dx^4} = \frac{d^2}{dx^2}\left(\frac{x}{l}\frac{d^2 w_0}{dx^2}\right)$$

has solution

$$w_1 = \frac{Px^2}{12EJ_0 l}(x^2 - 4lx + 6l^2).$$

3.5.2. After evaluating the constants A, B, C, and D in (3.5.15) with the help of the boundary conditions

$$w_0 = w_0'' = 0 \quad \text{for} \quad x = 0, \quad x = l,$$

one gets

$$w_0 = -\frac{qx}{24EJ_0}(x^3 - 2x^2 l + l^3).$$

In this case the solution of equation (3.5.16) is

$$w_1 = -\frac{qx}{120EJ_0 l}\left(3x^4 - 5x^3 l + 2l^4\right).$$

3.5.3. Let $\varepsilon = E S_0/(cl) \ll 1$ and seek a solution of the eigenvalue problem

$$u'' + \Lambda u = 0, \qquad u(0) = 0, \quad u(l) = -\varepsilon l u'(l)$$

in the form (3.5.19). The lowest eigenvalue, Λ_0, for the problem of first approximation,

$$u_0'' + \Lambda_0 u = 0, \qquad u_0(0) = 0, \quad u_0(l) = 0,$$

is determined by the formula $\Lambda_0 = \alpha^2$, where $\alpha = \pi/l$, and the corresponding
vibrations mode has the form $u_0 = \sin \alpha x$.

Taking the boundary conditions for the first approximation

$$u_1(0) = 0, \quad u_1(l) = -l u_0'(l)$$

into account, we obtain

$$\Lambda_1 = -\frac{2}{l} \left[u_1' u_0 - u_1 u_0' \right]_0^l = -\frac{2\pi^2}{l^2}.$$

3.5.4. The equations describing the longitudinal vibrations of bars are

$$\frac{d^2 u^{(k)}}{dx^2} + \Lambda u^{(k)} = 0, \quad k = 1, 2,$$

where $u^{(k)}$ are the displacements of points of the kth bar. We seek the solution of
these equations satisfying the boundary conditions

$$u^{(1)}(0) = 0, \quad u^{(1)}(l) = u^{(2)}(l), \quad \frac{du^{(1)}}{dx}(l) = (1+\varepsilon)\frac{du^{(2)}}{dx}(l), \quad \frac{du^{(2)}}{dx}(2l) = 0$$

in the form

$$u^{(k)} = u_0^{(k)} + u_1^{(k)}\varepsilon + \cdots, \quad \Lambda_0 + \Lambda_1\varepsilon + \cdots.$$

The eigenvalue problem of the zeroth approximation,

$$\frac{d^2 u_0^{(k)}}{dx^2} + \Lambda u_0^{(k)} = 0, \quad k = 1, 2,$$

$$u_0^{(1)}(0) = 0, \quad u_0^{(1)}(l) = u_0^{(2)}(l), \quad \frac{du_0^{(1)}}{dx}(l) = \frac{du_0^{(2)}}{dx}(l), \quad \frac{du_0^{(2)}}{dx}(2l) = 0,$$

has the solution

$$\Lambda_0 = \alpha^2, \quad u_0^{(1)} = u_0^{(2)} = \sin \alpha x.$$

The first eigenvalue is $\alpha = \pi/(4l)$.

Multiplying the equation of first approximation,

$$\frac{d^2 u_1^{(k)}}{dx^2} + \Lambda_0 u_1^{(k)} = -\Lambda_1 u_0^{(k)}, \quad k = 1, 2,$$

by $u_0^{(k)}$ and integrating by parts over the intervals $[0, l]$ for $k = 1$ and $[l, 2l]$ for $k = 2$ we obtain the equalities

$$\Lambda_1 I_1 = -\left[\frac{du_1^{(1)}}{dx} u_0^{(1)} - \frac{du_0^{(1)}}{dx} u_1^{(1)}\right]_0^l, \quad \Lambda_1 I_2 = -\left[\frac{du_1^{(2)}}{dx} u_0^{(2)} - \frac{du_0^{(2)}}{dx} u_1^{(2)}\right]_l^{2l},$$

where

$$I_1 = \int_0^l \sin^2 \frac{\pi x}{4l} \, dx, \quad I_2 = \int_l^{2l} \sin^2 \frac{\pi x}{4l} \, dx.$$

Adding the obtained equalities and taking the boundary conditions of the first approximation

$$u_1^{(1)}(0) = 0, \quad u_1^{(1)}(l) = u_1^{(2)}(l), \quad \frac{du_1^{(1)}}{dx}(l) = \frac{du_1^{(2)}}{dx}(l) + \frac{du_0^{(2)}}{dx}(l), \quad \frac{du_1^{(2)}}{dx}(2l) = 0$$

into account, we get the formula

$$\Lambda_1(I_1 + I_2) = -\frac{du_0^{(2)}}{dx}(l) \, u_0^{(2)}(l),$$

from which we have

$$\Lambda_1 = -\frac{\pi}{8l^2}.$$

3.5.5. In the zeroth approximation we get a problem for the vibrations of a beam with constant cross-section

$$w_0'''' - \Lambda_0 w_0 = 0, \quad w_0 = w_0'' = 0 \quad \text{for} \quad x = 0, l.$$

The vibrations mode $w_0 = A \sin \alpha_0 x$ corresponds to the first eigenvalue $\Lambda_0 = \alpha_0^4$, where $\alpha_0 = \pi/l$. From the equation of first approximation,

$$w_1'''' - \Lambda_0 w_1 = \frac{2}{l} w_0''' + \Lambda_1 w_0,$$

we find that $\Lambda_1 = 0$.

3.5.6. After separating the variables

$$w(x, y) = w(x) \sin \beta y, \quad \beta = \frac{\pi n}{b}, \quad n = 1, 2, \ldots,$$

we get the ordinary differential equation

$$\frac{d^2w}{dx^2} + \left[\Lambda(1 + \varepsilon g) - \beta^2\right]w = 0, \quad \Lambda = c_0\omega^2$$

for the function $w(x)$ and the boundary conditions

$$w(0) = w(a) = 0.$$

We seek a solution of the eigenvalue problem in the form (3.5.19). The zeroth approximation problem,

$$\frac{d^2w_0}{dx^2} + (\Lambda_0 - \beta^2)w_0 = 0, \quad w_0(0) = w_0(a) = 0,$$

has the nontrivial solution

$$w_0 = A\sin\alpha x, \quad \alpha = \frac{\pi m}{a}, \quad m = 1, 2, \ldots,$$

for

$$\Lambda_0 = \alpha^2 + \beta^2 = \pi^2\left(\frac{m^2}{a^2} + \frac{n^2}{b^2}\right).$$

From the solvability condition for the equation of first approximation,

$$\frac{d^2w_1}{dx^2} + \left(\Lambda_0 - \beta^2\right)w_1 + \Lambda_1 w_0 + \Lambda_0 g w_0 = 0,$$

we obtain

$$\Lambda_1 = -\frac{2\Lambda_0}{a}\int_0^a g\sin^2\alpha x\,dx.$$

3.5.7. We expand the coefficients of the system of equations in series in powers of β and omit the terms containing β^m for $m > 1$. We come to the system

$$u'' - \beta u' - \nu(1 + s\beta)w' - \beta w = -\Lambda u,$$
$$\nu(1 + s\beta)u' - \beta u - (1 + 2\beta s)w = -\Lambda w.$$

We represent its solutions in the form

$$u = u_0 + \beta u_1, \quad w = w_0 + \beta w_1, \quad \Lambda = \Lambda_0 + \beta\Lambda_1.$$

In the zeroth approximation we have the system of equations

$$u_0'' - \nu w_0' = -\Lambda_0 u_0, \quad \nu u_0' - w_0 = -\Lambda_0 w_0,$$

which describe the membrane vibrations of the cylindrical shell. This system is equivalent to the equation

$$u_0'' + \alpha^2 u_0 = 0, \quad \alpha^2 = \frac{(1 - \Lambda_0)\Lambda_0}{1 - \nu^2 - \Lambda_0}.$$

Taking the boundary conditions $u_0(0) = u_0(l) = 0$ into account, we find $u_{0n} = \sin[\alpha_n(s - l)]$, $\alpha_n = \pi n/l$, $n = 1, 2, \ldots$ The lowest eigenvalue,

$$\Lambda_0 = \frac{1}{2}\left[1 + \alpha_1^2 - \sqrt{\left(1 + \alpha_1^2\right)^2 - 4\left(1 - \nu^2\right)\alpha_1^2}\right],$$

corresponds to the vibrations mode

$$u_0 = \sin[\alpha_1(s - l)], \quad w_0 = \frac{\nu\alpha_1}{1 - \Lambda_0}\cos[\alpha_1(s - l)].$$

Consider the system of equations of first approximation

$$u_1'' - \nu w_1' + \Lambda_0 u_1 = u_0' + \nu s w_0' - \Lambda_1 u_0 + w_0,$$
$$\nu u_1' - w_1 + \Lambda_0 w_1 = -\nu s u_0' + u_0 + 2s w_0 - \Lambda_1 w_0.$$

Multiply the first equation by u_0, the second by w_0, add together and integrate the obtained equality by parts over the segment $[0, l]$. Taking the boundary conditions and the system of equations of zeroth approximation into account, we find the equation

$$\Lambda_1 I_0 = \int_0^l \left(u_0'' u_0 + \nu s w_0' u_0 + 2w_0 u_0 - \nu s u_0' w_0 + 2s w_0^2\right) ds,$$

where

$$I_0 = \int_0^l (u_0^2 + w_0^2)\, ds = \frac{l}{2}\left[1 + \left(\frac{\nu\alpha_1}{1 - \Lambda_0}\right)^2\right].$$

After evaluating the integral, we find

$$\Lambda_1 = \frac{\Lambda_0^2 \nu^2 l^2}{2I_0(1 - \nu^2 - \Lambda_0)(1 - \Lambda_0)}.$$

3.5.8. To construct the asymptotic expansion we apply the method that was used in Sect. 3.5.3 to solve Eq. (3.5.31). Representing the solution in the form (3.5.34) instead of Eq. (3.5.36) we get an equation for v_0,

$$-v_0'' + cn v_0 = \sin \pi x.$$

Its solution, $v_0 = \sin \pi x / (\pi^2 + cn)$, satisfies the boundary conditions $v_0(0) = v_0(1) = 0$. For the function $w_1(x, \xi)$ we again obtain expression (3.5.38). The equation

$$-v_1'' + cnv_1 = c^2 n^2 v_0/12$$

admits the solution

$$v_1 = \frac{c^2 n^2 v_0}{12(\pi^2 + cn)}.$$

Therefore

$$u(x) \simeq \frac{\sin \pi x}{\pi^2 + cn} \left\{ 1 - \frac{c}{2n} \left[(\xi - i)(\xi - i - 1) + \frac{1}{6} \right] + \frac{c^2}{12(\pi^2 + cn)} \right\}.$$

$$i \le \xi \le i + 1, \quad \xi = nx.$$

3.5.9. Represent the solution of the eigenvalue problem in the following form

$$u = v_0(x) + n^{-4}[v_1(x) + w_1(x, \xi)] + \cdots, \quad \langle w_1(x, \xi) \rangle = 0,$$

$$\Lambda = \Lambda_0 + n^{-4}\Lambda_1 + \cdots, \quad \xi = nx.$$

After averaging the equation of first approximation we obtain the equation

$$\frac{d^4 v_0}{dx^4} + (cn - \Lambda_0)v_0 = 0,$$

which describes the vibrations of a beam on an elastic foundation. Taking the boundary conditions

$$v_0 = \frac{d^2 v_0}{dx^2} = 0 \quad \text{for} \quad x = 0, \quad x = 1,$$

into account, we find the eigenvalues $\Lambda_{0k} = (\pi k)^4 + cn$ and the eigenfunctions $v_{0k} = \sin k\pi x$ for the eigenvalue problem of first approximation.

The equation for evaluating $w_1(x, \xi)$ is

$$\frac{\partial^4 w_1}{\partial \xi^4} = cnv_0 \left[1 - \sum_{i=1}^{n-1} \delta (\xi - i) \right].$$

Integrating this equation with respect to ξ, one gets

$$\frac{\partial^3 w_1}{\partial \xi^3} = cn(i - \xi)v_0 + C_{1i}.$$

After averaging the last equation we find $C_{1i} = 1/2$. Continuing integration with respect to ξ in combination with averaging we obtain

$$w_4(x, \xi) = cnv_0 \left[(\xi - i)^2 (\xi - i - 1)^2 - 1/720 \right], \quad i \leq \xi \leq i + 1.$$

An application of the averaging operator to the equation of second approximations produces the equation

$$\frac{d^4 v_1}{dx^4} + cnv_1 - \frac{c^2 n^2}{720} v_0 = \Lambda_0 v_1 + \Lambda_1 v_0,$$

from which we can find $\Lambda_1 = -c^2 n^2 / 720$.

Chapter 4
Singularly Perturbed Linear Ordinary Differential Equations

In this chapter, we study systems of linear differential equations with variable coefficients containing a small parameter μ in the derivative terms [10, 25, 49, 50, 57, 62, 63, 65]. Singular perturbation is characterized by the fact that for $\mu = 0$ the initial system transforms to a system of differential equations of lower order or even sometimes to a system of algebraic equations. In the absence of turning points, we construct an asymptotic expansion of the fundamental system of solutions as $\mu \to 0$.

We discuss methods of asymptotic solutions of linear boundary value problems. As examples, we analyze one-dimensional problems for equilibrium, dynamics and stability of solids.

4.1 Solutions of Linear Ordinary Differential Equations of the nth Order

Consider the linear differential equation of order n

$$M_\mu y = \sum_{k=0}^{n} \mu^k a_k(x, \mu) \frac{d^k y}{dx^k} = 0, \quad \mu > 0, \tag{4.1.1}$$

where

$$a_k(x, \mu) \simeq \sum_{j=0}^{\infty} \mu^j a_{kj}(x), \quad \mu \to 0, \quad x \in S,$$

in the real or complex domain S under the assumption that the coefficients $a_{kj}(x)$ are real analytic or complex analytic (holomorphic) in S, respectively. Later in the book, domains are assumed to be real unless specifically stated. However most of the

© Springer International Publishing Switzerland 2015
S.M. Bauer et al., *Asymptotic Methods in Mechanics of Solids*,
International Series of Numerical Mathematics 167,
DOI 10.1007/978-3-319-18311-4_4

results hold also for complex domains. After the introduction of additional unknown functions

$$y_j = \mu^j \frac{d^j y}{dx^j}, \quad j = 0, 1, \ldots, n - 1,$$

Equation (4.1.1) can be transformed into a system of equations. However for the sake of illustration the equation is studied separately.

4.1.1 Simple Roots of the Characteristic Equation

We seek a solution of equation (4.1.1) in the form

$$y(x, \mu) = U(x, \mu) \exp\left(\frac{1}{\mu} \int_{x_0}^{x} \lambda(x) \, dx\right), \quad U(x, \mu) = \sum_{k=0}^{\infty} \mu^k u_k(x). \qquad (4.1.2)$$

After substitution in (4.1.1) and equating the coefficients of μ^k to zero. we obtain a system of equations in the unknowns $\lambda(x)$ and $u_k(x)$:

$$P_0 u_0 = 0, \qquad (4.1.3)$$

$$P_0 u_1 + P_0^{(1)} u_0' + \frac{1}{2} P_0^{(2)} \lambda' u_0 + P_1 u_0 = 0, \quad ()' = \frac{d}{dx} \qquad (4.1.4)$$

$$P_0 u_2 + P_0^{(1)} u_1' + \frac{1}{2} P_0^{(2)} \lambda' u_1 + P_1 u_1 + \frac{1}{2} P_0^{(2)} u_0'' + \frac{1}{2} P_0^{(3)} \lambda' u_0' + \frac{1}{6} P_0^{(3)} \lambda'' u_0$$

$$+ \frac{1}{8} P_0^{(4)} (\lambda')_2 u_0 + P_1^{(1)} u_0' + \frac{1}{2} P_1^{(2)} \lambda' u_0 + P_2 u_0 = 0, \ldots, \qquad (4.1.5)$$

where the functions $P_k^{(m)}$ are polynomials in λ with coefficients depending on x and their derivatives in λ of order m:

$$P_j = P_j(x, \lambda) = \sum_{k=0}^{n} a_{kj}(x) \lambda^k, \quad j = 0, 1, \ldots,$$

$$P_j^{(m)} = P_j^{(m)}(x, \lambda) = \frac{d^m P_j}{d\lambda^m}, \quad m = 1, 2, \ldots \qquad (4.1.6)$$

To derive formulas (4.1.3)–(4.1.5) one should use the expansion

$$\frac{1}{\mu^k}\frac{d^k}{dx^k}\left[U\exp\left(\frac{1}{\mu}\int_{x_0}^{x}\lambda(x)\,dx\right)\right] = \left[U\lambda^k + \mu\left(k\lambda^{k-1}U' + \frac{k(k-1)}{2}\lambda^{k-2}\lambda'U\right)\right.$$
$$+ \mu^2\left(\frac{k(k-1)}{2}\lambda^{k-2}U'' + k(k-1)(k-2)\lambda^{k-3}\left(\frac{\lambda'U'}{2} + \frac{\lambda''U}{6}\right)\right.$$
$$+ \left.\left.\frac{k(k-1))k-2)(k-3)}{8}(\lambda')^2 U\right) + O(\mu^3)\right]\exp\left(\frac{1}{\mu}\int_{x_0}^{x}\lambda(x)\,dx\right),$$

which may be checked by means of mathematical induction. After that this expansion and also expansions for coefficients $a_k(x,\mu)$ and function $U(x,\mu)$ are substituted in Eq. (4.1.1).

We are interested only in nontrivial solutions of equation (4.1.1). Thus, Eq. (4.1.3) produces the characteristic equation in λ,

$$P_0(x,\lambda) = \sum_{k=0}^{n} a_{k0}(x)\lambda^k = 0. \qquad (4.1.7)$$

For $a_{n0}(x) \neq 0$, Eq. (4.1.7) has n roots

$$\lambda_1(x), \ \lambda_2(x), \ \ldots, \ \lambda_n(x). \qquad (4.1.8)$$

Let $\lambda(x)$ be a simple root of equation (4.1.7), i.e. $P_0^{(1)}(x,\lambda(x)) \neq 0$ for $x \in S$. Then Eqs. (4.1.4), (4.1.5),...have solutions $u_0(x)$, $u_1(x)$, ... analytic in S. Therefore, series (4.1.2), with analytic coefficients $u_k(x)$, transform equation (4.1.1) into an identity in S. Such series is called a *formal asymptotic solution*. Further, we limit ourselves to the construction of such solutions leaving aside the question of existence of exact solutions for which the obtained solutions are asymptotic expansions.

If all n roots (4.1.8) of Eq. (4.1.7) are simple, one can construct n linearly independent solutions of equation (4.1.1) in the following way.

Find the asymptotic expansion (4.1.2) of a linearly independent solution of the equation

$$\mu^2\frac{d^2y}{dx^2} + \rho(x)y = 0, \quad \rho(x) > 0, \qquad (4.1.9)$$

where $\mu > 0$ is a small parameter and the function $\rho(x)$ is analytic. The problem of free vibrations of a string with variable linear density along the length of the string is reduced to this equation (see Sect. 4.4.3).

The characteristic equation (4.1.7) takes the form $\lambda^2 + \rho(x) = 0$ and has solutions $\lambda(x) = \pm iq(x)$, $q(x) = (\rho(x))^{1/2}$. Construct a solution corresponding to the root $\lambda(x) = iq(x)$. The second solution is the complex conjugate of the first one. Equations (4.1.4), (4.1.5), ... give:

$$2qu'_n + u_nq' = iu''_{n-1}, \quad n = 0, 1, \ldots, \quad u_{-1} = 0, \quad ()' = \frac{d}{dx}. \tag{4.1.10}$$

Solving these equations we obtain

$$u_0 = [\rho(x)]^{-1/4}, \quad u_{n+1} = \frac{i}{2}u_0(x)\int_{x_0}^x u_0(\xi)u''_n(\xi)\,d\xi, \quad n = 0, 1, \ldots, \tag{4.1.11}$$

where the lower limit of integration, x_0, is arbitrary and may be chosen for convenience. In the particular case $\rho(x) = 1 + \alpha x$ we get

$$u_n = (-i\alpha)^n b_n(\rho(x))^{-\beta_n}, \tag{4.1.12}$$

where

$$b_0 = 1, \quad b_n = \frac{(6n-5)(6n-1)}{48n}b_{n-1}, \quad \beta_n = \frac{1}{4} + \frac{3n}{2}.$$

It is clear that series (4.1.2) converges since the ratio of two consequent terms goes to infinity with n.

As a second example, we construct the asymptotic expansions (4.1.2) for the solutions of the equation

$$\mu^4\frac{d^2}{dx^2}\left(p(x)\frac{d^2y}{dx^2}\right) - \rho(x)y = 0, \quad p(x), \rho(x) > 0, \tag{4.1.13}$$

where the functions $p(x)$ and $\rho(x)$ are analytic. The problem of transverse vibrations of a beam with variable cross-section can be reduced to this equation (see Sect. 4.4.4). In this case

$$P_0(\lambda) = p\lambda^4 - \rho, \quad P_1(\lambda) = 2p'\lambda^3, \quad P_2(\lambda) = p''\lambda^2.$$

The characteristic equation (4.1.7) has four roots:

$$\lambda_k = qr_k, \quad k = 1, 2, 3, 4, \quad q = \left(\frac{\rho}{p}\right)^{1/4}, \quad r_1 = i, \ r_2 = -i, \ r_3 = -1, \ r_4 = 1.$$
$$\tag{4.1.14}$$

Equation (4.1.4) gives

$$L_0u_0 = 4\lambda^3 pu'_0 + 6\lambda^2\lambda' pu_0 + 2\lambda^3 p'u_0 = 0,$$

from which we find $u_0 = p^{-1/8}\rho^{-3/8}$ for all λ_k.

Equation (4.1.5) has the form

$$L_0u_1^{(k)} + r_k^2 L_1u_0 = 0,$$

where

$$L_1 u_0 = 6pq^2 u_0'' + 12 pqq' u_0' + 4 pqq'' u_0 + 3 p(q')^2 u_0$$
$$+ 6 p' q^2 u_0' + 6 p' qq' u_0 + p'' q^2 u_0.$$

From the last relation, we obtain

$$u_1^{(k)} = -\frac{u_0}{4 r_k} \int u_0 L_1 u_0 \, dx, \quad k = 1, 2, 3, 4.$$

For the important case $p = \text{const}$, we get, after simplification,

$$u_1^{(k)} = -\frac{v_1}{r_k}, \quad v_1 = \frac{5 u_0}{8} \int \frac{3(q')^2 - 2 qq''}{q^3} \, dx. \tag{4.1.15}$$

For the general case, see Sect. 4.4.4.

The integrals $y^{(1)}$ and $y^{(2)}$ oscillate and the integrals $y^{(3)}$ and $y^{(4)}$ are edge effect integrals.

4.1.2 Multiple Roots of the Characteristic Equation

The case of multiple roots is a matter of significant difficulties. If the multiplicity of the roots (4.1.8) changes at some points $x = x_*$, such points are called *turning points* or *transition points*. They are considered in Chap. 5. Besides that, the roots may be multiple identically. In this case, fractional powers of μ can appear in expansion (4.1.2). Here we limit ourselves to study the case of the root zero of multiplicity m, that is widely met in applications.

Consider the linear differential equation of order $n = l + m$,

$$L_\mu y = \sum_{k=0}^{l} \mu^k a_{k+m}(x, \mu) \frac{d^{k+m} y}{dx^{k+m}} + \sum_{k=0}^{m-1} a_k(x, \mu) \frac{d^k y}{dx^k} = 0, \tag{4.1.16}$$

under the same assumptions on the coefficients $a_k(x, \mu)$ as for Eq. (4.1.1). For $\mu = 0$, Eq. (4.1.16) degenerates into the following equation of order m:

$$L_0 y = \sum_{k=0}^{m} a_k(x, 0) \frac{d^k y}{dx^k} = 0. \tag{4.1.17}$$

Multiply equation (4.1.16) by μ^m to get an equation of the form (4.1.1). The corresponding characteristic equation

$$\sum_{k=0}^{l} a_{k+m}(x,0)\lambda^{k+m} = 0, \tag{4.1.18}$$

has the root zero of multiplicity m. Let $a_n(x,0) \neq 0$ and $a_m(x,0) \neq 0$, and suppose that all roots of the equation

$$\sum_{k=0}^{l} a_{k+m}(x,0)\lambda^{k} = 0 \tag{4.1.19}$$

are simple. Then Eq. (4.1.16) has l solutions of the form (4.1.2), the other m solutions are not fast oscillating functions in x and they admit the expansions

$$y(x,\mu) = \sum_{k=0}^{\infty} \mu^{k} v_{k}(x), \tag{4.1.20}$$

where $v_0(x)$ satisfies Eq. (4.1.17), and the functions $v_k(x)$ satisfy non-homogeneous equations, with left sides the same as in (4.1.17).

Find asymptotic expansions for the solutions of the second-order equation

$$\mu \frac{d^2 y}{dx^2} + a_1(x)\frac{dy}{dx} + a_0(x)y = 0, \quad a_0(x) \neq 0, \quad a_1(x) \neq 0, \tag{4.1.21}$$

where μ is a small parameter, and the functions $a_0(x)$ and $a_1(x)$ are analytic.

The characteristic equation (4.1.18) has two roots: $\lambda_1(x) = -a_1(x)$ and $\lambda_2(x) = 0$. The root $\lambda_1(x)$ provides a solution of the form (4.1.2) and by (4.1.5) the functions $u_n(x)$ satisfy the equations

$$(\lambda u_n)' + a_0 u_n + u''_{n-1} = 0, \quad n = 0, 1, \ldots, \quad u_{-1} = 0.$$

When solving these equation we obtain

$$u_0 = \frac{1}{a_1(x)} \exp\left(\int \frac{a_0}{a_1} dx\right), \quad u_n = u_0 \int \frac{u''_{n-1}}{a_1 u_0} dx, \quad n = 1, 2, \ldots$$

In the particular case $a_1(x) = 1 + \alpha x$, we get

$$u_n = \alpha^n b_n (1 + \alpha x)^{\beta_n}, \quad \beta_n = \frac{1}{\alpha} - 2n - 1, \quad b_0 = 1, \quad b_{n+1} = -\frac{\beta_n(\beta_n - 1)}{2n+3} b_n.$$

The series (4.1.2) diverges.

Now, find the slowly varying solution (4.1.20) of Eq. (4.1.21) corresponding to the root $\lambda_2(x) = 0$. So, we have

$$a_1 v_0' + a_0 v_0 = 0, \quad v_0 = \exp\left(-\int \frac{a_0}{a_1} dx\right),$$

$$a_1 v_n' + a_0 v_n + v_{n-1}'' = 0, \quad v_n = -v_0 \int \frac{v_{n-1}''}{a_1 v_0} dx, \quad n = 1, 2, \dots$$

In the particular case $a_1(x) = 1 + \alpha x$, we have

$$v_n = \alpha^n c_n (1 + \alpha x)^{\gamma_n}, \quad \gamma_n = -\frac{1}{\alpha} - 2n, \quad c_0 = 1, \quad c_{n+1} = \frac{\gamma_n(\gamma_n - 1)}{2n + 3} c_n.$$

The series (4.1.20) is also divergent.

4.1.3 Asymptotic Solutions of Parameter-Free Equations

The above algorithm for the construction of solutions may be applied to some linear equations not containing the parameter μ as $\xi \to \infty$. Consider the equation

$$\sum_{k=0}^{n} d_k(\xi) \frac{d^k y}{d\xi^k} = 0 \tag{4.1.22}$$

in the domain $S = [\xi_0, \infty)$ for

$$d_k(\xi) = \sum_{j=0}^{\infty} d_{kj} \xi^{-j}. \tag{4.1.23}$$

After the change of variable $x = \mu\xi$, Eq. (4.1.22) takes the form (4.1.1) where $a_{kj} = d_{kj} x^{-j}$.

As before, we seek a solution in the form (4.1.2). Since the coefficients a_{k0} are constant, the roots (4.1.8) of Eq. (4.1.7) are also constant. For the simple root λ, Eqs. (4.1.4), (4.1.5), ... provide the system

$$P_0^{(1)} u_0' + b_1 x^{-1} u_0 = 0, \quad b_1 = \sum_{k=0}^{n} d_{k1} \lambda^k,$$

$$P_0^{(1)} u_1' + b_1 x^{-1} u_1 + \frac{1}{2} P_0^{(2)} u_0'' + b_1^{(1)} x^{-1} u_0' + b_2 x^{-2} u_0 = 0, \tag{4.1.24}$$

$$b_1^{(1)} = \sum_{k=0}^{n} k d_{k1} \lambda^{k-1}, \quad b_2 = \sum_{k=0}^{n} d_{k2} \lambda^k, \dots$$

Solving system (4.1.24), we obtain

$$u_k = c_k x^{\alpha - k}, \quad k = 0, 1, \ldots, \tag{4.1.25}$$

where c_0 is an arbitrary constant and

$$\alpha = -\frac{b_1}{P_0^{(1)}}, \quad c_1 = c_0 \left[\frac{1}{2} \alpha (\alpha - 1) P_0^{(2)} + b_1^{(1)} \alpha + b_2 \right] \left(P_0^{(1)} \right)^{-1}, \ldots$$

Returning to the original variable ξ we find an asymptotic solution of equation (4.1.22) in the form

$$y \simeq \sum_{k=0}^{\infty} c_k \xi^{\alpha - k} e^{\lambda \xi}, \quad \text{as} \ \xi \to \infty. \tag{4.1.26}$$

In this manner, if all n roots of equation (4.1.7) are simple we find all n solutions. Naturally, it is more convenient to seek solutions of equation (4.1.22) in the form (4.1.26) without introducing the auxiliary variable x. These manipulations were made to show the relation between asymptotic expansions in the parameter μ and asymptotic expansions of solutions of linear differential equations in a neighborhood of the singular point $\xi = \infty$.

Find the asymptotic expansion of the Bessel function $J_\nu(x)$ (see [1]) satisfying Bessel's equation

$$\frac{d^2 y}{dx^2} + \frac{1}{x} \frac{dy}{dx} + \left(1 - \frac{\nu^2}{x^2} \right) y = 0, \tag{4.1.27}$$

as $x \to +\infty$.

This equation is a particular case of equation (4.1.22) and its solutions have the form (4.1.26):

$$y = e^{\lambda x} \sum_{k=0}^{\infty} c_k x^{\alpha - k}.$$

Substituting this solution in Eq. (4.1.27) we have $\lambda_{1,2} = \pm i$. For $\lambda_1 = i$ we obtain

$$\alpha = -\frac{1}{2}, \quad c_0 = 1, \quad c_{k+1} = i \frac{(k + 1/2)^2 - \nu^2}{2(k + 1)} c_k, \quad k = 0, 1, \ldots$$

So, we have $J_\nu(x) \simeq \Re(Cy)$, where C is a complex constant, which can be evaluated by the integral representation of $J_\nu(x)$,

$$J_\nu(x) = \frac{(x/2)^\nu}{\sqrt{\pi}\, \Gamma(\nu + 1/2)} \int_0^\pi \cos(x \cos \theta) \sin^{2\nu} \theta \, d\theta.$$

Using the method of stationary phase (see Sect. 2.4) we find the main term in the asymptotic expansion

$$J_\nu(x) \simeq \sqrt{\frac{2}{\pi x}} \left[\cos\left(x - \frac{\pi\nu}{2} - \frac{\pi}{4} \right) + O\left(\frac{1}{x}\right) \right].$$

Comparing this formula with the first term of the series for $y(x)$, we obtain

$$C = \sqrt{\frac{2}{\pi}} \exp\left(-\frac{i\pi}{2}\left(\nu + \frac{1}{2} \right) \right).$$

4.1.4 Asymptotic Solutions of Non-homogeneous Equations

Now, we construct particular solutions for non-homogeneous equations. Consider the equation

$$M_\mu y = f(x, \mu) \exp\left(\frac{1}{\mu} \int_{x_0}^x \gamma(x)\, dx \right), \quad f(x, \mu) \simeq \sum_{k=0}^\infty \mu^k f_k(x), \quad (4.1.28)$$

where the functions $\gamma(x)$ and $f_k(x)$ are analytic in S and $M_\mu y$ is the operator in the left side of (4.1.1). If

$$\gamma(x) \neq \lambda_k(x), \quad k = 1, \ldots, n, \quad x \in S, \quad (4.1.29)$$

where $\lambda_k(x)$ are roots of equation (4.1.7), then a particular solution $y^*(x, \mu)$ of Eq. (4.1.28) has the form

$$y^*(x, \mu) \simeq \sum_{k=0}^\infty \mu^k v_k(x) \exp\left(\frac{1}{\mu} \int_{x_0}^x \gamma(x)\, dx \right), \quad \text{as} \quad \mu \to 0. \quad (4.1.30)$$

The functions $v_k(x)$ are analytic in S and they are evaluated after substitution of (4.1.30) in (4.1.28). In particular,

$$v_0(x) = f_0(x)(P_0(\gamma(x), x))^{-1}. \quad (4.1.31)$$

If condition (4.1.29) fails at distinct points x, then the solution differs from (4.1.30). This is the so-called *resonance case*, which is discussed, for example, in [25].

The non-homogeneous equation (4.1.16)

$$L_\mu y = f(x, \mu) \simeq \sum_{k=0}^\infty \mu^k f_k(x), \quad \text{as} \quad \mu \to 0, \quad (4.1.32)$$

where the operator $L_\mu y$ is as in (4.1.16) with $a_m(x, 0) \neq 0$, has particular solution

$$y^*(x, \mu) \simeq \sum_{k=0}^{\infty} \mu^k y_k^*(x), \quad \text{as} \quad \mu \to 0. \tag{4.1.33}$$

Here $y_0^*(x)$ is one of the particular solutions of the equation

$$L_0 y = f_0(x) \tag{4.1.34}$$

and $L_0 y$ is the operator in the left side of Eq. (4.1.17).

4.1.5 Exercises

4.1.1. Find the asymptotic expansion (4.1.2) for the solutions of the equation

$$\mu^2 \frac{d}{dx}\left(p(x)\frac{dy}{dx}\right) - \mu^2 r(x)y + \rho(x)y = 0, \quad p(x), \rho(x) > 0,$$

where the functions $p(x)$, $r(x)$ and $\rho(x)$ are analytic. This equations appears when studying asymptotic solutions of Sturm–Liouville problems. Consider the particular case $p = \rho = 1 + \alpha x$ and $r = 0$, corresponding to longitudinal vibrations of a bar with a linearly varying cross-section.

4.1.2. Find the asymptotic expansion (4.1.2) for the solutions of the equation

$$\mu^2 \frac{d^2 y}{dx^2} - c(x)y = 0, \quad c(x) > 0,$$

where $c(x)$ is analytic.

4.1.3. Find u_0 and $u_1^{(k)}$ for Eq. (4.1.13) for $p(x) = \rho(x) = 1 + \alpha x$. In this case, the equation describes the vibrations of a beam with a linearly varying width.

4.1.4. Under the conditions of Exercise 4.1.3 find u_0 and $u_1^{(k)}$ for a beam with a linearly varying thickness, i.e. $p(x) = \rho(x)^3$, $\rho(x) = 1 + \alpha x$.

4.1.5. Find the first two terms of the asymptotic expansions (4.1.2) for the solutions of the equation

$$\mu^4 \frac{d^2}{dx^2}\left(p(x)\frac{d^2 y}{dx^2}\right) + c(x)y = 0, \quad p(x), \ c(x) > 0,$$

describing the deflection of a beam on an elastic foundation.

4.1.6. Find the first terms of the asymptotic expansions for the solutions of the following equation with constant coefficients

$$-\mu^2 \frac{d^4 y}{dx^4} + \frac{d^2 y}{dx^2} + \Lambda y = 0,$$

which describes the vibrations of a non-absolutely flexible string (see Sect. 4.4.2).

4.1.7. The modified Bessel functions $I_\nu(x)$ and $K_\nu(x)$ satisfy the equation

$$\frac{d^2 y}{dx^2} + \frac{1}{x} \frac{dy}{dx} - \left(1 + \frac{\nu^2}{x^2}\right) y = 0.$$

Find the asymptotic expansions of the functions $I_\nu(x)$ and $K_\nu(x)$ for fixed ν as $x \to +\infty$. To evaluate the constant multipliers use the integral representations:

$$I_\nu(x) = \frac{1}{\pi} \int_0^\pi \exp(x \cos \theta) \cos(\nu\theta) \, d\theta - \frac{\sin(\pi\nu)}{\pi} \int_0^\infty \exp(-x \cosh t - \nu t) \, dt,$$

$$K_\nu(x) = \int_0^\infty \exp(-x \cosh t) \cosh(\nu t) \, dt.$$

and apply the Laplace method (see Sect. 2.3).

4.2 Solutions of Systems of Linear Ordinary Differential Equations

The results obtained in this section are largely similar to those found in Sect. 4.1.

Consider the system of equations

$$\mu \frac{dy}{dx} = A(x, \mu) y, \quad \mu > 0, \tag{4.2.1}$$

where y is an n-dimensional vector and A is a square matrix of order n with real or complex analytic coefficients in the complex domain S, in the form

$$A(x, \mu) \simeq \sum_{k=0}^\infty \mu^k A_k(x), \quad \text{as} \quad \mu \to 0, \tag{4.2.2}$$

4.2.1 Simple Roots of the Characteristic Equation

We seek a formal asymptotic solution of system (4.2.1) in the form

$$y(x, \mu) \simeq \sum_{k=0}^\infty U_k(x) \mu^k \exp\left(\frac{1}{\mu} \int_{x_0}^x \lambda(x) \, dx\right). \tag{4.2.3}$$

The function $\lambda(x)$ satisfies the characteristic equation

$$\det(A_0(x) - \lambda(x)I_n) = 0, \tag{4.2.4}$$

where I_n is the identity matrix of order n. If $\lambda(x)$ is a simple root of equation (4.2.4) for all $x \in S$, then all vector-functions $U_k(x)$ are recursively evaluated after substitution of (4.2.3) into (4.2.1) and they are analytic in domain S.

We consider in detail the process of evaluating a vector-function $U_0(x)$. After substitution of (4.2.3) in system (4.2.1) and equating coefficients at μ^0 and μ^1 we obtain the equations

$$[A_0 - \lambda(x)I_n]U_0 = 0,$$
$$[A_0 - \lambda(x)I_n]U_1 + A_1 U_0 = \frac{dU_0}{dx}. \tag{4.2.5}$$

From the first of these equations it follows that

$$U_0(x) = \varphi_0(x)V(x), \tag{4.2.6}$$

where $\varphi_0(x)$ is a scalar function and $V(x)$ is an eigenvector of the matrix $A_0(x)$ corresponding to the simple root $\lambda(x)$ of Eq. (4.2.4).

The function $\varphi_0(x)$ is evaluated only at the next approximation. The second equation in (4.2.5) is considered as a system of linear non-homogeneous equations in the components of the vector U_1. The determinant of this system is equal to zero and the compatibility condition for the system is

$$W^T A_1 U_0 = W^T \frac{dU_0}{dx}, \tag{4.2.7}$$

where $W(x)$ is an eigenvector of matrix $A_0^T(x)$ and the symbol T means matrix transpose,

$$\left[A_0^T(x) - \lambda(x)I_n\right] W = 0. \tag{4.2.8}$$

From (4.2.7) we obtain a differential equation of first order for the function $\varphi_0(x)$:

$$\frac{d\varphi_0(x)}{dx} = b_0(x)\varphi_0(x), \quad \text{where} \quad b_0 = \frac{W^T(A_1 V - V'_x)}{W^T V}, \tag{4.2.9}$$

which can be integrated by quadratures.

For a simple root $\lambda(x)$ all functions $\lambda(x)$ and $U_k(x)$ are analytic in S. If for all $x \in S$ all roots of equation (4.2.4) are simple, then formula (4.2.6) defines the fundamental matrix of formal asymptotic solutions.

Find the asymptotic expansion of the solutions of the system of equations

$$\mu \frac{dy}{dx} = y \cos x + z \sin x, \quad \mu \frac{dz}{dx} = y \sin x + 3z \cos x \qquad (4.2.10)$$

as $\mu \to 0$. We seek a solution in the form (4.2.3):

$$\{y(x, \mu), z(x, \mu)\} = \sum_{k=0}^{\infty} \mu^k \{y_k(x), z_k(x)\} \exp\left(\frac{1}{\mu} \int \lambda(x) \, dx\right). \qquad (4.2.11)$$

The characteristic equation (4.2.4)

$$[\lambda(x)]^2 - 4\lambda(x) \cos x + 3 \cos^2 x - \sin^2 x = 0$$

has the roots $\lambda_{1,2}(x) = 2 \cos x \pm 1$. These roots are simple for all x and thus the required solution exists.

We now consider only $\lambda_1(x) = 2 \cos x + 1$. Substituting (4.2.11) in system (4.2.10) and equating the coefficients of μ^k we get the equations

$$\begin{aligned} -y_k(1 + \cos x) + z_k \sin x &= y'_{k-1}, \\ y_k \sin x + z_k(\cos x - 1) &= z'_{k-1}, \quad k = 0, 1, \ldots, \end{aligned} \qquad (4.2.12)$$

and $y_{-1} = z_{-1} \equiv 0$.

For $k = 0$, we obtain

$$y_0 = \varphi_0(x) \sin \frac{x}{2}, \quad z_0 = \varphi_0(x) \cos \frac{x}{2},$$

where the function $\varphi_0(x)$ is evaluated at the next approximation.

For $k \geq 1$, system (4.2.12) in y_k and z_k has zero determinant and the compatibility conditions for the system are

$$y'_{k-1} \sin \frac{x}{2} + z'_{k-1} \cos \frac{x}{2} = 0, \quad k = 1, 2, \ldots$$

For $k = 1$, the compatibility condition implies $\varphi'_0 = 0$.

Assume that $\varphi_0(x) = 1$. Then the solution of system (4.2.12) for $k \geq 1$ has the form

$$y_k = \varphi_k(x) \sin \frac{x}{2} - \frac{1}{2} y'_{k-1}, \quad z_k = \varphi_k(x) \cos \frac{x}{2} - \frac{1}{2} z'_{k-1}, \qquad (4.2.13)$$

Substituting this solution into the compatibility condition we get

$$\varphi'_k = \frac{1}{2} y''_{k-1} \sin \frac{x}{2} + \frac{1}{2} z''_{k-1} \cos \frac{x}{2}, \quad k = 1, 2, \ldots, \qquad (4.2.14)$$

from which, for $k = 1$, we obtain

$$\varphi_1' = -\frac{1}{8}\left(\sin^2\frac{x}{2} + \cos^2\frac{x}{2}\right) = -\frac{1}{8}, \quad \varphi_1 = -\frac{x}{8},$$

$$y_1 = \varphi_1 \sin\frac{x}{2} - \frac{1}{4}\cos\frac{x}{2}, \quad z_1 = \varphi_1 \cos\frac{x}{2} + \frac{1}{4}\sin\frac{x}{2}.$$

Thus, the first two terms of series (4.2.11) are obtained. Using the recursive formulas (4.2.13) and (4.2.14) one can find any number of terms; however one could hardly expect to find a general formula.

4.2.2 Multiple Roots of the Characteristic Equation

As in the case of one equation of order n, we should separately consider the cases where the roots $\lambda_k(x)$ of Eq. (4.2.4) change multiplicity at some points (turning points) and when the roots are identically equal to each other. The first of these cases is discussed in Chap. 5.

Here, we consider the system of equations

$$\frac{dy}{dx} = A_{11}y + A_{12}z, \quad \mu\frac{dz}{dx} = A_{21}y + A_{22}z, \tag{4.2.15}$$

where y and z are vectors of dimensions m and l, respectively, and A_{ij} are matrices of the corresponding sizes regularly depending on μ:

$$A_{ij} = A_{ij}(x, \mu) = \sum_{k=0}^{\infty} \mu^k A_{ij}^{(k)}(x). \tag{4.2.16}$$

Multiplying the first equation in (4.2.15) by μ we get a system of characteristic equations of the form (4.2.1) which has m roots identically equal to zero. The other l roots satisfy the equation

$$\det\left(A_{22}^{(0)}(x) - \lambda(x)I_l\right) = 0. \tag{4.2.17}$$

A root $\lambda(x)$ of Eq. (4.2.17), which is simple and vanishes nowhere, provides a solution of the form (4.2.3). If all l roots of equation (4.2.17) are of this type then we get l solutions of the form (4.2.3).

For $\mu = 0$, $\det(A_{22}^{(0)}(x)) \neq 0$ and system (4.2.15) degenerates into a system of equations in y_0 of order m:

$$\frac{dy_0}{dx} = \left(A_{11}^{(0)} - A_{12}^{(0)}A_{22}^{(0)-1}A_{21}^{(0)}\right)y_0, \quad z_0 = -A_{22}^{(0)-1}A_{21}^{(0)}y_0, \tag{4.2.18}$$

which has m linearly independent solutions.

For $\mu \neq 0$, the original system (4.2.15) has m solutions of the form

$$y(x, \mu) \simeq \sum_{k=0}^{\infty} \mu^k y_k(x), \quad z(x, \mu) \simeq \sum_{k=0}^{\infty} \mu^k z_k(x), \quad \text{as} \quad \mu \to \infty, \quad (4.2.19)$$

where $y_0(x)$ and $z_0(x)$ are the same as in (4.2.18).

We examine the behavior of solutions of the gyroscopic system

$$\frac{d}{dx}\left(A(x)\frac{dy}{dx}\right) + HG(x)\frac{dy}{dx} + B(x)y = 0 \quad (4.2.20)$$

under the assumption that y is an n-dimensional vector, $A(x)$ is a positive definite symmetric matrix, $G(x)$ is a skew-symmetric matrix with det $G(x) \neq 0$, the elements of the matrices A, G, and B are analytic, and H is a large parameter. System (4.2.20) is analyzed in [46]. Note that n is even since, for odd n, det $G(x) \equiv 0$.

System (4.2.20) is reduced to a system of the form (4.2.15) by introducing the auxiliary variables $z = A\,dy/dx$:

$$\frac{dy}{dx} = A^{-1}z, \quad \mu\frac{dz}{dx} = -GA^{-1}z + \mu By, \quad (4.2.21)$$

where $\mu = H^{-1}$ is a small parameter. This system has n rapidly oscillating solutions of the form (4.2.3) and Eq. (4.2.17) for $\lambda(x)$ is

$$f(\lambda(x)) = \det(G + \lambda(x)A) = 0. \quad (4.2.22)$$

We prove that all roots of this equation are pure imaginary by considering the auxiliary system of equations

$$A(x)\frac{du}{dt} + G(x)u = 0,$$

where x is a parameter. By virtue of Lyapunov's Theorem [45] a trivial solution of this system is stable since the quadratic form $V = u^T Au$ is positive definite and its derivative $dV/dt = -2u^T Gu \equiv 0$ because the matrix G is skew-symmetric. Therefore, Eq. (4.2.22) does not have any roots $\lambda(x)$ with $\Re\,\lambda(x) > 0$. But $f(-\lambda(x)) \equiv f(\lambda(x))$, and also roots with $\Re\,\lambda(x) < 0$ do not exist. There are no trivial roots because det $G \neq 0$.

Thus, the system has n rapidly oscillating integrals. The behavior of the other n integrals differs from those of type (4.2.19) since they are very slow varying integrals $(dy/dx \sim \mu y)$. To find the integrals we replace system (4.2.20) with the following equivalent system of Volterra integral equation of the first kind

$$y = y_0 - \mu \int_0^x G^{-1} \left[B y + \frac{d}{dx} \left(A \frac{dy}{dx} \right) \right] dx \equiv L(y),$$ (4.2.23)

where y_0 is an arbitrary vector. The solution of this system obtained by the iterative scheme $y_{n+1} = L(y_n)$ results in the asymptotic expansion (4.2.19), in which

$$y_0 = \text{const}, \quad y_1 = - \int_0^x G^{-1} B y_0 \, dx,$$

$$y_2 = \int_0^x \left[G^{-1} B \int_0^x G^{-1} B y_0 \, dx + G^{-1} \frac{d}{dx} (A G^{-1} B) y_0 \right] dx.$$

4.2.3 Asymptotic Solutions of Parameter-Free Systems

Similar to Sect. 4.1, we consider asymptotic equations for solutions of the system of equations

$$\frac{dy}{d\xi} = A(\xi) y$$ (4.2.24)

as $\xi \to \infty$. Assume that

$$A(\xi) \simeq \sum_{k=0}^{\infty} A_k \xi^{-k}, \quad \text{as} \quad \xi \to \infty,$$ (4.2.25)

where A_k are constant matrices.

The change of variable $x = \mu \xi$ transforms system (4.2.24) into one of the form (4.2.1). A simple root $\lambda(\xi)$ of the characteristic equation (4.2.4) defines the solution

$$y(\xi) \simeq \sum_{k=0}^{\infty} U_k \xi^{\alpha-k} e^{\lambda \xi}, \quad \text{as} \quad \xi \to \infty,$$ (4.2.26)

where U_k are constant and the factor α is equal to

$$\alpha = - \frac{V_0^T A_1 U_0}{V_0^T U_0}.$$ (4.2.27)

Here U_0 is the eigenvector of the matrix A_0 corresponding to the eigenvalue $\lambda(\xi)$ and V_0 is the eigenvector of the transposed matrix A_0^T.

4.2.4 Asymptotic Solutions of Non-homogeneous Systems

Consider the non-homogeneous system of equations

$$\mu \frac{dy}{dx} = A(x, \mu)y + F(x, \mu) \exp\left(\frac{1}{\mu} \int_{x_0}^{x} \gamma(x)dx\right), \qquad (4.2.28)$$

where

$$F(x, \mu) \simeq \sum_{k=0}^{\infty} \mu^k F_k(x), \quad \text{as} \quad \mu \to 0,$$

the matrix $A(x, \mu)$ is the same as in (4.2.1), and the vector-functions $F_k(x)$ and the function $\gamma(x)$ are analytic in S.

If condition (4.1.29) holds, where now $\lambda_k(x)$ are the roots of equation (4.2.4), then the solution of equation (4.2.28) has the asymptotic expansion

$$y^*(x, \mu) \simeq \sum_{k=0}^{\infty} y_k^*(x)\mu^k \exp\left(\frac{1}{\mu} \int_{x_0}^{x} \gamma(x)\,dx\right), \qquad (4.2.29)$$

The analytic coefficients, $y_k^*(x)$, are found recursively by substituting (4.2.29) into (4.2.28). In particular,

$$y_0^*(x) = [A_0(x) - I_n\gamma(x)]^{-1} F_0(x). \qquad (4.2.30)$$

The case $\gamma(x) = 0$ is not excluded from our consideration. However condition (4.1.29) requires that none of the roots of equation (4.2.4) vanishes at some points.

4.2.5 Equations of the Theory of Shells

The system of Donnell's equations [18]

$$\frac{Eh^3}{12(1 - \nu^2)} \Delta^1\Delta^1 w - \Delta_k^1\Phi^1 - \rho h\omega^2 w = 0,$$

$$\frac{1}{Eh} \Delta^1\Delta^1\Phi^1 + \Delta_k^1 w = 0, \qquad (4.2.31)$$

describes the free vibrations of a shallow shell, where w is deflection, Φ^1 is the stress function, E is Young's modulus, ν is Poisson's ratio, ρ is the density, h is the shell thickness, ω is the vibrations frequency, Δ^1 and Δ_k^1 are linear differential operators of the second order. We apply system (4.2.31) to describe non-axisymmetric vibrations of a shell of revolution with m waves in the circumferential direction. On the surface

Fig. 4.1 A shell of
revolution with curvilinear
coordinates s^1 and φ

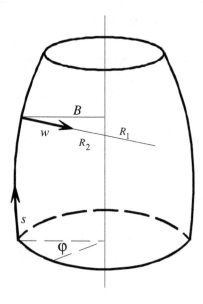

of the shell we introduce the curvilinear coordinates s^1, φ, where s^1 is the arc length
of the meridian and φ is the angle in the circumferential direction (see Fig. 4.1).

After separation of variables,

$$w(s^1, \varphi) = w(s^1) \cos m\varphi, \quad \Phi(s^1, \varphi) = \Phi(s^1) \cos m\varphi, \tag{4.2.32}$$

and the operators Δ^1 and Δ_k^1 take the form

$$\Delta^1 w = \frac{1}{B}(Bw')' - \frac{m^2}{B^2}w, \quad \Delta_k^1 w = \frac{1}{B}\left(\frac{Bw'}{R_2}\right)' - \frac{m^2}{R_1 B^2}w, \quad ' = \frac{d()}{ds^1}, \tag{4.2.33}$$

where $B(s^1)$ is the distance from the axis of revolution, and $R_1(s^1)$ and $R_2(s^1)$ are
the main radii of curvature (see Fig. 4.1).

Let R be the characteristic size of the neutral surface. We turn to non-dimensional
variables:

$$s = \frac{s^1}{R}, \quad b = \frac{B}{R}, \quad k_1 = \frac{R}{R_1}, \quad k_2 = \frac{R}{R_2},$$

$$\Phi = \frac{\Phi^1}{EhR\mu^2}, \quad \Lambda = \frac{\rho\omega^2 R^2}{E}, \quad \mu^4 = \frac{h^2}{12(1 - \nu^2)R^2}, \tag{4.2.34}$$

where Λ is the frequency parameter and $\mu > 0$ is a small parameter.

System (4.2.31) contains three main parameters, μ, m, and Λ, and the asymptotic
expansions of the solutions of the system depend on the relations between the orders
of the parameters [30].

4.2.6 The Case $m \sim \mu^{-1}$, $\Lambda \sim 1$

Consider non-axisymmetric vibrations with m waves in the circumferential direction, where m is large. Assume that $\Lambda = O(\mu^0)$, $m \sim \mu^{-1}$. Let $r = \mu m \sim 1$. Then system (4.2.31) can be rewritten in the form

$$\Delta \Delta w - \Delta_k \Phi - \Lambda w = 0, \quad \Delta \Delta \Phi + \Delta_k w = 0, \quad (4.2.35)$$

where

$$\Delta w = \mu^2 \frac{1}{b} (bw')' - \frac{r^2}{b^2} w, \quad \Delta_k w = \mu^2 \frac{1}{b} (bk_2 w')' - \frac{k_1 r^2}{b^2} w, \quad ()' = \frac{d()}{ds}.$$

System (4.2.35) contains a small parameter μ in the derivative terms. This system can be reduced to the standard form (4.2.1) by introducing the auxiliary functions

$$y_k = \mu^{k-1} \frac{d^{k-1} w}{ds^{k-1}}, \quad y_{k+4} = \mu^{k-1} \frac{d^{k-1} \Phi}{ds^{k-1}}, \quad k = 1, 2, 3, 4, \quad (4.2.36)$$

Hence, one should seek a solution of system (4.2.35) in the form

$$w(s, \mu) \simeq \sum_{k=0}^{\infty} \mu^k w_k(s) \exp\left(\frac{1}{\mu} \int_{s_0}^{s} \lambda(s) \, ds\right),$$

$$\Phi(s, \mu) \simeq \sum_{k=0}^{\infty} \mu^k \Phi_k(s) \exp\left(\frac{1}{\mu} \int_{s_0}^{s} \lambda(s) \, ds\right), \quad (4.2.37)$$

Substituting series (4.2.37) in system (4.2.35) and equating the coefficients of equal powers of μ we find that the function $\lambda(s)$ satisfies the characteristic equation, which is an algebraic equation of the degree eight:

$$f(\lambda, s) = \left(\lambda^2 - \frac{r^2}{b^2}\right)^4 - \Lambda \left(\lambda^2 - \frac{r^2}{b^2}\right)^2 + \left(k_2 \lambda^2 - \frac{k_1 r^2}{b^2}\right)^2 = 0, \quad (4.2.38)$$

and the leading coefficients of series (4.2.37) are equal to

$$w_0 = \left(\lambda^2 - \frac{r^2}{b^2}\right) \left(b \frac{\partial f}{\partial \lambda}\right)^{-1/2},$$

$$\Phi_0 = -\left(k_2 \lambda^2 - \frac{k_1 r^2}{b^2}\right) \left(\lambda^2 - \frac{r^2}{b^2}\right)^{-1} \left(b \frac{\partial f}{\partial \lambda}\right)^{-1/2}. \quad (4.2.39)$$

See Example 4.2.3.

Each simple root $\lambda(s)$ of Eq. (4.2.38) provides solution (4.2.37) with analytic coefficients $w_k(s)$ and $\Phi_k(s)$. The exceptions are shells in the shape of a cupola. At the apex of the cupola $b = 0$ and therefore expansions (4.2.37) are not applicable in a neighborhood of the apex.

With this method we get a general solution of system (4.2.35) if all eight roots of equation (4.2.38) are simple.

At the turning points $s = s_*$ the roots $\lambda(s)$ become multiple, $\partial f/\partial\lambda = 0$ and expansions (4.2.37) are inapplicable. This case is considered in Sect. 5.3.

4.2.7 The Case $m \sim \mu^{-1/2}$, $\Lambda \sim 1$

We obtain asymptotic expansions for solutions of system (4.2.31) under the assumption that $m \sim \mu^{-1/2}$ and $\Lambda \sim 1$. We now take system (4.2.35) as the original system:

$$\Delta w = \mu^2 \frac{1}{b}(bw')' - \mu \frac{r_0^2}{b^2} w, \quad \Delta_k w = \mu^2 \frac{1}{b}(bk_2 w')' - \mu \frac{k_1 r_0^2}{b^2} w, \quad (4.2.40)$$

where $r_0 = \mu^{1/2} m \sim 1$. In this case, one cannot assume that all solutions have asymptotic expansions of type (4.2.37). To clarify the structure of asymptotic expansions we start by seeking solutions in the form

$$w(s, \mu) \simeq w_0 \exp\left(\int_{s_0}^s p\, ds\right), \quad \Phi(s, \mu) \simeq \Phi_0 \exp\left(\int_{s_0}^s p\, ds\right), \quad |p| \gg 1.$$

For the function $p(s)$, we get an algebraic equation similar to (4.2.38)

$$\mu^2 \left(\mu p^2 - \frac{r_0^2}{b^2}\right)^4 - \Lambda\left(\mu p^2 - \frac{r_0^2}{b^2}\right)^2 + \left(\mu k_2 p^2 - \frac{k_1 r_0^2}{b^2}\right)^2 = 0. \quad (4.2.41)$$

This equation has four roots $p_j \sim \mu^{-1}$:

$$p_j = \mu^{-1}\lambda_j + O(1), \quad \lambda_j^4 + k_2^2 - \Lambda = 0, \quad j = 1, 2, 3, 4, \quad (4.2.42)$$

and four roots $p_j \sim \mu^{-1/2}$:

$$p_j = \mu^{-1/2} q_j + O(1), \quad \Lambda\left(q_j^2 - \frac{r_0^2}{b^2}\right)^2 = \left(k_2 q_j^2 - \frac{k_1 r_0^2}{b^2}\right)^2, \quad j = 5, 6, 7, 8,$$

$$q_{5,6} = \pm\frac{r_0}{b}\sqrt{\frac{\sqrt{\Lambda} + k_1}{\sqrt{\Lambda} + k_2}}, \quad q_{7,8} = \pm\frac{r_0}{b}\sqrt{\frac{\sqrt{\Lambda} - k_1}{\sqrt{\Lambda} - k_2}}. \quad (4.2.43)$$

Thus, the four solutions of system (4.2.35), (4.2.40) have expansions (4.2.37), where

$$w_0(s) = b^{-1/2}\lambda^{-3/2}\exp\left(-r_0^2\int_{s_0}^s \frac{b(\Lambda - 2\lambda^4) + b''}{2b^3\lambda^5}\,ds\right), \quad \Phi_0 = -\frac{k_2 w_0}{\lambda^2}.$$

$$(4.2.44)$$

The variation index (see Sect. 1.4) of these solutions is equal to 1. Four other solutions have variation index 1/2 and may be represented as

$$w(s,\mu) \simeq \sum_{k=0}^{\infty} \mu^{k/2} w_k(s)\exp\left(\mu^{-1/2}\int_{s_0}^s q(s)\,ds\right),$$

$$\Phi(s,\mu) \simeq \mu^{-1}\sum_{k=0}^{\infty}\mu^{k/2}k\Phi_k(s)\exp\left(\mu^{-1/2}\int_{s_0}^s q(s)\,ds\right),$$

$$(4.2.45)$$

and

$$w_0 = \frac{1}{\sqrt{\Lambda}\pm k_2}\left(\frac{k_1 - k_2}{qb^3}\right)^{1/2}, \quad \Phi_0 = -\frac{\Lambda w_0}{k_2 q^2 - k_1 r_0^2 b^{-2}},$$

where we take the plus sign for $q = q_{5,6}$, and the minus sign for $q = q_{7,8}$.

As it follows from the above formulas for w_0 and Φ_0, the obtained solutions are inapplicable in neighborhoods of the cupola apex ($b = 0$) and of the turning points $s = s_*$, where

$$\Lambda = k_1^2\left(s_*^{(1)}\right) \quad \text{or} \quad \Lambda = k_2^2\left(s_*^{(2)}\right).$$

For $s = s_*^{(1)}$, two of the roots q_j vanish as hey become equal to each other and for $s = s_*^{(2)}$ four roots λ_j vanish and two of the roots q_j become infinite.

4.2.8 Low Frequency Vibrations of Shells of Revolution of Zero Gaussian Curvature

Now consider non-axisymmetric low frequency vibrations of shells of revolution of zero Gaussian curvature for the following values of the parameters: $k_1 = 0$, $m \sim \mu^{-1/2}$, $\Lambda \sim \mu^2$. As in Sect. 4.2.7, we proceed from system (4.2.35) and (4.2.40), in which we assume that $k_1 = 0$, $\Lambda = \mu^2\Lambda_0$, and $\Lambda_0 \sim 1$.

Under the above assumptions, the characteristic equation (4.2.41) is

$$\left(\mu p^2 - \frac{r_0^2}{b^2}\right)^4 - \Lambda_0\left(\mu p^2 - \frac{r_0^2}{b^2}\right)^2 + k_2 p^4 = 0. \qquad (4.2.46)$$

This equation has four roots $p_j \sim \mu^{-1}$, $j = 1, 2, 3, 4$, which correspond to solutions (4.2.37), (4.2.42) and (4.2.44) for $\Lambda = 0$.

Four other roots $p_j \sim 1$, $j = 5, 6, 7, 8$, have small absolute values; thus, the solutions corresponding to them are

$$w(s, \mu) \simeq \sum_{k=0}^{\infty} \mu^k w_k(s), \quad \Phi(s, \mu) \simeq \sum_{k=0}^{\infty} \mu^k \Phi_k(s). \tag{4.2.47}$$

Asymptotic expansions (4.2.47) are used to find approximately the lower part of the frequency spectrum for free vibrations of shells of revolution of zero Gaussian curvature [30].

4.2.9 Low Frequency Vibrations of Shells of Revolution of Negative Gaussian Curvature

Consider non-axisymmetric vibrations of a shell of revolution under the assumptions that $k_1 k_2 < 0$, $m \sim \mu^{-2/3}$, $\Lambda \sim \mu^{4/3}$. Introduce the small parameter $\mu_1 = \mu^{2/3}$ and set $r_1 = m\mu_1 \sim 1$, $\Lambda = \mu_1^2 \Lambda_1$, and $\Lambda_1 \sim 1$. Then system (4.2.35) may be written as

$$\mu_1 \Delta_1 \Delta_1 w - \Delta_{k1} \Phi - \mu_1 \Lambda_1 w = 0, \quad \mu_1 \Delta_1 \Delta_1 \Phi + \Delta_{k1} w = 0, \tag{4.2.48}$$

where

$$\Delta_1 w = \mu_1^2 \frac{1}{b} (bw')' - \frac{r_1^2}{b^2} w, \quad \Delta_{k1} w = \mu_1^2 \frac{1}{b} (bk_2 w')' - \frac{k_1 r_1^2}{b^2} w.$$

The characteristic equation for system (4.2.48),

$$\mu_1^2 \left(\mu_1^2 p^2 - \frac{r_1^2}{b^2} \right)^4 - \mu_1^2 \Lambda_1 \left(\mu_1^2 p^2 - \frac{r_1^2}{b^2} \right)^2 + \left(\mu_1^2 k_2 p^2 - \frac{k_1 r_1^2}{b^2} \right)^2 = 0,$$

has the roots

$$p_j = \mu_1^{-3/2} \lambda_j + O\left(\mu_1^{-1/2} \right), \quad \lambda_j^4 + k_2^2 = 0, \quad j = 1, 2, 3, 4, \tag{4.2.49}$$

$$p_j = \mu^{-1} q_j + O(1), \quad \left(k_2 q_j^2 - \frac{k_1 r_1^2}{b^2} \right)^2 = 0, \quad j = 5, 6, 7, 8. \tag{4.2.50}$$

The roots (4.2.49) define the solutions

$$w(s, \mu_1) \simeq \sum_{k=0}^{\infty} \mu_1^{k/2} w_k(s) \exp\left(\mu^{-3/2} \int_{s_0}^{s} \lambda(s)\,ds + \mu^{-1/2} \int_{s_0}^{s} \lambda^{(1)}(s)\,ds\right),$$

$$\lambda^{(1)} = -\frac{r_1^2}{2\lambda b^2}\left(1 + \frac{b'}{bk_2}\right), \qquad w_0 = b^{-1/2}\lambda^{-3/2}.$$

and the roots (4.2.50) provide the solutions

$$w(s, \mu_1) \simeq \sum_{k=0}^{\infty} \mu_1^k w_k(s) \exp\left(\frac{1}{\mu_1} \int_{s_0}^{s} q(s)\,ds\right),$$

$$\Phi(s, \mu_1) \simeq \sum_{k=0}^{\infty} \mu_1^k \Phi_k(s) \exp\left(\frac{1}{\mu_1} \int_{s_0}^{s} q(s)\,ds\right). \tag{4.2.51}$$

Solutions (4.2.51) are used to find approximately the lower part of the frequency spectrum for free vibrations of shells of revolution of negative Gaussian curvature [30].

4.2.10 Exercises

4.2.1. Find the first two terms of the asymptotic expansions of the solutions of system (4.2.10) for $\lambda_1(x) = 2\cos x - 1$.

4.2.2. Small vibrations of a dynamically symmetric top are described by the system of equations

$$\frac{d^2 y_1}{dt^2} - H \frac{dy_2}{dt} - k^2 y_1 = 0, \quad \frac{d^2 y_2}{dt^2} + H \frac{dy_1}{dt} - k^2 y_2 = 0, \tag{4.2.52}$$

where y_1 and y_2 are small deviations of the top axis from the vertical, t is time, $H = C\omega/A$ and $k^2 = Wz_c/A$. Here C is the moment of inertia of the top about the axis of symmetry, A is the moment of inertia of the top about the orthogonal axis passing through the point of support, W is the top weight, z_c is the distance between the center of gravity and the point of support, ω is the angular velocity of the top.

Under the assumption $H \to \infty$, find the asymptotic expansion of the solutions of system (4.2.52) and compare it with the exact solution.

4.2.3. Determine formulas (4.2.39) for w_0, Φ_0.

4.2.4. The stability of a membrane axisymmetric stress state in a shell of revolution is described by the following system of equations [10, 31, 56]:

$$\Delta\Delta w + \Lambda\Delta_t w - \Delta_k \Phi = 0, \quad \Delta\Delta\Phi + \Delta_k w = 0, \tag{4.2.53}$$

where

$$\Delta_t w = \mu^2 \frac{1}{b}(bt_1 w')' - \frac{t_2 r^2}{b^2} w, \quad t_k = -\frac{T_k^0}{\Lambda E h \mu^2}, \ k = 1, 2.$$

Here $T_k^0(s)$ are the membrane initial stress-resultants and $\Lambda > 0$ is the loading parameter. The other notations are the same as for system (4.2.35).

Find the asymptotic expansions of the solutions of system (4.2.53).

4.2.5. Establish the equation for evaluating the functions w_0 and Φ_0 in the asymptotic expansions (4.2.47).

4.2.6. Derive the system of equations for evaluating the main terms of series (4.2.51).

4.3 Non-homogeneous Boundary Value Problems

4.3.1 Statement of Boundary Value Problems

Let the domain S contain a segment $x_1 \le x \le x_2$ of the real axis. Here we consider both the differential equation (4.1.28) and the system of first order equations (4.2.28). We suppose that the conditions of Sects. 4.1 and 4.2 on the solutions hold, the main condition being the absence of turning points. Then, the general solution may be written in the form

$$y(x, \mu) = \sum_{k=1}^{n} C_k y^{(k)}(x, \mu) + y^*(x, \mu), \tag{4.3.1}$$

where C_k are arbitrary constants, $y^{(k)}(x, \mu)$ are particular solutions of the homogeneous equations (4.1.1) or (4.2.1), $y^*(x, \mu)$ is a particular solution of the non-homogeneous equation (4.1.28) or (4.2.28).

For brevity's sake, we write formula (4.3.1) as

$$y(x, \mu) = Y(x, \mu)C + y^*(x, \mu), \quad C = (C_1, \ldots, C_n)^T, \tag{4.3.2}$$

where $Y(x, \mu)$ is a fundamental system, i.e. nonsingular matrix of order n consisting of particular solutions $y^{(k)}(x, \mu)$ of the homogeneous equation.

For system (4.2.28) the boundary conditions are introduced in the form

$$B_j(\mu)y = b_j(\mu) \quad \text{for } x = x_j, \quad j = 1, 2, \tag{4.3.3}$$

where y is an n-dimensional vector, B_j is a matrix of size $n_j \times n$, b_j is a vector of dimension n_j, and $n_1 + n_2 = n$. The vector y and the matrix B_j are represented in powers of μ as

$$B_j(\mu) \simeq \sum_{k=0}^{\infty} B_{jk}\mu^k, \quad b_j(\mu) \simeq \sum_{k=0}^{\infty} b_{jk}\mu^k, \quad \text{as} \quad \mu \to 0. \tag{4.3.4}$$

We assume that the boundary conditions are linearly independent in the zeroth approximation, i.e.

$$\text{rank}\{B_{0j}\} = n_j, \quad j = 1, 2. \tag{4.3.5}$$

The general form of the boundary conditions of Eq. (4.1.28) is also of the form (4.3.3), if for y we use the n-dimensional vector

$$y = \left[y, \ \mu \frac{dy}{dx}, \ldots, \mu^{n-1} \frac{d^{n-1}y}{dx^{n-1}} \right]^T. \tag{4.3.6}$$

Here the condition on the matrices B_{j0} is the same as above.

Substituting the general solution (4.3.1) in the boundary conditions (4.3.3) results in a system of n linear non-homogeneous equations in the constants C_k:

$$D(\mu)C = d(\mu), \tag{4.3.7}$$

where

$$D = \begin{bmatrix} D_1 \\ D_2 \end{bmatrix}, \quad d = \begin{bmatrix} d_1 \\ d_2 \end{bmatrix},$$

$$D_j(\mu) = B_j(\mu)y(x_j, \mu), \quad d_j(\mu) = b_j(\mu) - B_j(\mu)y^*(x_j, \mu), \quad j = 1, 2. \tag{4.3.8}$$

If

$$\det D(\mu) \neq 0, \tag{4.3.9}$$

then system (4.3.7) with the boundary value problem at hand has a unique solution. If

$$\det D(\mu) = 0, \tag{4.3.10}$$

the boundary value problem does not have a solution for all right sides $f(x, \mu)$ and $b(\mu)$. In this case we deal with *the spectrum problem*.

The difficulty with the roots of equation (4.3.10) is that the limit

$$D_0 = \lim_{\mu \to 0} D(\mu). \tag{4.3.11}$$

does not always exists

We now study some particular cases (see also Sects. 4.4 and 4.5).

4.3.2 Classification of Solution Types

Firstly, consider all possible variants of the behavior of integrals (4.1.2) and (4.2.3) depending on the roots $\lambda_k(x)$. For $\Re(\lambda_k(x)) > 0$ the integral grows exponentially, for $\Re(\lambda_k(x)) < 0$ it decreases, and for $\Re(\lambda_k(x)) = 0$ and $\Im(\lambda_k(x)) \neq 0$ it oscillates. If $\lambda_k(x) \equiv 0$, then integrals (4.1.20) and (4.2.19) vary slowly.

If the conditions

$$\Re(\lambda_k(x_1)) < 0, \quad \Re \int_{x_1}^x \lambda_k(x))dx < 0, \quad x_1 < x < x_2, \tag{4.3.12}$$

hold, then the integrals (4.1.2) or (4.2.3) are called *edge effect integral* or *boundary layer integral* at the left edge $x = x_1$. In this, case we assume that $x_0 = x_1$ in (4.1.2) or (4.2.3). Then the given integral decreases exponentially away form the edge $x = x_1$ and remains exponentially small for all $x > x_1$. In computing the corresponding entries of the matrix $Y(x_2, \mu)$ in (4.3.8) with an error of order $e^{-c/\mu}$, $c > 0$ we may assume them to be zero.

To describe an edge effect integral, besides (4.1.2) or (4.2.3) one may use the representation

$$y_k(x, \mu) \simeq \sum_{m=0}^{\infty} \mu^m P_m^{(k)}(\xi)e^{\lambda_k^0 \xi}, \quad \xi = \frac{x - x_1}{\mu}, \tag{4.3.13}$$

where $\lambda_k^0 = \lambda_k(x_1) < 0$, $P_m^{(k)}(\xi)$ are polynomials in ξ of degree smaller than $2m$.

Edge effect integrals at the right edge $x = x_2$ are introduced the same way.

If for all x, $\Re(\lambda_k(x)) = 0$ and $\Im(\lambda_k(x)) \neq 0$, then the integral is called *oscillating*. There may exist integrals belonging to none of the mentioned types. We note among them the integrals oscillating in $S_0 \subset [x_1, x_2]$ (see Chap. 5). If the coefficients of equations (4.1.1) or (4.2.1) are real, then only the turning points or terminal points, x_1 or x_2, can be the ends of S_0.

4.3.3 The Simplest Case

Now consider the system of equations (4.3.7). We start with the simplest case when the characteristic equation (4.2.4) or (4.1.7) does not have zero or pure imaginary roots for all x. Let

$$\begin{aligned} \Re(\lambda_k(x)) < 0, \quad k = 1, 2, \ldots, l_1, \\ \Re(\lambda_k(x)) > 0, \quad k = l_1 + 1, l_1 + 2, \ldots, l_1 + l_2, \end{aligned} \tag{4.3.14}$$

where $l_1 + l_2 = l = n$, $l_1 = n_1$, $l_2 = n_2$ and n_1 and n_2 are the same as in (4.3.3), i.e. they are equal to the numbers of boundary conditions at $x = x_1$ and at $x = x_2$.

Then with an error of order $e^{-c/\mu}$, $c > 0$, the matrix (4.3.8) is block-diagonal

$$D = \begin{bmatrix} D_1 & 0 \\ 0 & D_2 \end{bmatrix}, \quad D(\mu) \simeq \sum_{k=0}^{\infty} \mu^k D^{(k)}, \tag{4.3.15}$$

where D_1 and D_2 are square matrices of dimensions n_1 and n_2, respectively.

For sufficiently small μ, condition (4.3.9) holds if

$$\Delta_j(0) = \det D_j(0) \neq 0, \quad j = 1, 2. \tag{4.3.16}$$

The solution of the boundary value problem (4.2.1) and (4.3.3) consists of a particular solution $y^*(x, \mu)$ corrected in the neighborhoods of the edges $x = x_1$ and $x = x_2$ with edge effect integrals.

Example 1

Consider the deflection, $y(x_1)$, of a string on an elastic foundation. The function $y(x_1)$ satisfies the equation

$$T \frac{d^2 y}{dx_1^2} - c_1(x_1)y + q_1(x_1) = 0, \quad y(0) = y(l) = 0, \tag{4.3.17}$$

where T is the tension, $c_1(x_1) > 0$ is the foundation stiffness, $q_1(x_1)$ is the intensity of external load. Under the assumption that the tension T is relatively small, find an approximate value of the deflection. The functions $c_1(x_1)$ and $q_1(x_1)$ are considered to be analytic.

In Eq. (4.3.17) we introduce the non-dimensional variables

$$x_1 = lx, \ 0 \leq x \leq 1, \quad c_1(x_1) = c_0 c(x), \quad c(x) \sim 1.$$

Then, Eq. (4.3.17) is transformed into

$$\mu^2 \frac{d^2 y}{dx^2} - c(x)y + q(x) = 0, \quad y(0) = y(1) = 0, \tag{4.3.18}$$

where $\mu^2 = T/(c_0 l^2)$ and $q(x) = q_1(x_1)/c_0$. Assume the parameter μ to be small. This is a formalization of the original assumption on the relative smallness of the tension.

Here and below, when introducing a small parameter we are guided by a matter of convenience. We also wish that the obtained asymptotic series will contain only integer powers of μ. For that, the variation index (see Sect. 1.4) of fast oscillating solutions is equal to 1 as for formulas (4.1.2).

The general solution of equation (4.3.18) has the form (4.3.1):

$$y(x, \mu) = C_1 y^{(1)}(x, \mu) + C_2 y^{(2)}(x, \mu) + y^*(x, \mu), \tag{4.3.19}$$

where C_1 and C_2 are arbitrary constants, and $y^{(1)}$ and $y^{(2)}$ are particular solutions of the corresponding homogeneous equation obtained in Exercise 4.1.1, which has characters of edge effect integrals. The particular solution $y^*(x, \mu)$ of the non-homogeneous equation (4.3.18) is of the form (4.1.33):

$$y^*(x, \mu) = \sum_{k=0}^{\infty} \mu^k y_k^*(x),$$

where

$$y_0^*(x) = \frac{q(x)}{c(x)}, \quad y_{2k+1}^*(x) \equiv 0, \quad y_{2k+2}^*(x) = \frac{1}{c(x)} \frac{d^2 y_{2k}^*}{dx^2}, \quad k = 0, 1, 2, \ldots$$

The constants C_1 and C_2 in (4.3.19) are obtained from the boundary conditions $y(0) = y(1) = 0$. The boundary conditions

$$y(0) = C_1 y^{(1)}(0, \mu) + y^*(0, \mu) = 0, \quad y(1) = C_2 y^{(2)}(1, \mu) + y^*(1, \mu) = 0,$$

can be satisfied separately within an error or order

$$\varepsilon = \exp\left(-\frac{1}{\mu} \int_0^1 \sqrt{c(x)}\, dx\right).$$

Now choose the lower limit of integration $x^{(1)} = 0$, $x^{(2)} = 1$ in the formulas for evaluating $y_k^{(j)}(x)$, $k > 0$ (see the answer to Exercise 4.1.1) is such a way that $u_k^{(1)}(0) = 0$, $u_k^{(2)}(1) = 0$ for $k > 0$. Then

$$C_1 = -c(0)^{1/4} y^*(0, \mu), \quad C_2 = -c(1)^{1/4} y^*(1, \mu).$$

4.3.4 Deflection of a Beam on an Elastic Foundation

The deflection of a beam on an elastic foundation is described by the equation

$$\frac{d^2}{dx_1^2}\left(EI\, \frac{d^2 y}{dx_1^2}\right) + c_1(x_1)y = g_1(x_1), \quad 0 \le x_1 \le l,$$

where EI is the beam bending stiffness, $c_1(x_1)$ is the foundation stiffness and $g_1(x_1)$ is the load intensity.

In terms of the non-dimensional variables

$$EI = E_0 I_0 p(x), \quad c_1(x_1) = c_0 c(x), \quad g_1(x_1) = E_0 I_0 g(x),$$

$$x_1 = lx, \qquad 0 \le x \le 1,$$

the above equation becomes

$$\mu^4 \frac{d^2}{dx^2}\left(p(x)\frac{d^2 y}{dx^2}\right) + c(x)y = g(x), \quad \mu^4 = \frac{E_0 I_0}{c_0 l^4}, \tag{4.3.20}$$

where μ is a small parameter and the functions $p(x)$, $c(x)$ and $g(x)$ are assumed to be analytic. Moreover $p(x) \sim 1$, $c(x) \sim 1$, $p(x) > 0$, $c(x) > 0$.

The small parameter μ allows us to find approximate asymptotic solutions and satisfy separately boundary conditions at $x = 0$ and $x = 1$. We seek an approximate solution of equation (4.3.20) which satisfies the boundary conditions of free support type

$$y = \frac{d^2 y}{dx^2} = 0 \quad \text{for} \quad x = 0, \ x = 1.$$

The general solution of equation (4.3.20) is a linear combination of the edge effect integrals and a particular solution (see Exercise 4.1.5):

$$y(x,\mu) = \sum_{k=1,2} C_k \left[u_0(x) + \mu u_1^{(k)}(x) + O(\mu^2)\right]\exp\left(\frac{r_k}{\mu}\int_0^x q(x)dx\right)$$

$$+ \sum_{k=3,4} C_k \left[u_0(x) + \mu u_1^{(k)}(x) + O\left(\mu^2\right)\right]\exp\left(\frac{r_k}{\mu}\int_1^x q(x)dx\right)$$

$$+ y_0^*(x) + O\left(\mu^4\right),$$

where

$$u_0(x) = (pq^3)^{-1/2}, \quad q(x) = \left(\frac{c}{p}\right)^{1/4}, \quad y_0^*(x) = \frac{g}{c},$$

$$r_{1,2} = -\frac{1}{\sqrt{2}} \pm \frac{i}{\sqrt{2}}, \quad r_{3,4} = \frac{1}{\sqrt{2}} \pm \frac{i}{\sqrt{2}},$$

and the functions $u_1^{(k)}(x)$ are the same as in Exercise 4.1.5. We find the constants C_k from the boundary conditions.

We take $u_j^{(k)}(x)$, $j > 0$ in such a way that the equalities

$$u_j^{(k)}(0) = 0, \ k = 1,2, \quad u_j^{(k)}(1) = 0, \ k = 3,4,$$

are satisfied. Then with an error of order

$$\varepsilon = \exp\left(-\frac{1}{\mu\sqrt{2}}\int_0^1 q(x)\,dx\right)$$

the boundary conditions at $x = 0$ and $x = 1$ can be satisfied separately. At $x = 0$, the boundary conditions provide the equations

$$\sum_{k=1,2} C_k u_0(0) + y_0^*(0) = 0,$$

$$\sum_{k=1,2} C_k \left[\frac{r_k^2 q^2(0) u_0(0)}{\mu^2} + \frac{r_k}{\mu} [2q(0)u_0'(0) + q'(0)u_0(0)] + O(1) \right] + y_0^{*\prime\prime}(0) = 0.$$

Similar equations for C_3 and C_4 are obtained from the boundary conditions at $x = 1$. Solving these equations, we get

$$C_{1,2} = -\frac{y_0^*(0)}{2u_0(0)} \left[1 \pm \frac{i\mu a(0)}{\sqrt{2}} + O\left(\mu^2\right) \right],$$

$$C_{3,4} = -\frac{y_0^*(1)}{2u_0(1)} \left[1 \pm \frac{i\mu a(1)}{\sqrt{2}} + O\left(\mu^2\right) \right],$$

$$a(x) = \frac{1}{q(x)} \frac{d}{dx} \ln(q u_0^2),$$

where the plus sign goes with C_1 and C_3 and the negative sign goes with C_2 and C_4.

Thus, the constants C_k are found within an error of order $O\left(\mu^2\right)$. We try to simplify the solution $y(x, \mu)$ in such a way that its error is not larger than $O\left(\mu^2\right)$. Taking

$$x e^{-bx/\mu} = O(\mu), \quad 0 \le x \le 1.$$

for $b \sim 1$ into account, we get

$$y(x, \mu) = 2u_0(x)\Re \left[C_1 \exp\left(\frac{r_1}{\mu} \int_0^x q(x)\,dx \right. \right.$$

$$\left. \left. + C_3 \exp\left(\frac{r_3}{\mu} \int_1^x q(x)dx \right) \right] + y_0^*(x) + O\left(\mu^2\right). \quad (4.3.21)$$

If we accept a larger error of order $O(\mu)$, then the solution can be simplified further,

$$y(x, \mu) = y_0^*(x) - \Re \left[y_0^*(0) \exp\left(\frac{r_1 q_0 x}{\mu} \right) + y_0^*(1) \exp\left(\frac{r_3 q_1 (x-1)}{\mu} \right) \right] + O(\mu),$$

$$(4.3.22)$$

where $q_0 = q(0)$ and $q_1 = q(1)$.

Example 2

Study the deflection of a beam on an elastic foundation in a neighborhood of the point of discontinuity of the load. As an example, we consider the equation with constant coefficients

$$\mu^4 \frac{d^4 y}{dx^4} + y = g(x), \quad \begin{cases} g(x) = 0, & x < 0, \\ g(x) = 1, & x \geq 0. \end{cases}$$

In a neighborhood of the point of discontinuity, $x = 0$, of the load, the solution has the form:

$$y = C_3\, e^{r_3 x/\mu} + C_4\, e^{r_4 x/\mu}, \quad r_{3,4} = \frac{1}{\sqrt{2}} \pm \frac{i}{\sqrt{2}}, \quad x < 0,$$

$$y = C_1\, e^{r_1 x/\mu} + C_2\, e^{r_2 x/\mu} + 1, \quad r_{1,2} = -\frac{1}{\sqrt{2}} \pm \frac{i}{\sqrt{2}}, \quad x > 0.$$

The constants C_k are found from the continuity conditions on the function y and its first three derivatives at $x = 0$,

$$C_3 r_3^m + C_4 r_4^m = \delta_{m0} + C_1 r_1^m + C_2 r_2^m, \quad m = 0, 1, 2, 3,$$

$$\delta_{00} = 1, \quad \delta_{m0} = 1, \ m > 0.$$

The solution of this system is

$$C_1 = C_2 = -1/4, \quad C_3 = C_4 = 1/4.$$

Now, we find

$$y(x_1) = \frac{1}{2} e^{x_1} \cos x_1, \quad x_1 = \frac{x}{\mu\sqrt{2}}, \quad x < 0;$$

$$y(x_1) = 1 - \frac{1}{2} e^{-x_1} \cos x_1, \quad x \geq 0.$$

The deflection, $y_1(x)$, of a clamped beam $(y(0) = y'(0) = 0)$ and the deflection, $y_2(x)$, of a freely supported beam $(y(0) = y''(0) = 0)$ are given by the expressions

$$y_1(x_1) = 1 - e^{-x_1}(\cos x_1 + \sin x_1), \quad x \geq 0.$$

$$y_2(x_1) = 1 - e^{-x_1} \cos x_1, \quad x \geq 0.$$

In Fig. 4.2, the functions $y(x)$, $y_1(x)$ and $y_2(x)$ are plotted as curves 0, 1 and 2, respectively.

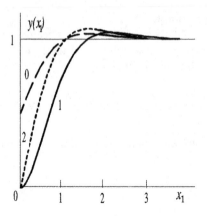

Fig. 4.2 Plot of $y(x)$, $y_1(x)$ and $y_2(x)$ as curves 0, 1 and 2, respectively

The maximal deflection values are

$$y^{\max} = y(3\pi/4) = 1.0335,$$
$$y_1^{\max} = y(\pi) \quad\;\, = 1.0432,$$
$$y_2^{\max} = y(3\pi/4) = 1.0670.$$

Away from the point $x = 0$, $y \simeq 1$. Therefore, in a neighborhood of the point of discontinuity of the load or the point of support of the beam, the deflection slightly increases. A similar effect also exists for a beam with variable parameters since, in this case, the main terms of the asymptotic expansions are the same as for a beam with constant parameters.

4.3.5 Regular Degeneracy

We consider now the case of a regular degeneracy firstly studied by Vishik and Lyusternik [62]. Let the characteristic equation (4.2.4) or (4.1.7) have l roots of the form (4.3.14) with $l < n$. The other $m = n - l$ roots are identically equal to zero. In other words, we consider Eq. (4.1.16) or system (4.2.15). We assume that

$$m_1 = n_1 - l_1 \geq 0, \quad m_2 = n_2 - l_2 \geq 0, \quad m = m_1 + m_2, \tag{4.3.23}$$

where n_1 and n_2 are the same as in (4.3.3).

As before, with an error of order $O\left(e^{-c/\mu}\right)$, where $c > 0$, matrix (4.3.8) is

$$\boldsymbol{D} = \begin{bmatrix} \boldsymbol{D}_1 & \boldsymbol{D}_{01} & 0 \\ 0 & \boldsymbol{D}_{02} & \boldsymbol{D}_2 \end{bmatrix}, \quad \boldsymbol{D}(\mu) \simeq \sum_{k=0}^{\infty} \mu^k \boldsymbol{D}^{(k)}, \tag{4.3.24}$$

where the rectangular matrices $\boldsymbol{D}_1$, $\boldsymbol{D}_{01}$, $\boldsymbol{D}_{02}$, and $\boldsymbol{D}_2$ are of dimensions $n_1 \times l_1$, $n_1 \times m$, $n_2 \times m$, and $n_2 \times l_2$, respectively.

We assume that

$$\text{rank } \boldsymbol{D}_j(0) = l_j, \quad j = 1, 2. \tag{4.3.25}$$

If

$$\det \boldsymbol{D}(0) \neq 0, \tag{4.3.26}$$

then, for a sufficiently small μ, conditions (4.3.9) holds. Then, the required solution exists and has the asymptotic expansion

$$\boldsymbol{y} \simeq \sum_{k=0}^{\infty} \mu^k \boldsymbol{y}_k(x) + \sum_{j=1,2} \sum_{k=0}^{\infty} \mu^k \boldsymbol{z}_k^{(j)}(\xi_j) + \boldsymbol{y}^*(x, \mu), \tag{4.3.27}$$

where the first term is a solution of type (4.1.20) with (4.2.19) corresponding to the zero roots of the characteristic equations (4.1.7) or (4.2.4). The second term consists of the edge effect integrals at the edges $x = x_1$ and $x = x_2$ and

$$\xi_1 = \frac{x - x_1}{\mu}, \quad \boldsymbol{z}_k^{(1)}(\xi_1) \to 0 \quad \text{as} \quad \xi_1 \to \infty,$$
$$\xi_2 = \frac{x - x_2}{\mu}, \quad \boldsymbol{z}_k^{(2)}(\xi_2) \to 0 \quad \text{as} \quad \xi_2 \to -\infty. \tag{4.3.28}$$

The last term in (4.3.27) is the particular solution (4.1.30) or (4.2.29). If $\gamma(x) \equiv 0$, then the particular solution $\boldsymbol{y}^*(x, \mu)$ is also slowly varying and can be included in the first sum in (4.3.27).

We consider Eq. (4.1.32) with boundary conditions of the special type

$$\frac{d^k y}{dx^k} = b_{kj} \quad \text{for} \quad x = x_j, \quad k = 0, 1, \ldots, m_j + l_j - 1, \quad j = 1, 2, \tag{4.3.29}$$

where b_{kj} do not depend on μ. If conditions (4.3.14) hold, then, as $\mu \to 0$, the boundary value problem (4.1.32), (4.3.29) degenerates to problem (4.1.34) with the boundary conditions

$$\frac{d^k y}{dx^k} = b_{kj} \quad \text{for} \quad x = x_j, \quad k = 0, 1, \ldots, m_j - 1, \quad j = 1, 2. \tag{4.3.30}$$

If the solution $y_0(x)$ of the degenerate problem exists and is unique, then, for sufficiently small μ, the solution $y_0(x, \mu)$ of the original problem exists and is unique. The solution has the form (4.3.27) where some of the first terms in the second sum are equal to zero,

$$y(x, \mu) \simeq \sum_{k=0}^{\infty} \mu^k y_k(x) + \sum_{j=1,2} \sum_{k=m_j}^{\infty} \mu^k z_k^{(j)}(\xi_j). \tag{4.3.31}$$

The passage to the limit

$$\lim_{\mu \to 0} y(x, \mu) = y_0(x) \tag{4.3.32}$$

in the segment $x_1 + \varepsilon \leq x \leq x_2 - \varepsilon$, $\varepsilon > 0$, is uniform in x together with all derivatives. In the segment $x_1 \leq x \leq x_2 - \varepsilon$ it is uniform in x together with $m_1 - 1$ derivatives, and in the segment $x_1 + \varepsilon \leq x \leq x_2$ it is uniform in x together with $m_2 - 1$ derivatives.

The appearance of pure imaginary roots, $\lambda_k(x)$, of the characteristic equation (4.1.19) or (4.2.4) makes the construction of the solution significantly more difficult since, in this case, condition (4.3.11) does not usually hold and the set of zeros of the function $D(\mu)$ has the accumulation point $\mu = 0$. Under such conditions, we consider the homogeneous problem of Sect. 4.4 and do not consider the non-homogeneous problem in this book. We find approximate asymptotic solutions of some boundary value problems of mechanics of solids for which regular degeneracy takes place.

4.3.6 Non-absolutely Flexible String

The deflection of a non-absolutely flexible string is described by the equation

$$\frac{d^2}{dx_1^2}\left(EI\frac{d^2y}{dx_1^2}\right) - T\frac{d^2y}{dx_1^2} = g_1(x_1), \quad 0 \leq x_1 \leq l,$$

where EI is the string bending stiffness, T is the tension and $g_1(x_1)$ is the intensity of the external load.

The introduction of the non-dimensional variables

$$x_1 = lx, \ 0 \leq x \leq 1, \quad EI = E_0 I_0 p(x) > 0, \quad g_1(x_1) = E_0 I_0 g(x)$$

transforms the equation into

$$\mu^2 \frac{d^2}{dx^2}\left(p(x)\frac{d^2y}{dx^2}\right) - \frac{d^2y}{dx^2} = g(x), \quad \mu^2 = \frac{E_0 I_0}{Tl^2}, \tag{4.3.33}$$

where $\mu > 0$ is a small parameter.

We consider the boundary conditions

$$y(x) = \frac{dy}{dx} = 0 \quad \text{for} \quad x = 0, \ x = 1, \tag{4.3.34}$$

corresponding to fixing the string edges.

Find the first terms of the asymptotic expansions of the solution of the boundary value problem (4.3.33) and (4.3.34) assuming for simplicity that $p(x) \equiv 1$.

The characteristic equation (4.1.19), $\lambda^2 - 1 = 0$, has one positive and one negative root. Therefore, as $\mu \to 0$, problem (4.3.33) and (4.3.34) degenerates into the regular boundary value problem

$$-\frac{d^2 y_0}{dx^2} = g(x), \quad y_0(0) = y_0(1) = 0. \tag{4.3.35}$$

The solution of this problem can be represented by three sums (4.3.27):

$$y \simeq \sum_{k=0}^{\infty} \mu^k y_k(x) + \sum_{j=1,2} \sum_{k=1}^{\infty} \mu^k z_k^{(j)}(\xi_j),$$

where

$$\xi_1 = \frac{x}{\mu}, \quad z_k^{(1)}(\xi_1) \to 0 \quad \text{as} \quad \xi_1 \to \infty,$$

$$\xi_2 = \frac{x-1}{\mu}, \quad z_k^{(2)}(\xi_2) \to 0 \quad \text{as} \quad \xi_2 \to -\infty.$$

The first sum provides the slow varying part of the solution and the other two sums (for $j = 1, 2$) provide the edge effect integrals. It is convenient to evaluate the terms in these sums step by step. We show how this can be done.

Let solution (4.3.35) be found

$$y_0(x) = xg_2(1) - g_2(x), \quad g_2(x) = \int_0^x \left(\int_0^t g(t_1)dt_1 \right) dt.$$

In general, this solution does not satisfies the boundary conditions $y'(x) = 0$ at $x = 0$ and $x = 1$, which were omitted in the transition from the original problem (4.3.33) and (4.3.34) to the degenerate problem (4.3.35). We consider the edge $x = 0$ and add to the solution y_0 the edge effect integral

$$y(x, \mu) = y_0(x) + C\mu e^{-x/\mu}.$$

For $C = y_0'(0)$, the condition $y'(x) = 0$ at $x = 0$ holds. If this condition is also satisfied at $x = 1$, we get an approximate solution of the form

$$y(x, \mu) = y_0(x) + \mu y_0'(0) e^{-x/\mu} - \mu y_0'(1) e^{(x-1)/\mu}.$$

As before, we neglect the mutual influence of the edge effects. This allows an error of order $\varepsilon = e^{-1/\mu}$, i.e. we approximately consider $\varepsilon = e^{-1/\mu} = 0$.

Now the obtained solution does not satisfies the conditions $y(0) = y(1) = 0$. The residual has order μ. We assume that

$$y^{(1)}(x, \mu) = y_0(x) + \mu y_1(x) + \mu y_0'(0) e^{-x/\mu} - \mu y_0'(1) e^{(x-1)/\mu}.$$

The function $y_1(x)$ can be found from the solution of the boundary value problem

$$\frac{d^2 y_1}{dx^2} = 0, \quad y_1(0) = -y_0'(0), \quad y_1(1) = y_0'(1).$$

Therefore $y_1(x) = -y_0'(0) + x(y_0'(0) + y_0'(1))$.

So far, the function $y^{(1)}(x, \mu)$ does not satisfy the conditions $y_1'(0) = y_1'(1) = 0$. The residual has order μ. The next step should be refining the edge effect integrals. In such a manner, one can find any number of terms of series (4.3.27).

4.3.7 Axisymmetric Deformation of a Shell of Revolution

The thin shell theory provides numerous problems to be solved by asymptotic methods [10, 28, 29, 30, 56]. We consider a system of equations describing the axisymmetric deformation of a thin shell of revolution. We write three equations of equilibrium of a shell in the form [31, 56]:

$$\frac{dT_1}{ds} + \frac{B'}{B}(T_1 - T_2) - \frac{Q_1}{R_1} + q_1^* = 0,$$

$$\frac{dQ_1}{ds} + \frac{B'}{B} \cdot Q_1 + T_1 \left(\frac{1}{R_1} + \kappa_1 \right) + T_2 \left(\frac{1}{R_2} + \kappa_2 \right) + q_n^* = 0, \qquad (4.3.36)$$

$$\frac{dM_1}{ds} + \frac{B'}{B}(M_1 - M_2) + Q_1 = 0, \quad ()' = \frac{d}{ds},$$

where s in the length of the meridian arc. The shape of the neutral surface is characterized by the functions $B(s)$, $R_1(s)$, $R_2(s)$, $\theta(s)$ (see Fig. 4.3), where $B(s)$ is the distance between a point on the neutral surface and the axis of revolution, $R_1(s)$ and $R_2(s)$ are the main radii of curvature and

$$R_1 = -\frac{d\theta}{ds}, \quad R_2 = \frac{B}{\sin \theta}, \quad B' = -\cos \theta.$$

Here, T_1 and T_2 are the stress-resultants in the neutral surface, Q_1 is the transverse shear force, M_1 and M_2 are the bending moments, q_1^* and q_n^* are the intensity of the external distributed loads in the projections on the tangent and on the normal to the shell.

The stress-resultants and the moments are related to the deformations of the neutral surface by the formulas:

Fig. 4.3 Functions characterizing the shape of the neutral surface

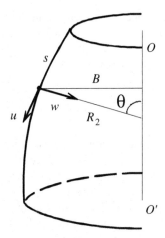

$$T_1 = K(\varepsilon_1 + \nu\varepsilon_2), \qquad T_2 = K(\varepsilon_2 + \nu\varepsilon_1), \qquad K = \frac{Eh}{1-\nu^2},$$

$$M_1 = D(\kappa_1 + \nu\kappa_2), \qquad M_2 = D(\kappa_2 + \nu\kappa_1), \qquad D = \frac{Eh^3}{12(1-\nu^2)}, \qquad (4.3.37)$$

where E, ν, h are Young's modulus, Poisson's ratio and the shell thickness, respectively. The tensile deformations of the neutral surface ε_1 and ε_2 and its bending deformations κ_1 and κ_2 are related to the displacements $u(s)$, $w(s)$ (see Fig. 4.3) as

$$\varepsilon_1 = \frac{du}{ds} - \frac{w}{R_1}, \qquad\qquad \varepsilon_2 = \frac{B'}{B}u - \frac{w}{R_2},$$

$$\kappa_1 = \frac{d^2w}{ds^2} + \frac{1}{R_1}\frac{du}{ds}, \qquad \kappa_2 = \frac{B'}{B}\left(\frac{dw}{ds} + \frac{u}{R_1}\right). \qquad (4.3.38)$$

Equations (4.3.36)–(4.3.38) is a closed system of order six. The system becomes linear if we neglect the nonlinear terms $T_1\kappa_1 + T_2\kappa_2$ in the second equation in (4.3.36). The role of these terms is clarified below.

At each edge of the shell, one should introduce three boundary conditions denoted by C_i, S_i, $i = 1, 2, 3, 4$,

$$
\begin{array}{llll}
u = w = 0, & \{w' = 0 \ \text{or} \ M_1 = 0\} & C_1 \ \text{or} \ S_1 \\
T_1 = w = 0, & \{w' = 0 \ \text{or} \ M_1 = 0\} & C_2 \ \text{or} \ S_2 \\
u = Q_1 = 0, & \{w' = 0 \ \text{or} \ M_1 = 0\} & C_3 \ \text{or} \ S_3 \\
T_1 = Q_1 = 0, & \{w' = 0 \ \text{or} \ M_1 = 0\} & C_4 \ \text{or} \ S_4 \qquad (4.3.39)
\end{array}
$$

In particular, the conditions of clamped type are denoted by C_1, freely support by S_2, and free edge by S_4. If nonzero displacements or forces are introduced at the edges then the boundary conditions are non-homogeneous.

One often uses a simplified approach for the analysis of shells, the so-called *membrane theory*, when a shell is assumed to be absolutely flexible, i.e. $M_1 = M_2 = Q_1 = 0$. As a result one gets only the first two equations from the three equations in (4.3.36)

$$\frac{dT_1^0}{ds} + \frac{B'}{B}\left(T_1^0 - T_2^0\right) + q_1^* = 0, \qquad \frac{T_1^0}{R_1} + \frac{T_2^0}{R_2} + q_n^* = 0. \qquad (4.3.40)$$

Together with relations (4.3.37) and (4.3.38), system (4.3.40) has second order. Therefore, from the three boundary conditions (4.3.39) one should keep only one, namely, the first condition $u = 0$ or $T_1 = 0$.

A sufficiently accurate approximate solution of the original system (4.3.36) can be obtained as a sum of the membrane solution and the edge effect integrals in neighborhoods of the shell edges $s = s_1$ and $s = s_2$.

Introduce the characteristic size, R, of the neutral surface and refer to it all linear sizes. Introduce a small parameter μ by the formula

$$\mu^4 = \frac{h^2}{12R^2(1 - \nu^2)}.$$

Then, system (4.3.36)–(4.3.38) transforms into two equations in the displacements $u(s)$ and $w(s)$:

$$\begin{aligned}
L_{11}u + \mu^4 N_{11}u + L_{13}w + \mu^4 N_{13}w + q_1 &= 0, \\
L_{31}u + \mu^4 N_{31}u + L_{33}w + \mu^4 N_{33}w + q_n &= 0,
\end{aligned} \qquad (4.3.41)$$

where

$$L_{11}u = \frac{1}{1 - \nu^2}\left(\frac{d^2u}{ds^2} + \frac{B'}{B}\frac{du}{ds} - \left(\frac{B'}{B}\right)^2 u + \frac{\nu B''}{B}\right),$$

$$L_{13}w = \frac{1}{1 - \nu^2}\left(\frac{B'c_2 w}{B} - \frac{(Bc_1 w)'}{B}\right),$$

$$L_{31}u = \frac{1}{1 - \nu^2}\left(c_1\frac{du}{ds} + \frac{B'c_2 u}{B}\right),$$

$$L_{33}w = -\frac{1}{1 - \nu^2}\left(\frac{w}{R_1^2} + \frac{2\nu w}{R_1 R_2} + \frac{w}{R_2^2}\right),$$

$$N_{33} = -\frac{d^4w}{ds^4} - \frac{2B'}{B}\frac{d^3w}{ds^3} + \cdots,$$

$$c_1 = \frac{1}{R_1} + \frac{\nu}{R_2}, \quad c_2 = \frac{\nu}{R_1} + \frac{1}{R_2}, \quad q_1 = \frac{q_1^*}{Eh}, \quad q_2 = \frac{q_2^*}{Eh}.$$

We keep only the main terms of the operator N_{33} and do not show the other operators N_{ij} since they are not used in the construction of the main terms of the solution.

We seek edge effect integrals in the form (4.2.3):

$$w^k(s, \mu) = \sum_{n=0}^{\infty} \mu^n w_n^k(s) \exp\left(\frac{1}{\mu} \int \lambda(s)\, ds\right),$$

$$u^k(s, \mu) = \mu \sum_{n=0}^{\infty} \mu^n u_n^k(s) \exp\left(\frac{1}{\mu} \int \lambda(s)\, ds\right), \quad k = 1, 2, 3, 4,$$

(4.3.42)

After substitution in system (4.3.41) we find

$$\lambda^4(s) + \frac{1}{R_2^2(s)} = 0, \quad w_0^k(s) = B^{-1/2} R_2^{3/4}, \quad u_0^k(s) = \frac{1}{\lambda}\left(\frac{1}{R_1} + \frac{\nu}{R_2}\right) w_0^k(s).$$

(4.3.43)

For the particular case of a circular cylindrical shell, $R_1 = 0$, $R_2 = B = 1$ and system (4.3.41) transforms into

$$\frac{1}{1-\nu^2}\left(\frac{d^2 u}{ds^2} - \nu\frac{dw}{ds}\right) + q_1 = 0,$$

$$\frac{1}{1-\nu^2}\left(\nu\frac{du}{ds} - w\right) - \mu^4\frac{d^4 w}{ds^4} + q_n = 0.$$

(4.3.44)

Eliminating the function u, we get a fourth order equation for $w(s)$,

$$\mu^4\frac{d^4 w(s)}{ds^4} + w(s) = q_n(s) - \nu I(s),$$

$$u(s) = \int_{s_0}^{s}\left[\nu w(s) - \left(1 - \nu^2\right) I(s)\right] ds, \quad I(s) = \int_{s_0}^{s} q_1(\xi)\, d\xi.$$

(4.3.45)

The homogeneous equation (4.3.45) describes the edge effect. Note that this equation is identical to the equation for the deflection of a beam on an elastic foundation (4.3.20).

Neglecting terms with multiplier μ^4 in (4.3.44), we come to the membrane system of equations:

$$\frac{d^2 u}{ds^2} = q_1 - \nu\frac{dq_n}{ds}, \quad w = \nu\frac{du}{ds} + \left(1 - \nu^2\right) q_n.$$

(4.3.46)

4.3.8 Shell Deformation Under External Pressure

We find approximate expressions for the displacements u and w of a cylindrical shell under an external uniform normal pressure, q_n. We introduce the boundary conditions of free support type S_2 in (4.3.39) at the edges $x = 0$ and $x = l$ in the form

$$\frac{du}{ds} = w = \frac{d^2w}{ds^2} = 0 \quad \text{for} \quad s = 0, \ s = l.$$

The membrane part of the solution is

$$u_0 = C + \nu q_n s, \quad w_0 = q_n,$$

where the constant C is unknown so far since the given boundary conditions do not resist axial displacement of the shell. When constructing a membrane solution, we satisfy the condition

$$T_1 = \frac{du}{ds} - \nu w = 0.$$

We add edge effect integrals to the obtained membrane solution in order to satisfy the boundary conditions $w = d^2w/ds^2 = 0$, the first of which does not hold for the membrane solution. Hence

$$u = C + \nu q_n s + \frac{\mu \nu q_n}{\sqrt{2}} \left[e^{-s_1}(\cos s_1 - \sin s_1) - e^{s_2}(\cos s_2 + \sin s_2) \right],$$

$$w = q_n(1 - e^{-s_1} \cos s_1 - e^{s_2} \cos s_2), \quad s_1 = \frac{s}{\mu\sqrt{2}}, \ s_2 = \frac{s - l}{\mu\sqrt{2}}.$$

This solution satisfies equations (4.3.44) exactly. At the same time, the boundary conditions S_2 in (4.3.39) hold with an error of order $\exp(-l/(\mu\sqrt{2}))$. Therefore, for short shells, for which $l = L/R \ll 1$, one should take the mutual influence of the edge effects into account. Here we assume that $l \sim 1$.

In Fig. 4.4 we plot the functions $u(s)$ and $w(s)$ for $l = 2$, $\mu = 0.1$, $\nu = 0.3$, $q_n = 1$, $C = -\nu l$.

Fig. 4.4 Plot of the functions $u(s)$ and $w(s)$

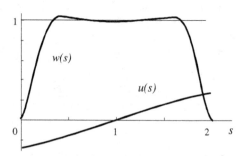

4.3.9 Shell Deformations Under Axial Force

Here we examine the displacements in a cylindrical shell under axial tension or compression under the assumption that the following boundary conditions are given at the shell edges:

$$T_1 = T_1^0, \quad w = \frac{d^2 w}{ds^2} = 0 \quad \text{for} \quad s = 0, \ s = l.$$

Due to the first equation in (4.3.36), $T_1(s) \equiv T_1^{(0)}$. Taking

$$T_1 = \frac{Eh}{(1 - \nu^2)R} \left(\frac{du}{ds} - \nu w \right)$$

into account, we obtain the membrane solution

$$u^{(0)} = C + as, \quad w^{(0)} = \nu a, \quad a = \frac{T_1^0 R}{Eh}, \tag{4.3.47}$$

where C is an arbitrary constant.

Find the edge effect integrals that should be added to the obtained membrane solution to satisfy the boundary conditions for the deflection w. Consider a neighborhood of the edge $s = 0$. Here, as in Sect. 4.3.8, the solution has the form

$$u = C + as + \frac{\mu \nu^2 a}{\sqrt{2}} e^{-s_1} (\cos s_1 - \sin s_1),$$

$$w = \nu a \left(1 - e^{-s_1} \cos s_1 \right), \quad s_1 = \frac{s}{\mu \sqrt{2}}. \tag{4.3.48}$$

Find the validity limits for the obtained solution resulting from an increase in the axial force $|T_1^0|$ (under compression $T_1^0 < 0$). For that purpose, we consider the second equation in (4.3.36) and keep in it the nonlinear term, $T_1 \kappa_1$, that was omitted when we passed to system (4.3.44). As a result, instead of (4.3.45) for the edge effect integrals, we get the equation

$$\mu^4 \frac{d^4 w}{ds^4} - 2\mu^2 \tau \frac{d^2 w}{ds^2} + w = 0, \quad \tau = \frac{T_1^0}{2Eh\mu^2}. \tag{4.3.49}$$

We consider its characteristic equation

$$\lambda^4 - 2\tau \lambda^2 + 1 = 0. \tag{4.3.50}$$

For $\tau > -1$, i.e. under the tension $T_1^0 > 0$ and for a not too heavy compression, $T_1^0 < 0$ and $|\tau| < 1$, Eq. (4.3.7) has two roots with positive real parts and two roots with negative real parts. In other words, one can find edge effect integrals. We construct the integral so that their sum and solution (4.3.47) satisfy the boundary conditions $w = d^2w/ds^2 = 0$ at $s = 0$. Thus, for $|\tau| < 1$, we get

$$u = C + as + O(\mu), \quad w = w_0 \left[1 - e^{-\alpha_1}(\cos \beta_1 + b \sin \beta_1) + O(\mu) \right],$$

$$\lambda_{1,2} = -\sqrt{\frac{1+\tau}{2}} \pm i\sqrt{\frac{1-\tau}{2}} = -\alpha \pm i\beta, \quad b = \frac{\tau}{\tau_1}, \quad \alpha_1 = \frac{as}{\mu}, \quad \beta_1 = \frac{\beta s}{\mu},$$

and for $\tau > 1$ we get

$$u = C + as + O(\mu), \quad w = w_0 \left[1 - \frac{1}{2\tau_1} \left(\lambda_2^2 e^{-\lambda_1 s/\mu} - \lambda_1^2 e^{-\lambda_2 s/\mu} \right) + O(\mu) \right],$$

$$\lambda_1 = \sqrt{\tau - \tau_1}, \quad \lambda_2 = \sqrt{\tau + \tau_1}, \quad w_0 = \nu a = \frac{\nu h \tau}{\sqrt{3(1 - \nu^2)}},$$

where $\tau_1 = \sqrt{|1 - \tau^2|}$.

We should note that $w \to \infty$ as $\tau \to -1 + 0$. For $\tau = -1$ or under the compressive force $T_1^0 = -2Eh\mu^2$, the shell buckles under axial compression. Note that for $\tau = +1$ the function $w(s)$ is nonsingular since the uncertainty of type 0/0 is expanded as

$$w = w_0 \left[1 - e^{-\xi}(1 + \xi/2) + O(\mu) \right], \quad \xi = s/\mu, \quad \tau = 1.$$

4.3.10 Exercises

4.3.1. Solve the boundary value problem (4.3.18) for the particular case $c(x) = 1 + x$ and $q(x) \equiv 1$ keeping in the solutions only the terms of orders up to μ^2 inclusively. Plot the function $y(x, \mu)$ for $\mu = 0.1$.

4.3.2. Find the asymptotic expansion of the solution of the equation $\mu^2 y'' - y + q(x) = 0$, $y \to 0$ as $x \to \pm\infty$, where $q(x) = x$ for $0 \le x \le 1$ and $q(x) \equiv 0$ for $x < 0$ or $x > 1$. This equation describes the deflection of a sufficiently long string on an elastic foundation under a variable load. Plot the function $y(x, \mu)$ for $\mu = 0.2$.

4.3.3. In the conditions of Exercise 4.3.2 consider $q(x) = 1$ and $0 \le x \le 1$. Find the exact and approximate solutions of the equation $\mu^2 y'' - y + q(x) = 0$, as $y \to 0$ and $x \to \pm\infty$.

4.3.4. Plot the approximate solutions (4.3.21) and (4.3.22) for a beam of variable thickness when $p(x) = (1 + \alpha x)^3$, $c(x) \equiv 1$ and $g(x) \equiv 1$. Consider $\alpha = 0.2$ and $\mu = 0.2$.

4.3.5. Find approximate solutions of equation (4.3.20) similar to solutions (4.3.21) and (4.3.22) for a beam with two clamped ends $y(x) = y'(x) = 0$ at $x = 0$ and $x = 1$.

4.3.6. Find the terms of order μ^2 of the asymptotic solution of the boundary value problem (4.3.33) and (4.3.34).

4.3.7. Substituting (4.3.42) in system (4.3.41) derive formulas (4.3.43).

4.3.8. In the conditions of Sect. 4.3.8, consider the boundary conditions C_2 in (4.3.39).

4.3.9. In the conditions of Sect. 4.3.8, consider the boundary conditions of a clamped edge type C_1 in (4.3.39).

4.3.10. Find the deflection, w, of a cylindrical shell under axial tension, or compression under the assumption that the following boundary conditions are introduced at the shell edges: $T_1 = T_1^0$, $w = w' = 0$ at $s = 0$ and $s = l$.

4.4 Eigenvalue Problems

In this section, we consider the same problems as in Sect. 4.3; however, the equations and the boundary conditions are assumed to be homogeneous, i.e. we study Eqs. (4.1.1), (4.1.16), (4.2.1) and (4.2.15) with boundary conditions of type (4.3.3) for $b_j(\mu) = 0$. Let the coefficients of the equations and perhaps also of the boundary conditions depend on a spectral parameter, Λ. We seek the eigenvalues $\Lambda = \Lambda^{(k)}(\mu)$ for which there exist nontrivial solutions of given eigenvalue problems. We limit ourselves to real Λ.

4.4.1 Asymptotics Solutions of Eigenvalue Problems

To evaluate Λ we use Eq. (4.3.10) written in the form

$$\Delta(\Lambda, \mu) = \det D(\Lambda, \mu) = 0. \tag{4.4.1}$$

Here two cases are possible. In the first case, the limit

$$\lim_{\mu \to 0} \Delta(\Lambda, \mu) = \Delta(\Lambda, 0) \tag{4.4.2}$$

exist, but not in the second case.

In the first case, under natural auxiliary conditions, the expansion in integer powers of μ,

$$\Lambda = \Lambda_0 + \mu \Lambda_1 + \mu^2 \Lambda_2 + \cdots, \tag{4.4.3}$$

exists. Here, Λ_0 is a root of the equation $\Delta(\Lambda, 0) = 0$. This case occurs for some eigenvalue problems (see Sect. 4.3) for which the characteristic equation does not have pure imaginary roots. In particular, if the characteristic equation does not have pure imaginary and zero roots and conditions (4.3.14) hold, then

$$\Delta(\Lambda, \mu) = \Delta_1(\Lambda, 0)\Delta_2(\Lambda, 0) + O(\mu) \tag{4.4.4}$$

and for each simple root, Λ_0, of the equations $\Delta_j(\Lambda, 0) = 0$, $j = 1, 2$, expansion (4.4.3) takes place. The eigenfunctions have the character of edge effect integrals and they are localized in a neighborhood of the corresponding edge of the segment $[x_1, x_2]$.

If the boundary conditions (4.3.29) for $b_{kj} = 0$ are introduced for Eq. (4.1.16) and the nontrivial roots of equation (4.1.18) satisfy conditions (4.3.14), then, as $\mu \to 0$, the original eigenvalue problem degenerates to a problem for equation (4.1.17) with boundary conditions (4.3.30) for $b_{kj} = 0$.

Now we write (4.1.17) and (4.3.30) in the form

$$L_0(\Lambda)y \equiv \sum_{k=0}^{m} a_{k0}(x, \Lambda)\frac{d^k y}{dx^k} = 0,$$

$$\left.\frac{d^k y}{dx^k}\right|_{x=x_j} = 0, \quad k = 0, 1, \ldots, m_j - 1, \quad j = 1, 2. \tag{4.4.5}$$

For each simple eigenvalue, Λ_0, of the degenerate problem (4.4.5), expansion (4.4.3) exists and the eigenfunction has expansion (4.3.31).

For the last problem, we describe the algorithm for the construction of the eigenvalue and the eigenfunction, limiting ourselves to the evaluation of Λ_1 in (4.4.3). Let Λ_0 and $y_0(x)$ be obtained for the degenerate problem and suppose that Λ_0 is simple, i.e. the function $y_0(x)$ is unique up to a multiple constant. We find functions $z_{m_j}^{(j)}(\xi_j)$ satisfying the equation

$$\sum_{k=0}^{n} a_{k+m}(x_j, 0)\frac{d^k z_{m_j}^{(j)}}{d\xi_j^k} = 0, \quad \xi_j = \frac{x - x_j}{\mu}, \quad j = 1, 2, \tag{4.4.6}$$

the damping conditions (4.3.28) and the boundary conditions at $\xi_j = 0$:

$$\frac{d^{m_j} z_{m_j}^{(j)}}{d\xi_j^{m_j}} = -\frac{d^{m_j} y_0(x_j)}{dx^{m_j}}, \quad j = 1, 2,$$

$$\frac{d^k z_{m_j}^{(j)}}{d\xi_j^k} = 0, \quad k = m_j + 1, m_j + 2, \ldots, m_j + l_j - 1. \tag{4.4.7}$$

The function $z_{m_j}^{(j)}(\xi_j)$ exists and is unique.

For the function $y_1(x)$ we obtain the eigenvalue problem

$$L_0(\Lambda_0)y_1 = \varphi_1(x),$$

$$\varphi_1(x) = \sum_{k=0}^{m} a_{k1}(x, \Lambda_0)\frac{d^k y_0}{dx^k} + c_1\Lambda_1, \quad c_1 = \sum_{k=0}^{m} \frac{\partial a_{k0}(x, \Lambda_0)}{\partial\Lambda_0}\frac{d^k y_0}{dx^k} = 0,$$

$$\frac{d^k y_1}{dx^k}\bigg|_{x=x_j} = 0, \quad k = 0, 1, \ldots, m_j - 2, \tag{4.4.8}$$

$$\frac{d^{m_j-1} y_1}{dx_j^{m_j-1}} = d_j = -\frac{d^{m_j-1} z_{m_j}^{(j)}}{d\xi_j^{m_j-1}}\bigg|_{\xi_j=0}, \quad j = 1, 2.$$

Since the homogeneous problem corresponding to (4.4.8) has a nontrivial solution, $y_0(x)$, the non-homogeneous problem (4.4.8) does not have a solution for all right sides. The compatibility condition for $c_1 \neq 0$, which is used for evaluating Λ_1, is written in the form

$$\int_{x_1}^{x_2} \varphi_1(x)z(x)\,dx - (-1)^{m_1} a_{m0}(x_2, \Lambda_0)d_2 \frac{d^{m_1} z(x_2)}{dx_2^{m_1}}$$

$$+ (-1)^{m_2} a_{m0}(x_1, \Lambda_0)d_1 \frac{d^{m_2} z(x_1)}{dx_1^{m_2}} = 0, \quad (4.4.9)$$

where $z(x)$ is a nontrivial solution of the eigenvalue problem conjugated to (4.4.5)

$$\sum_{k=0}^{m}(-1)^k \frac{d^k}{dx^k}\left[a_{k0}(x, \Lambda_0)z\right] = 0,$$

$$\frac{d^k z}{dx^k}\bigg|_{x=x_j} = 0, \quad k = 0, 1, \ldots, m - m_j - 1, \quad j = 1, 2. \tag{4.4.10}$$

If problem (4.4.5) is self-adjoint, then $z(x) = y_0(x)$ in (4.4.9).

If the characteristic equation has one or several pairs of pure imaginary roots, then limit (4.4.2) may not exist. For example, if Eq. (4.1.7) has only one pair of pure imaginary roots, $\lambda_{1,2} = \pm iq(x, \Lambda)$, and have no zero roots, then the determinant (4.4.1) is given in the form

$$\Delta(\Lambda, \mu) = a(\Lambda, \mu)\sin(\mu^{-1}\varphi(\Lambda) + \Psi(\Lambda, \mu)) = 0, \quad \varphi(\Lambda) = \int_{x_1}^{x_2} q(x, \Lambda)\,dx. \tag{4.4.11}$$

If $r = \partial\varphi/\partial\Lambda \neq 0$, then Eq. (4.4.11) has dense roots because its small multiplier μ makes it fast oscillating. The distance between neighboring roots has order μ and is equivalent to

$$\Lambda^{(k+1)} - \Lambda^{(k)} \simeq \pi\mu r^{-1}. \tag{4.4.12}$$

Later, we shall find asymptotic solutions for some vibrations problems.

4.4.2 Vibrations of Non-absolutely Flexible Strings

The free vibrations of a non-absolutely flexible string are described by the equation

$$-EJ \frac{d^4y}{dx_1^4} + T \frac{d^2y}{dx_1^2} + \rho\omega^2 y = 0, \quad 0 \le x_1 \le l,$$

where EJ is the bending stiffness, T is the tension, ρ is the linear density, ω is the vibrations frequency, and l is the string length. We write this equation in non-dimensional variables as

$$-\mu^2 \frac{d^4y}{dx^4} + \frac{d^2y}{dx^2} + \Lambda y = 0, \quad 0 \le x \le 1, \tag{4.4.13}$$

where

$$\mu^2 = \frac{EJ}{Tl^2}, \quad \Lambda = \frac{\rho\omega^2 l^2}{T}, \quad x_1 = lx.$$

As $\mu \to +0$, we obtain asymptotic expansions of the eigenvalues Λ of (4.4.13) with the boundary conditions

$$y = \frac{dy}{dx} = 0 \quad \text{at} \quad x = 0, \ x = 1. \tag{4.4.14}$$

The asymptotic expansions of the integrals of equation (4.4.13) are found in Exercise 4.1.5 and they can be used to satisfy the boundary conditions (4.4.14). However, now we face difficulties since $\Lambda = \Lambda_0 + \mu\Lambda_1 + \cdots$. With this in mind, we apply an iteration method (see Sect. 4.3.6).

For the zeroth approximation we use the degenerate problem,

$$\frac{d^2y_0}{dx^2} + \Lambda_0 y_0 = 0, \quad y_0(0) = y_0(1) = 0, \tag{4.4.15}$$

which describes an absolutely flexible string. Here, regular degeneracy, due to Vishik–Lyusternik, takes place. The solution of the problem is

$$\Lambda_0 = n^2\pi^2, \quad y_0 = \sin n\pi x, \quad n = 1, 2, \ldots,$$

and the solution of the original problem has the form

$$y(x, \mu) \simeq \sum_{k=0}^{\infty} \mu^k y_k(x) + \sum_{j=1,2} \sum_{k=1}^{\infty} \mu^k z_k^{(j)}(\xi_j), \tag{4.4.16}$$

where the notation of Sect. 4.3.6 is used.

To satisfy the boundary conditions $y'(0) = y'(1) = 0$, we find the edge effect integrals

$$z_1^{(1)} = n\pi\,e^{-\xi_1}, \ \xi_1 = \frac{x}{\mu}, \quad z_1^{(2)} = (-1)^{n+1} n\pi\,e^{\xi_2}, \ \xi_2 = \frac{x-1}{\mu}.$$

Therefore,

$$y(x,\mu) \simeq \sin n\pi x + \mu n\pi \left[e^{-x/\mu} + (-1)^{n+1}\, e^{(x-1)/\mu} \right].$$

Then we evaluate Λ_1 and $y_1(x)$ by means of the eigenproblem

$$\frac{d^2 y_1}{dx^2} + \Lambda_0 y_1 + \Lambda_1 y_0 = 0, \quad y_1(0) = -n\pi, \ y_1(1) = (-1)^n n\pi.$$

We obtain

$$\Lambda_1 = 4n^2\pi^2, \quad y_1(x) = n\pi(2x-1)\cos n\pi x.$$

Thus, the approximate solution of the problem is

$$\Lambda = n^2\pi^2 + 4\mu n^2\pi^2 + O(\mu^2), \quad n = 1, 2, \ldots,$$

$$y(x,\mu) = \sin n\pi x + \mu n\pi \left[e^{-x/\mu} + (-1)^{n+1} e^{(1-x)/\mu} + (2x-1)\cos n\pi x \right] + O\left(\mu^2\right).$$

In Fig. 4.5, the first vibrations mode of an absolutely flexible (curve 1) and non-absolutely flexible strings (curve 2) with clamped ends are plotted. As it may be expected, the string frequencies increase when the string bending stiffness is taken into account ($\mu \neq 0$).

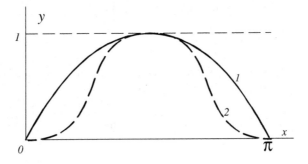

Fig. 4.5 Absolutely flexible string (*curve 1*) and non-absolutely flexible string (*curve 2*) with clamped ends

4.4.3 Vibrations of Strings with Variable Density

The free vibrations of a string with variable thickness are described by the equation

$$T \frac{d^2 y}{dx_1^2} + \rho_1(x_1)\omega^2 y = 0, \quad y(0) = y(l) = 0,$$

where T is the tension, $\rho_1(x_1)$ is the linear density, ω is the vibrations frequency, and l is the string length.

We write this equation in non-dimensional form

$$\frac{d^2 y}{dx^2} + \Lambda\rho(x)y = 0, \quad y(0) = y(1) = 0,$$

where

$$x_1 = lx, \quad \rho_1(x_1) = \rho_0\rho(x), \quad \rho_0 \sim 1, \quad \Lambda = \frac{\rho_0\omega^2}{T}.$$

To find asymptotics for the eigenvalues Λ_n as $n \to \infty$ we introduce the small parameter $\mu = \Lambda^{-1/2}$. Then, the equation takes the form

$$\mu^2 \frac{d^2 y}{dx^2} + \rho(x)y = 0, \quad y(0) = y(1) = 0, \tag{4.4.17}$$

similar to the equation in Exercise 4.1.1 and we use the solution obtained in that exercise:

$$y(x, \mu) \simeq \sum_{n=0}^{\infty} \mu^n u_n(x)\, e^{i\varphi(x)/\mu}, \quad \varphi(x) = \int_0^x \sqrt{\rho(x)}\, dx. \tag{4.4.18}$$

As it was shown in that exercise, the functions $u_n(x)$ with even n are real and with odd n are pure imaginary. Assume that

$$U = u_0 + \mu^2 u_2 + \cdots, \quad V = \mu v_1 + \mu^3 v_3 + \cdots, \quad v_{2n+1} = -i u_{2n+1}. \tag{4.4.19}$$

Then, separating the real and imaginary parts of solution (4.4.18), we find the general solution

$$y(x, \mu) = C_1 \left[U \cos\left(\frac{\varphi}{\mu}\right) - V \sin\left(\frac{\varphi}{\mu}\right) \right] + C_2 \left[U \sin\left(\frac{\varphi}{\mu}\right) + V \cos\left(\frac{\varphi}{\mu}\right) \right], \tag{4.4.20}$$

where C_1 and C_2 are arbitrary constants. After substituting this solution in the boundary conditions (4.4.17) and equating to zero the determinant of second order we get the following equations for μ:

$$\tan\frac{\varphi_1}{\mu} = \frac{V(0)U(1) - U(0)V(1)}{U(0)U(1) + V(0)V(1)} \simeq \sum_{k=0}^{\infty} a_k \mu^{2k+1}, \quad \varphi_1 = \int_0^1 \sqrt{\rho(x)}\, dx.$$

(4.4.21)

The asymptotic expansion of the nth solution of this equation has the form

$$\mu_n \simeq \sum_{k=0}^{\infty} \frac{c_k}{(n\pi)^{2k+1}}, \quad \text{as } n \to \infty,$$

where

$$c_0 = \varphi_1, \quad c_1 = a_0 \varphi_1^2, \quad a_0 = \frac{v_1(1)}{u_0(1)} - \frac{v_1(0)}{u_0(0)}.$$

Evaluating the coefficients c_k for large k is troublesome since one needs to divide series by series and substitute one series into another. For an approximate evaluation of the roots μ_n it is more convenient to solve Eq. (4.4.21) directly with an iterative method while keeping a particular number of terms in expansions (4.4.19).

4.4.4 Vibrations of Beams with Variable Cross-Section

Free vibrations of a beam with variable cross-section are described by the equation

$$\frac{d^2}{dx_1^2}\left(EJ(x_1)\frac{d^2 y}{dx_1^2}\right) - \omega^2 \rho_0 S_1(x_1)y = 0, \quad 0 \le x_1 \le l,$$

where EJ in the bending stiffness, ρ_0 is the density, $S_1(x_1)$ in the cross-sectional area, l is the beam length, and ω is the vibrations frequency.

In the non-dimensional variables

$$x_1 = lx, \quad EJ(x_1) = EJ_0 p(x), \quad S_1(x_1) = S_0 \rho(x), \quad \mu^4 = \frac{EJ_0}{\rho_0 S_0 \omega^2 l^4},$$

$$\omega = \sqrt{\frac{EJ_0}{\rho_0 S_0 l^4}}\,\Omega, \quad \Omega = \frac{1}{\mu^2}.$$

(4.4.22)

the equation of vibrations of a beam is

$$\mu^4 \frac{d^2}{dx^2}\left(p(x)\frac{d^2 y}{dx^2}\right) - \rho(x)y = 0, \quad 0 \le x \le 1.$$

We find the first terms of the asymptotic expansions of the frequencies and vibrations modes for a beam with clamped ends

$$y = \frac{dy}{dx} = 0 \quad \text{at} \quad x = 0, \ x = 1.$$

The general solution of equation (4.4.22) is obtained in Sect. 4.1.1. Here we write the first two terms of its asymptotic expansion:

$$y(x, \mu) = C_1(u_0 \cos z - \mu v_1 \sin z) + C_2(u_0 \sin z + \mu v_1 \cos z)$$
$$+ C_3 e^{-z}(u_0 + \mu v_1) + C_4 e^{z_1}(u_0 - \mu v_1),$$

where C_k are arbitrary constants,

$$z = \frac{1}{\mu} \int_0^x q(x)\, dx, \quad z_1 = \frac{1}{\mu} \int_1^x q(x)\, dx, \quad q(x) = \left(\frac{p}{p}\right)^{1/4},$$

$$u_0 = p^{-1/2} q^{-3/2}, \quad v_1 = \frac{u_0}{8} \int_0^x \left[\frac{15(q')^2 - 10qq''}{q^3} + \frac{3(p')^2 - 4pp''}{p^2 q} \right] dx.$$

After substitution into the boundary conditions and setting the determinant of fourth order to zero we get, to an accuracy of order μ^2,

$$\cos z(1) - \mu b_1 \sin z(1) + O\left(\mu^2\right) = 0, \quad b_1 = \frac{v_1(1)}{u_0(1)},$$

whence we find (see Example 1 in Sect. 1.5)

$$\mu_n = \frac{f_1 c_n}{c_n^2 - f_1/b_1}, \quad c_n = \frac{\pi}{2}(2n + 1), \quad f_1 = \int_0^1 q(x)\, dx.$$

The value of Ω_n, which is proportional to the vibrations frequency ω_n, is defined by the formula

$$\Omega_n = \left(\frac{c_n}{f_1} - \frac{1}{b_1 c_n}\right)^2 + O\left(n^{-1}\right), \quad \text{as} \quad n \to \infty.$$

4.4.5 Axisymmetric Vibrations of Cylindrical Shells

Now, we study the spectrum of free axisymmetric vibrations of a circular cylindrical thin shell with freely supported edges.

Add to Eq. (4.3.44) the inertial terms:

$$\frac{1}{1 - v^2} \left(\frac{d^2 u}{ds^2} - v \frac{dw}{ds} \right) + \Lambda u = 0,$$

$$\frac{1}{1 - v^2} \left(v \frac{du}{ds} - w \right) - \mu^4 \frac{d^4 w}{ds^4} + \Lambda w = 0,$$

(4.4.23)

where

$$\Lambda = \frac{\rho \omega^2 R^2}{E}, \quad \mu^4 = \frac{h^2}{12(1 - \nu^2)R^2}.$$

Here, u and w are the projections of the displacement in the directions of the generatrix and normal, ω is the vibrations frequency, ρ is the density, E is Young's modulus, ν is Poisson's ratio, μ is a small parameter, and Λ is the frequency parameter. The boundary conditions of free support type have the form

$$\frac{du}{ds} = w = \frac{d^2 w}{ds^2} = 0 \quad \text{at} \quad s = 0, \ s = l = \frac{L}{R}, \qquad (4.4.24)$$

where L is the shell length and R is the shell radius.

For $\mu = 0$, we get the equations for the membrane vibrations of a shell. Since the order of the system of equations reduces from 6 to 2 we should keep only one of the three boundary conditions in (4.3.39), namely S_2,

$$T_1 = 0 \quad \text{or} \quad \frac{du}{ds} - \nu w = 0 \quad \text{for} \quad s = 0, \ s = l.$$

The solution of both the original and the membrane problems have the same form:

$$u = u_0 \cos \frac{n \pi s}{l}, \quad w = w_0 \sin \frac{n \pi s}{l}, \quad n = 0, 1, \ldots, \qquad (4.4.25)$$

Substituting this in system (4.4.23) leads to an equation for evaluating the frequency parameter Λ:

$$\left(\Lambda - \frac{k_n^2}{1 - \nu^2} \right) \left(\Lambda - \frac{1}{1 - \nu^2} - \mu^4 k_n^4 \right) - \frac{\nu^2 k_n^2}{(1 - \nu^2)^2} = 0, \quad k_n = \frac{n \pi}{l}. \quad (4.4.26)$$

For each n this equation has two roots: Λ_{n1} and Λ_{n2}. For fixed μ and $n \to \infty$,

$$\Lambda_{n1} \sim \frac{k_n^2}{1 - \nu^2}, \quad \Lambda_{n2} \sim \mu^4 k_n^4,$$

i.e. both frequency series have accumulation point $\Lambda = \infty$.

For $\mu = 0$ we denote the roots of equation (4.4.26) as Λ_{n1}^0 and Λ_{n2}^0:

$$\Lambda_{n1}^0 = \frac{k_n^2}{1 - \nu^2} + O(1), \quad \Lambda_{n2}^0 = 1 - \frac{\nu^2}{(1 - \nu^2)k_n^2} + O\left(n^{-4}\right), \quad n \to \infty, \quad (4.4.27)$$

i.e. the membrane shell vibrations have two frequency series. The series with Λ_{n1}^0 has accumulation point $\Lambda = \infty$, and the other with Λ_{n2}^0 has accumulation point $\Lambda = 1$.

Comparing the roots Λ_{nj} and Λ^0_{nj} of Eq. (4.4.26) for $\mu \neq 0$ and $\mu = 0$ we get

$$\Lambda_{nj} = \Lambda^0_{nj} + O\left(\mu^4\right), \quad j = 1, 2.$$

The residual term in the last formula for $j = 2$ is non-uniform in n since, for $n \sim \mu^{-1}$, it has the order of the main term.

We now consider the spectrum of free axisymmetric vibrations of a thin circular cylindrical shell with clamped edges. The complete eigenproblem consists in the system of equation (4.4.23) and the boundary conditions C_1 in (4.3.39),

$$u = w = \frac{dw}{ds} = 0 \quad \text{for} \quad s = 0, \ s = l. \tag{4.4.28}$$

We compare the solution of the complete eigenproblem with the solution of the degenerate problem

$$\frac{d^2 u^0}{ds^2} + a(\Lambda^0)u^0 = 0, \quad u^0 = 0 \quad \text{for} \quad s = 0, \ s = l, \tag{4.4.29}$$

$$a\left(\Lambda^0\right) = \frac{\Lambda^0\left[1 - (1 - \nu^2)\Lambda^0\right]}{1 - \Lambda^0}, \quad w^0 = \frac{\nu}{1 - (1 - \nu^2)\Lambda^0} \frac{du^0}{ds},$$

which is obtained for $\mu = 0$ after eliminating w from system (4.4.23). This gives the frequencies and modes for the membrane vibrations of a shell. The frequencies of the free membrane vibrations are evaluated by the equation

$$a\left(\Lambda^0_{nj}\right) = \frac{n^2 \pi^2}{l^2}, \quad n = 1, 2, \ldots, \quad j = 1, 2$$

and coincide with the frequencies of a freely supported shell for $n > 0$ [see (4.4.26)]. The vibrations mode is the following:

$$u^0 = \sin \frac{n\pi s}{l}, \quad w^0 = \frac{\nu n \pi}{l[1 - (1 - \nu^2)\Lambda^0]} \cos \frac{n\pi s}{l}. \tag{4.4.30}$$

In the present case, the complete eigenproblem does not have a simple solution of type (4.4.25) satisfying all the boundary conditions (4.4.28). So, firstly we find the general solution of system (4.4.23),

$$u = \sum_{k=1}^{6} C_k u_k \, e^{p_k s}, \quad w = \sum_{k=1}^{6} C_k w_k \, e^{p_k s}, \tag{4.4.31}$$

where C_k are arbitrary constants and u_k and w_k satisfy the equation

$$\left[p_k^2 + \left(1 - \nu^2\right) \Lambda \right] u_k - \nu p_k w_k = 0,$$

where p_k are obtained from the polynomial equation of degree 6:

$$\mu^4 p^6 + \left(1 - \nu^2\right) \Lambda \mu^4 p^4 + (1 - \Lambda) p^2 + \Lambda \left[1 - \left(1 - \nu^2\right) \Lambda \right] = 0. \quad (4.4.32)$$

Let, firstly, $\Lambda \neq 1$. Then (see Sect. 1.3) this equation has two roots $p_k \sim 1$ ($k = 1, 2$). For the other four roots $p_k \sim \mu^{-1}$, $k = 3, 4, 5, 6$, and if $\Lambda < 1$, then all roots have nontrivial real parts. The edge effect integrals correspond to these roots. If $\Lambda > 1$, then two of the roots, p_k, $k = 3, 4$, are pure imaginary and they provide oscillating integrals. In the discussion below, the specific features of the spectrum are explained with the different behavior of the integrals for $\Lambda < 1$ and for $\Lambda > 1$.

For $\Lambda < 1$, the problem A_μ degenerates regularly into the problem A_0 and its solution can be obtained by an iterative method (see Exercise 4.4.8).

For $\Lambda > 1$, we represent the approximate expression for the roots (4.4.32):

$$p_{1,2} = \pm ib + O\left(\mu^2\right), \quad p_{3,4} = \pm \frac{ic}{\mu} + O(\mu), \quad p_{5,6} = \pm \frac{c}{\mu} + O(\mu), \quad (4.4.33)$$

where

$$b^2 = \frac{\Lambda \left[1 - \left(1 - \nu^2\right) \Lambda\right]}{1 - \Lambda}, \quad c^4 = \Lambda - 1$$

and substitute the general solution (4.4.31) into the boundary conditions (4.4.28). Equating to zero the obtained sixth-order determinant and after transformations, we find the relation

$$\sin bl \cos \frac{cl}{\mu} + O(\mu) = 0. \quad (4.4.34)$$

With the error of $O(\mu)$, the left part of this equation is transformed into the product of two multipliers. Therefore, for $\Lambda > 1$ the spectrum consists of two parts. The equation $\sin bl = 0$ provides $b = n\pi/l$ and due to (4.4.33) has roots coinciding with the eigenvalues of the membrane vibrations problem A_0 [see (4.4.29)].

From the equation $\cos(cl/\mu) = 0$ we calculate

$$\Lambda_n = 1 + \left[\frac{(2n + 1)\pi\mu}{2l}\right]^4, \quad n = 1, 2, \ldots \quad (4.4.35)$$

These are additional eigenvalues which do not have analogs in the membrane vibrations problem.

It should be underlined that the above consideration have a sense only if the value of $1 - \Lambda$ is not close to zero. Indeed, in a neighborhood of $\Lambda = 1$ the estimates (4.4.33) become non-uniform. An analysis of the residual terms in (4.4.33) shows that

$$p_{1,2} = \pm ib \left\{1 + O\left[\frac{\mu^2}{(\Lambda - 1)^{3/2}}\right]\right\}, \quad p_k = \frac{r_k c}{\mu}\left\{1 + O\left[\frac{\mu^2}{(\Lambda - 1)^{3/2}}\right]\right\},$$

where $k = 3, 4, 5, 6$ and $r_k^4 = 1$. Therefore for $\Lambda - 1 \sim \mu^{4/3}$ the above formulas are not valid any more.

Find the spectrum of the eigenvalues Λ for the eigenproblem A_μ [see (4.4.23) and (4.4.28)] in a neighborhood of the point $\Lambda = 1$ for the free axisymmetric vibrations of a circular cylindrical thin shell with clamped edges. The point $\Lambda = 1$ is an accumulation point for the frequencies of the membrane vibrations of the shell.

Assume that $\Lambda = 1 + \mu^{4/3}z$ and represent Eq. (4.4.32) in the form

$$q^6 - z_1 q^2 + 1 = 0, \tag{4.4.36}$$

where

$$p = \mu^{-2/3}a^{1/6}q\left[1 + O\left(\mu^{4/3}\right)\right], \quad a = \nu^2, \quad z = z_1 a^{2/3}.$$

For any z_1 $(-\infty < z_1 < \infty)$, Eq. (4.4.36) has a pair of pure imaginary roots $q_{1,2} = \pm ix$ and $z_1 = (x^6 - 1)/x^2$. We find the other roots with nonzero real parts from the equation

$$q^4 - x^2 q^2 + \frac{1}{x^2} = 0.$$

Substituting the general solution (4.4.31) into the boundary conditions (4.4.28) and after cumbersome transformations we obtain the equation for evaluating the unknown x:

$$\tan\left(\mu^{-2/3}a^{1/6}lx\right) = \frac{2\left(x^3 - 1\right)\sqrt{x^6 + 2x^3}}{4x^3 - 1}. \tag{4.4.37}$$

The parameter Λ is related to x by the formula

$$\Lambda = 1 + \mu^{4/3}a^{2/3}\frac{x^6 - 1}{x^2}. \tag{4.4.38}$$

4.4.6 Exercises

4.4.1. Find the terms of order μ^2 in the expansions for Λ and $y(x)$ from Sect. 4.4.2.

4.4.2. Find the asymptotic expansions of the eigenvalue $\Lambda(\mu)$ and eigenfunction $y(x, \mu)$, as $\mu \to 0$, for the free vibrations of a non-absolutely flexible string [see Eq. (4.4.13)] with freely supported edges $y = y'' = 0$ for $x = 0$, $x = 1$.

4.4.3. Find the asymptotic expansions of the eigenvalues μ_n, as $n \to \infty$, of the eigenproblem $\mu^2 y'' + \rho(x)y = 0$, $y(0) = y'(1) = 0$.

4.4.4. In the conditions of Sect. 4.4.3, consider the particular case $\rho(x) = 1 + x$ corresponding to a string with a linearly varying density being twice larger at one end than at the other. Compare the exact and approximate values of μ_n obtained from Eq. (4.4.21) with U and V of different accuracy. Plot the graphs of the exact and approximate eigenfunctions for $n = 1, 2, 3$.

4.4.5. The vibrations of a bar with a weight attached at one end are described by the eigenproblem $\mu^2 \left(S(x)y'\right)' + S(x)y = 0$, $y(0) = 0$, $\mu^2 S(x)y' = my$ for $x = 1$. Find the asymptotic approximations for the first eight values of μ_n for $\alpha = 1$, $m = 1$ if $S(x) = 1 + \alpha x$.

4.4.6. Find the asymptotic expansions of the frequencies and modes of the free vibrations of a clamped beam with constant thickness and linearly varying width.

4.4.7. Compare the frequencies of a complete and a membrane eigenproblems for the axisymmetric vibrations of a freely supported circular cylindrical thin shell for $R/h = 100$, $l = 3$ and $\nu = 0.3$.

4.4.8. Find a two-term asymptotic expansion for the eigenvalues $\Lambda < 1$ of eigenproblems (4.4.23) and (4.4.28).

4.4.9. For a clamped cylindrical shell with $R/h = 100$, $l = 3$ and $\nu = 0.3$, compare the exact values of the frequency parameter Λ with the values obtained by the asymptotic formulas (4.4.8), (4.4.26), (4.4.35) and (4.4.38).

4.5 Eigenfunctions Localized in a Neighborhood of One End of the Interval

In Sect. 4.4, the spectrum points are associated with oscillating integrals of the system for which edge effects integrals are auxiliary. In this section, we consider some examples where the eigenfunctions are linear combinations of edge effect integrals exponentially decreasing away from the edge. The study of such examples is important since for such eigenfunctions we obtain the lower eigenvalues.

4.5.1 Vibrations of Rectangular Plates

The free vibrations of a rectangular plate are described by the equation

$$D\Delta^2 w - \rho\omega^2 w = 0, \quad 0 \le x \le a, \quad 0 \le y \le b,$$

$$\Delta = \frac{\partial^2}{\partial x^2} + \frac{\partial^2}{\partial y^2}, \quad D = \frac{Eh^3}{12(1 - \nu^2)},$$

where $w(x, y)$ is the plate deflection, a and b are the lengths of the plate sides, ω is the vibrations frequency, E, ν, h, ρ are Young's modulus, Poisson's ratio, plate thickness and material density, respectively.

Let the edges $y = 0$ and $y = b$ be freely supported

$$w = \frac{\partial^2 w}{\partial y^2} = 0 \quad \text{for} \quad y = 0, \quad y = b,$$

and the edge $x = 0$ be free

$$\frac{\partial^2 w}{\partial x^2} + \nu \frac{\partial^2 w}{\partial y^2} = 0, \quad \frac{\partial^3 w}{\partial x^3} + (2 - \nu) \frac{\partial^3 w}{\partial x \partial y^2} = 0 \quad \text{for} \quad x = 0.$$

So far, we did not define concretely the boundary conditions at the end $x = a$ since, for sufficiently large a, they do not affect the result. We now find the lowest frequency for plate vibrations as $a \to \infty$ [56].

After separating variables and using the new variable x_1,

$$w(x, y) = w(x_1) \sin \frac{\pi y}{b}, \quad x_1 = \frac{\pi x}{b} \tag{4.5.1}$$

we obtain the following ordinary differential equation for the function $w(x_1)$:

$$\frac{d^4 w}{dx_1^4} - 2 \frac{d^2 w}{dx_1^2} + w - \Lambda w = 0, \quad \Lambda = \frac{\rho h \omega^2 b^4}{D \pi^4} \tag{4.5.2}$$

with the boundary conditions at the free edge

$$\frac{d^2 w}{dx_1^2} - \nu w = 0, \quad \frac{d^3 w}{dx_1^3} - (2 - \nu) \frac{dw}{dx_1} = 0 \quad \text{for} \quad x_1 = 0. \tag{4.5.3}$$

If we consider the cylindrical bending of a plate which is infinite in the direction x_1, then $\Lambda = 1$ and the function w does not depend on x_1.

If the edges $x = 0$ and $x = a$ are freely supported, then

$$w(x_1) = \sin \frac{bx_1}{a}, \quad \Lambda = \left(1 + \frac{b^2}{a^2}\right)^2 > 1$$

and the parameter $\Lambda \to 1$ as $b/a \to 0$.

Let $\Lambda < 1$. We seek a nontrivial solution of equation (4.5.2) in the form

$$w(x_1) = C_1 e^{-sx_1} + C_2 e^{-rx_1}, \quad s > 0, r > 0,$$

which satisfies conditions (4.5.3) and $w(x_1) \to 0$ as $x_1 \to \infty$, where s and r are the roots of the equation

$$\left(z^2 - 1\right)^2 - \Lambda = 0,$$

and C_1 and C_2 are arbitrary constants. So, we have

$$s = \sqrt{1 - \sqrt{\Lambda}}, \quad r = \sqrt{2 - s^2}.$$

Substituting s and r into the boundary conditions (4.5.3), we obtain the following equation for Λ:

$$s\left(r^2 - \nu\right)^2 = r\left(\nu - s^2\right)^2, \tag{4.5.4}$$

which has only one root $\Lambda < 1$.

Taking $\nu < 1/2$ into account, we approximately obtain

$$\Lambda = 1 - \frac{4\nu^4}{(2 - \nu)^2}, \quad s = \frac{\sqrt{2}\nu^2}{(2 - \nu)^2}.$$

In particular, for $\nu = 0.3$ we get $\Lambda = 0.9962$, $s = 0.0436$, $r = 1.4135$.

The existence of the free edge $x = 0$ leads to an insignificant decrease of the parameter Λ compared to $\Lambda = 1$, and the vibrations mode (assuming $w(0) = 1$)

$$w(x_1) = 0.8507\, e^{-sx_1} + 0.1493\, e^{-rx_1}$$

slowly goes down as x_1 increases. For $\nu = 0.5$ (rubber), we have $\Lambda = 0.9571$, $s = 0.1472$, $r = 1.4065$, and

$$w(x_1) = 0.7555\, e^{-sx_1} + 0.2445\, e^{-rx_1},$$

and for $\nu = 0.1$ (foam), we have $\Lambda = 0.99997$, $s = 0.0039$, $r = 1.4142$, and

$$w(x_1) = 0.95\, e^{-sx_1} + 0.05\, e^{-rx_1}.$$

4.5.2 Vibrations and Buckling of Shells

In Sect. 4.5.1, a decrease of the eigenvalue due to a free edge was insignificant and less than 1 %. Now, in the problems of free vibrations and buckling considered below the existence of a free or weakly supported edge can reduce the eigenvalue several times.

We write the linear system of equation for shallow shells [31, 56] in a non-dimensional form [see also (4.2.53)]

$$\mu^2 \Delta^2 w - \Delta_k \Phi + Z = 0, \quad \mu^2 \Delta^2 \Phi + \Delta_k w = 0, \tag{4.5.5}$$

where

$$\Delta w = \frac{\partial^2 w}{\partial x^2} + \frac{\partial^2 w}{\partial y^2}, \quad \Delta_k w = k_2 \frac{\partial^2 w}{\partial x^2} + k_1 \frac{\partial^2 w}{\partial y^2}, \quad 0 \le x \le a, \ 0 \le y \le b.$$

Here $w(x, y)$ and $\Phi(x, y)$ are unknown deflection and stress functions, k_1 and k_2 are non-dimensional variables, $\mu > 0$ is a small parameter and

$$k_1 = \frac{R}{R_1}, \quad k_2 = \frac{R}{R_2}, \quad \mu^4 = \frac{h^2}{12(1 - \nu^2)R^2},$$

where R_1 and R_2 are the radii of curvature, E, ν and h are Young's modulus, Poisson's ratio and shell thickness, respectively. We let R denote the characteristic linear size of the coordinates x and y, and Z be the loading term. Further, we consider free vibrations and buckling of shells. For the free vibrations with frequency ω we have

$$Z = -\frac{\rho \omega^2 R^2}{E \mu^2} w = \frac{c \Lambda}{\mu^2} w, \quad \Lambda = \frac{\rho \omega^2 R^2}{Ec}, \tag{4.5.6}$$

and, for the stability of the membrane, the stress state is given with the initial stress-resultants T_1^0 and T_2^0 (under tension $T_i^0 < 0$),

$$Z = 2\Lambda \left(t_1 \frac{\partial^2 w}{\partial x^2} + t_2 \frac{\partial^2 w}{\partial y^2} \right), \quad T_i^0 = -2\Lambda E h \mu^2 t_i, \ i = 1, 2. \tag{4.5.7}$$

Here Λ, which is either a frequency parameter or a loading parameter depending on the problem under consideration, is the required parameter. We normalize Λ and also the parameter c in such a manner to make the characteristic value of Λ equal to 1.

Assume that, at the edges $y = 0$ and $y = b$, the following boundary conditions of free support type are:

$$w = \frac{\partial^2 w}{\partial y^2} = \Phi = \frac{\partial^2 \Phi}{\partial y^2} = 0 \quad \text{for} \ \ y = 0, \ y = b, \tag{4.5.8}$$

allowing separation of variables (for k_i, $t_i = \text{const}$):

$$w(x, y) = w(x) \sin \frac{n \pi y}{b}, \quad \Phi(x, y) = \Phi(x) \sin \frac{n \pi y}{b}, \quad n = 1, 2, \ldots \tag{4.5.9}$$

At the edge $x = 0$, we consider 16 standard variants of the boundary conditions denoted by C_i, S_i, $i = 1, 2, \ldots, 8$.

$$u = v = w = 0, \qquad \{w' = 0 \text{ or } M_1 = 0\} \qquad C_1 \text{ or } S_1$$
$$T_1 = v = w = 0, \qquad \{w' = 0 \text{ or } M_1 = 0\} \qquad C_2 \text{ or } S_2$$
$$u = S = w = 0, \qquad \{w' = 0 \text{ or } M_1 = 0\} \qquad C_3 \text{ or } S_3$$
$$T_1 = S = w = 0, \qquad \{w' = 0 \text{ or } M_1 = 0\} \qquad C_4 \text{ or } S_4$$
$$u = v = Q_1^* = 0, \qquad \{w' = 0 \text{ or } M_1 = 0\} \qquad C_5 \text{ or } S_5 \qquad (4.5.10)$$
$$T_1 = v = Q_1^* = 0, \qquad \{w' = 0 \text{ or } M_1 = 0\} \qquad C_6 \text{ or } S_6$$
$$u = S = Q_1^* = 0, \qquad \{w' = 0 \text{ or } M_1 = 0\} \qquad C_7 \text{ or } S_7$$
$$T_1 = S = Q_1^* = 0, \qquad \{w' = 0 \text{ or } M_1 = 0\} \qquad C_8 \text{ or } S_8$$

where u and v are the projections of displacements on the x and y axes, T_1 and S are stress-resultants, Q_1^* is the generalized transverse force, M_1 is the bending moment. In particular, the conditions of clamped type are denoted by C_1, free support (4.5.8) by S_2, and free edge by S_8. Compare with the boundary conditions (4.3.39).

We seek eigenfunctions, localized at the edge $x = 0$ and exponentially decreasing away from it, in the form

$$w(x, y) = \sum_{k=1}^{4} C_k w_k \, e^{p_k x / \mu} \sin \frac{qy}{\mu}, \quad q = \frac{n \pi \mu}{b}, \qquad (4.5.11)$$

where C_k are arbitrary constants, p_k are the roots of the eighth order equation satisfying condition $\Re \, p_k < 0$ that guarantees decay of solution (4.5.11). For the case of vibrations, this equation has the form

$$\left(p^2 - q^2\right)^4 + \left(k_2 p^2 - k_1 q^2\right)^2 - c\Lambda \left(p^2 - q^2\right)^2 = 0 \qquad (4.5.12)$$

and for buckling,

$$\left(p^2 - q^2\right)^4 + \left(k_2 p^2 - k_1 q^2\right)^2 + 2\Lambda \left(t_1 p^2 - t_2 q^2\right) \left(p^2 - q^2\right)^2 = 0. \quad (4.5.13)$$

The functions u, T_1, Q_1^*, M_1 have the form (4.5.11) and for the functions v, S one should take $\cos(qy/\mu)$ instead of $\sin(qy/\mu)$. Here, we represent expressions for u_k, $v_k, \ldots$ in w_k (with the index k omitted):

$$u = \mu p^{-1} \left(k_1 w - \left(q^2 + \nu p^2\right) \Phi\right), \quad T_1 = -EhR^{-1} q^2 \Phi,$$
$$v = \mu q^{-1} \left(k_2 w + \left(p^2 + \nu q^2\right) \Phi\right), \quad S = EhR^{-1} pq\Phi,$$
$$Q_1^* = -EhR^{-1} \mu \left(p^3 - (2 - \nu)pq^2 + 2\Lambda t_1 p\right) w, \qquad (4.5.14)$$
$$M_1 = Eh\mu^2 \left(p^2 - \nu q^2\right) w, \quad \Phi = -\left(k_2 p^2 - k_1 q^2\right) \left(p^2 - q^2\right)^{-2} w.$$

We denote the generalized transverse force Q_1^* as

$$Q_1^* = Q_1 - \frac{1}{R}\frac{\partial H}{\partial y} + T_1^0\frac{\partial w}{\partial x},$$

where H is the torsion.

A substitution of solutions (4.5.11) into the boundary conditions at $x = 0$ yields a system of four equations in C_k. If its determinant vanishes one gets an equation for evaluating Λ,

$$\Delta(\Lambda, q) = 0. \tag{4.5.15}$$

We analyze solution (4.5.11) and Eq. (4.5.15) in the domain of the parameters Λ, q, k_i, t_i, where Eqs. (4.5.12) or (4.5.13) do not have pure imaginary roots since, in this case, only four of its roots satisfy condition $\Re\, p_k < 0$.

Depending on the boundary conditions at $x = 0$ and also on the parameters q, k_i and t_i, Eq. (4.5.15) may have or not roots we are interested in. If the root Λ exists, the corresponding boundary condition at the edge $x = 0$ is called *weak support*.

Following the above scheme, we shall consider a series of particular problems of free vibrations and buckling of thin shells.

4.5.3 Vibrations of Cylindrical Panels

Consider the free vibrations of a cylindrical panel (curved plate) with free rectangular edge $x = 0$ and free supported curvilinear edges $y = 0$ and $y = b$. Find the vibrarion frequencies of a panel for which the vibrations modes are localized in a neighborhood of the edge $x = 0$ under the assumption that $b \sim 1$.

In this case, we should take $k_1 = 1$ and $k_2 = 0$ in the formulas of Sect. 4.5.2. Equation (4.5.12) has the form

$$\left(p^2 - q^2\right)^4 + q^4 - 2\Lambda q^2\left(p^2 - q^2\right)^2 = 0, \quad q = \frac{n\pi\mu}{b}. \tag{4.5.16}$$

Consider, firstly, a shell closed in the circumferential direction. The periodicity condition in the direction x provides $p = im\mu$, where $m = 1, 2, \ldots$ is the wavenumber in the circumferential direction. Representing Eq. (4.5.16) in the form

$$\Lambda = \frac{\left(m^2\mu^2 + q^2\right)^4 + q^4}{c\left(m^2\mu^2 + q^2\right)^2},$$

we find

$$\min_{m}\,\Lambda = 1$$

for $c = 2q^2$. Because of that, we compare the value of the required parameter Λ with the value $\Lambda = 1$ corresponding to the lowest vibrations frequency for a cylindrical shell closed in the circumferential direction. For $\Lambda < 1$, Eq. (4.5.16) does not have pure imaginary roots.

Coming back to a panel with free edge $x = 0$ [see boundary conditions S_8, (4.5.10)], we find the roots of equation (4.5.16) for $\Lambda < 1$. By (4.5.16), $q = n\pi\mu/b$. We assume that $n = 1$ or $n \sim 1$. Then q and μ are small parameters and the roots of equation (4.5.16) can be represented in the form

$$p_{1,2} = -\sqrt{q}\,e^{\pm i\theta} + O\left(\sqrt{q^3}\right), \quad p_{3,4} = -\sqrt{q}\,e^{\pm i(\pi/2-\theta)} + O\left(\sqrt{q^3}\right), \quad (4.5.17)$$

where $\Lambda = \cos 4\theta$ and

$$0 < \theta < \frac{\pi}{8}. \tag{4.5.18}$$

In (4.5.17), we write only the roots p_k with $\Re\, p_k < 0$.

For the boundary conditions S_8, due to (4.5.14) and (4.5.17) with accuracy of order q, Eq. (4.5.15) is transformed into

$$\begin{vmatrix} p_1^3 & p_2^3 & p_3^3 & p_4^3 \\ p_1^2 & p_2^2 & p_3^2 & p_4^2 \\ p_1^{-4} & p_2^{-4} & p_3^{-4} & p_4^{-4} \\ p_1^{-3} & p_2^{-3} & p_3^{-3} & p_4^{-3} \end{vmatrix} = 0.$$

After transformations and neglecting the values of order q compared to 1 we get

$$\sin 2\theta - \sin^2 6\theta = 0. \tag{4.5.19}$$

This equation has two roots, $\theta^{(1)} = 0.36448$ and $\theta^{(2)} = 0.05770$, which satisfy conditions (4.5.18) and, correspondingly, $\Lambda^{(1)} = 0.113$ and $\Lambda^{(2)} = 0.973$.

In Fig. 4.6 we plot the corresponding eigenfunctions $w^{(1)}(x)$ and $w^{(2)}(x)$ given by the formulas

$$w^{(k)}(x) = D^{(k)}\Re\left[C_1^{(k)}\exp\left(-\xi\,e^{i\theta^{(k)}}\right) + C_3^{(k)}\exp\left(i\xi\,e^{-i\theta^{(k)}}\right)\right], \quad k = 1, 2,$$

$$C_1^{(k)} = i\,e^{2i\theta^{(k)}} + i\,e^{-2i\theta^{(k)}}\left[\cos 6\theta^{(k)} - \sin 6\theta^{(k)}\right],$$

$$C_3^{(k)} = -i\,e^{-2i\theta^{(k)}} + i\,e^{2i\theta^{(k)}}\left[\cos 6\theta^{(k)} + \sin 6\theta^{(k)}\right],$$

where $\xi = x\sqrt{q}$.

Here, $D^{(k)}$ are the normalization factors evaluated from the condition $\max_x w^{(k)} = 1$. The damping rate of $w^{(k)}(x)$ as $x \to \infty$ decreases with q. The

Fig. 4.6 Eigenfunctions $w^{(1)}(x)$ and $w^{(2)}(x)$

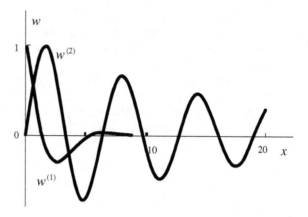

relatively slow damping of the function $w^{(2)}(x)$ (see Fig. 4.6) is explained by the closeness of $\Lambda^{(2)} = 0.973$ to $\Lambda = 1$ for which system (4.5.5) has oscillating integrals.

Thus, we obtain the values of the roots $\Lambda^{(k)}$ for $\Lambda(q)$ as $q \to 0$ for Eq. (4.5.15). Roots $\Lambda(q)$ for $q > 0$ are found in Exercise 4.5.4.

4.5.4 Buckling of Cylindrical Panels

Consider the buckling of a circular cylindrical panel (curved plate) with a weakly supported curvilinear edge $x = 0$ under axial compression (see Fig. 4.7). The straight edges $y = 0$ and $y = b$ are freely supported.

Now we obtain the type of week support of the edge $x = 0$. Using the notation accepted in this subsection we have

Fig. 4.7 Buckling of a *circular cylindrical* panel under axial compression

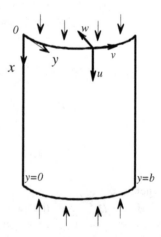

$$T_1^{(0)} = -2Eh\mu^2\Lambda, \quad k_1 = 0, \ k_2 = 1, \ t_1 = 1, \ t_2 = 0,$$

and Eq. (4.5.13) becomes

$$\left(p^2 - q^2\right)^4 + p^4 + 2\Lambda p^2 \left(p^2 - q^2\right)^2 = 0, \quad q = \frac{n\pi\mu}{b}, \ n = 1, 2, \ldots \quad (4.5.20)$$

For $b \sim 1$ and $n = 1$, the parameter q is small. For $\Lambda < 1$, Eq. (4.5.20) has two roots, $p_k \sim 1$ and $(\Re p_k < 0)$, and two roots $p_k \sim q^2$ and $(\Re p_k < 0)$:

$$p_{1,2} = \pm i e^{\pm i\psi} + O\left(q^2\right), \quad p_{3,4} = \pm i q^2 e^{\pm i\psi} + O\left(q^4\right), \quad (4.5.21)$$

where $\Lambda = \cos 2\psi$.

Consider, for example, the boundary conditions of clamped type $u = v = w = w' = 0$, i.e. conditions C_1 [see (4.5.10)]. Substituting the values of the roots (4.5.21) into (4.5.14) we obtain Eq. (4.5.15) in the form

$$\begin{vmatrix} \nu p_1^{-1} & \nu p_2^{-1} & p_1 & p_2 \\ -(2+\nu)q^2 p_1^{-2} & -(2+\nu)q^2 p_2^{-2} & 1 & 1 \\ 1 & 1 & 1 & 1 \\ p_1 & p_2 & q^2 p_1 & q^2 p_2 \end{vmatrix} = 0. \quad (4.5.22)$$

In the determinant (4.5.22) we cancel common multipliers in the rows and neglect terms of order q^2 compared to 1. With an accuracy of order q^2, Eq. (4.5.22) can be represented as

$$-(p_2 - p_1)^2 = 4\cos^2\psi = 0.$$

It does not have any roots satisfying the condition

$$0 < \psi < \frac{\pi}{2}, \quad (4.5.23)$$

i.e. clamped edge is not a weak support.

4.5.5 Exercises

4.5.1. Under the conditions of Sect. 4.5.1, study the effect of the ratio a/b on the lowest vibrations frequency assuming that the edge $x = 0$ is free and the edge $x = a$ is clamped. Assume that $\nu = 0.3$.

4.5.2. The buckling of a rectangular plate $0 \leq x \leq a$, $0 \leq y \leq b$, compressed with force T in the direction y is described by the equation [56]:

Fig. 4.8 Buckling of
cylindrical panel with
weakly supported
rectangular edge $x = 0$
under axial compression

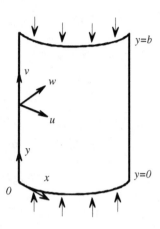

$$DΔ^2 w + T \frac{\partial^2 w}{\partial y^2} = 0,$$

where the notation of Sect. 4.5.1 is used. Find the critical force T as $a/b \to \infty$ if
the edges $y = 0$ and $y = b$ are freely supported and the edge $x = 0$ is free.

4.5.3. For all 16 types of boundary conditions (4.5.10) develop equations similar to
(4.5.19) and find their roots satisfying condition (4.5.18) [56].

4.5.4. In Sect. 4.5.3 and Exercise 4.5.3, the roots of equation (4.5.15) are found as
$q \to 0$. We recall that $q = n\pi\mu/b, n = 1, 2, \ldots$. For large n and/or for small b, the
assumption $q \ll 1$ does not hold any more. Find the roots $Λ(q)$ of Eq. (4.5.15) for
$q \sim 1$. Assume that $\nu = 0.3$.

4.5.5. Consider the buckling of a cylindrical panel with weakly supported rectangular
edge $x = 0$ under axial compression (see Fig. 4.8). The curvilinear edges $y = 0$ and
$y = b$ are freely supported.

For different types of boundary conditions at $x = 0$, analyze the existence of
buckling modes localized near the edge $x = 0$ and find the critical values of the
force $T_2^{(0)}$.

4.5.6. Under the conditions of Sect. 4.5.4, study the dependence of the loading parameter $Λ(q)$ on the value of q assuming that $q \sim 1$ and $\nu = 0.3$.

4.6 Answers and Solutions

4.1.1 In the accepted notation for formula (4.1.2), we have

$$\lambda = iq(x), \quad q(x) = \left(\frac{\rho}{p}\right)^{1/2}, \quad u_0 = (p\rho)^{-1/4},$$

$$u_{n+1} = \frac{iu_0}{2} \int u_0((pu_n')' - ru_n)\, dx.$$

In the particular case $p = \rho = 1 + \alpha x, r = 0$ we get

$$u_0 = p^{-1/2}, \quad u_n = (-i\alpha)^n p^{-\beta_n} b_n,$$

where $b_0 = 1$, $b_n = (2n - 1)^2 b_{n-1}/(8n)$, $\beta_n = n + 1/2$, $n = 1, 2, \ldots$
Series (4.1.2) diverges.

4.1.2 In formula (4.1.2), $u_0^{(1,2)}(x) = (c(x))^{-1/4}$,

$$\lambda_1(x) = -\sqrt{c(x)}, \quad u_{n+1}^{(1)}(x) = \frac{u_0}{2} \int_{x^{(1)}}^{x} u_0(\xi)\, \frac{d^2 u_n^{(1)}}{d\xi^2}\, d\xi,$$

$$\lambda_2(x) = \sqrt{c(x)}, \quad u_{n+1}^{(2)}(x) = -\frac{u_0}{2} \int_{x^{(2)}}^{x} u_0(\xi)\, \frac{d^2 u_n^{(2)}}{d\xi^2}\, d\xi,$$

and, as with (4.1.11), the lower limit of integration $x^{(1)}$ or $x^{(2)}$ is arbitrary.

4.1.3. $q(x) = 1$, $u_0 = p^{-1/2}$,

$$u_1^{(k)}(x) = \frac{3\alpha u_0(x)}{8r_k}\left(\frac{1}{p(x)} - \frac{1}{p(0)}\right), \quad k = 1, 2, 3, 4.$$

4.1.4. $q(x) = (\rho(x))^{-1/2}$, $u_0 = (\rho(x))^{-3/4}$,

$$u_1^{(k)}(x) = \frac{3\alpha u_0(x)}{16r_k}(q(0) - q(x)), \quad k = 1, 2, 3, 4.$$

4.1.5. We have

$$y^{(k)}(x, \mu) = \left[u_0(x) + \mu u_1(x) + O\left(\mu^2\right)\right]\exp\left(\frac{1}{\mu}\int \lambda_k(x)\, dx\right), \quad k = 1, 2, 3, 4,$$

$$\lambda_k = qr_k, \quad q = \left(\frac{c}{p}\right)^{1/4}, \quad r_k^4 + 1 = 0, \quad r_{1,2} = -\frac{1}{\sqrt{2}} \pm \frac{i}{\sqrt{2}}, \quad r_{3,4} = \frac{1}{\sqrt{2}} \pm \frac{i}{\sqrt{2}},$$

and the expressions for u_0 and $u_1^{(k)}$ coincide with those obtained for Eq. (4.1.13). All integrals $y^{(k)}(x, \mu)$ are edge effect integrals.

4.1.6. Two solutions, y_1 and y_2, of the given equation are slowly oscillating. The expansion of y_1 is of the form

$$y_1(x, \mu) = \left\{1 - \mu^2 \frac{(i\omega)^3 x}{2} + \mu^4\left[\frac{7(i\omega)^5 x}{8} - \frac{(i\omega)^6 x^2}{8}\right] + O\left(\mu^6\right)\right\}\exp(i\omega x),$$

where $\omega = \sqrt{\Lambda}$. The second solution, $y_2(x, \mu)$, is the complex conjugate of $y_1(x, \mu)$.
The edge effect integrals have the form

$$y_{3,4}(x, \mu) = \left[1 \pm \mu \frac{\Lambda x}{2} + \mu^2 \frac{\Lambda^2 x^2}{8} + O\left(\mu^3\right)\right] e^{\pm x/\mu}.$$

4.1.7. $K_\nu(x) \simeq \sqrt{\dfrac{\pi}{2x}} e^{-x} \displaystyle\sum_{n=0}^{\infty} (-1)^n a_n, \quad I_\nu(x) \simeq \dfrac{e^x}{\sqrt{2\pi x}} \displaystyle\sum_{n=0}^{\infty} a_n,$ where $a_0 = 1$,

$$a_n = \frac{(\nu_1 - 1) \cdots (\nu_1 - (2n-1)^2)}{n!(8x)^n}, \quad \nu_1 = 4\nu^2.$$

4.2.1. We seek an asymptotic expansion in the form (4.2.3):

$$y(x, \mu) = \left\{\cos \frac{x}{2} - \mu \left[\varphi_1(x) \cos \frac{x}{2} - \frac{1}{4} \sin \frac{x}{2}\right] + O\left(\mu^2\right)\right\} e^{q/\mu},$$

$$z(x, \mu) = \left\{-\sin \frac{x}{2} + \mu \left[\varphi_1(x) \sin \frac{x}{2} + \frac{1}{4} \cos \frac{x}{2}\right] + O\left(\mu^2\right)\right\} e^{q/\mu},$$

where $q(x) = 2 \sin x - x$, φ_1 is the same as in Sect. 4.2.1.

4.2.2. The asymptotic expansions are

$$y_1 = D\left[\cos z + \mu k^2 t \sin z + O\left(\mu^4\right)\right] + y_1^0 + \mu k^2 y_2^0 t + O\left(\mu^2\right),$$

$$y_2 = D\left[-\sin z + \mu k^2 t \cos z + O\left(\mu^4\right)\right] + y_2^0 - \mu k^2 y_1^0 t + O\left(\mu^2\right),$$

where $z = \mu^{-1} t + \alpha$, and D, α, y_1^0 and y_2^0 are arbitrary constants.
   The exact solution of system (4.2.52) has the form

$$y_1 = D_1 \cos(\nu_1 t + \alpha_1) + D_2 \cos(\nu_2 t + \alpha_2),$$

$$y_2 = -\frac{\nu_1 H D_1}{k^2 + \nu_1^2} \sin(\nu_1 t + \alpha_1) + \frac{\nu_2 H D_2}{k^2 + \nu_2^2} D_2 \sin(\nu_2 t + \alpha_2),$$

where $D_1, \alpha_1, D_2, \alpha_2$ are arbitrary constants, and

$$\nu_{1,2}^2 = \frac{H}{2} - k^2 \pm \sqrt{\frac{H^2}{4} - Hk^2}, \quad \nu_1 \sim H, \quad \nu_2 \sim \frac{1}{H}.$$

For $H > 4k^2$, the values of ν_1 and ν_2 are real and the vertical position of a top is
stable. A comparison shows that the estimates of the residual terms in the asymptotic
expansion are non-uniform as $t \to \infty$.

4.2.3. Substituting (4.2.37) in system (4.2.35), at μ^0 we get

$$(F - \Lambda)w_0 - G\Phi_0 = 0, \quad Gw_0 + F\Phi_0 = 0,$$

$$F(\lambda, s) = \left(\lambda^2 - \frac{r^2}{b^2}\right)^2, \quad G(\lambda, s) = k_2\lambda^2 - \frac{k_1 r^2}{b^2}, \tag{4.6.1}$$

from where follow Eq. (4.2.38) and the relation between w_0 and Φ_0.
   For μ^1 we obtain

$$(F - \Lambda)w_1 - G\Phi_1 + L_1 w_0 - L_2\Phi_0 = 0,$$
$$Gw_1 + F\Phi_1 + L_2 w_0 + L_1\Phi_0 = 0, \tag{4.6.2}$$

where

$$L_1 w_0 = \left[4\lambda w_0' + 2\left(\lambda' + \frac{\lambda b'}{b}\right)w_0\right]\left[\lambda^2 - \frac{r^2}{b^2}\right] + 2\lambda\left[\lambda^2 - \frac{r^2}{b^2}\right]' w_0$$

$$= \sqrt{\frac{F_\lambda}{b}}\left(w_0\sqrt{bF_\lambda}\right)',$$

$$L_2 w_0 = 2\lambda k_2 w_0' + \lambda' k_2 w_0 + \frac{\lambda(k_2 b)'}{b}w_0 = \sqrt{\frac{G_\lambda}{b}}\left(w_0\sqrt{bG_\lambda}\right)',$$

$$F_\lambda = \frac{\partial F}{\partial\lambda}, \quad G_\lambda = \frac{\partial G}{\partial\lambda}, \quad ()' = \frac{d()}{ds}.$$

From the first equation in (4.6.2) multiplied by bw_0 we subtract the second equation multiplied by $b\Phi_0$. This eliminates the variables w_1 and Φ_1 and we obtain

$$w_0\sqrt{bF_\lambda}\left(w_0\sqrt{bF_\lambda}\right)' - w_0\sqrt{bG_\lambda}\left(\Phi_0\sqrt{bG_\lambda}\right)'$$

$$\Phi_0\sqrt{bG_\lambda}\left(w_0\sqrt{bG_\lambda}\right)' - \Phi_0\sqrt{bF_\lambda}\left(\Phi_0\sqrt{bF_\lambda}\right)' = 0. \tag{4.6.3}$$

The second equation in (4.6.1) is satisfied if we assume

$$w_0 = \sqrt{F}z, \quad \Phi_0 = -\frac{G}{\sqrt{F}}z. \tag{4.6.4}$$

Then taking (4.2.38) into account, Eq. (4.6.3) is transformed into the equation

$$\left(z^2 b\frac{\partial f}{\partial\lambda}\right)' = 0, \quad f = F(F - \Lambda) + G^2,$$

where the function f is as in (4.2.38). For $z = \left(b\dfrac{\partial f}{\partial \lambda}\right)^{-1/2}$, formulas (4.6.4) and (4.2.39) are equivalent.

4.2.4. The asymptotic expansions of the integrals of the system of equation (4.2.53) have the form (4.2.37), where the function $\lambda(s)$ satisfies the equation

$$f(\lambda, s) = \left(\lambda^2 - \frac{r^2}{b^2}\right)^4 + \Lambda\left(t_1\lambda^2 - \frac{t_2 r^2}{b^2}\right)\left(\lambda^2 - \frac{r^2}{b^2}\right)^2 + \left(k_2\lambda^2 - \frac{k_1 r^2}{b^2}\right)^2 = 0,$$
(4.6.5)

and the first coefficients w_0 and Φ_0 are the same as in (4.2.39).

4.2.5. Substituting (4.2.47) into system (4.2.35) we find that the functions w_0 and Φ_0 satisfy the system of equations:

$$\frac{d^2 w_0}{ds^2} + \frac{r_0^4}{b^3\cos\alpha}\Phi_0 = 0, \quad \frac{d^2\Phi_0}{ds^2} - \frac{r_0^4}{b^3\cos\alpha}w_0 + \frac{\Lambda b}{\cos\alpha}w_0 = 0,$$

or the fourth order equation

$$\frac{d^2}{ds^2}\left(b^3\frac{d^2 w_0}{ds^2}\right) + \left(\frac{r_0^8}{b^3\cos^2\alpha} - \frac{\Lambda b r_0^4}{\cos^2\alpha}\right)w_0 = 0. \qquad (4.6.6)$$

For a cylindrical shell of radius R one may assume that $b = 1$ and $\alpha = 0$, and for a conic shell α is the vertex angle and $b = b(s)$ is a linear function of s.

4.2.6. Substituting (4.2.51) into system (4.2.48) and equating the coefficients of μ_1 we obtain the following system of two differential equations for w_0 and Φ_0:

$$2\sqrt{\frac{k_2 q}{b}}\left(\sqrt{k_2 q b}\, w_0\right)' + \frac{r_1^4}{b^4}\Phi_0 = 0,$$

$$2\sqrt{\frac{k_2 q}{b}}\left(\sqrt{k_2 q b}\, \Phi_0\right)' + \left(\Lambda_1 - \frac{r_1^4}{b^4}\right)w_0 = 0.$$

Note that in the case of a simple root of the characteristic equation, the coefficients of the asymptotic series can be obtained by quadratures since, in contrast to the case under consideration, they are evaluated from first order differential equations [see (4.1.4), (4.2.9), (4.2.43)].

4.3.1.

$$y(x, \mu) = -\frac{1 + 2\mu^2}{c^{1/4}}\left[1 + \frac{5\mu}{48}\left(1 - c^{-3/2}\right)\right]\exp\left(-\frac{2}{3\mu}\left(c^{3/2} - 1\right)\right)$$

$$-\frac{4 + \mu^2}{(2^{11}c)^{1/4}}\left[1 - \frac{5\mu}{48}\left(2^{-3/2} - c^{-3/2}\right)\right]\exp\left(\frac{2}{3\mu}\left(c^{3/2} - 2^{3/2}\right)\right)$$

Fig. 4.9 Asymptotic
solution (*solid line*) and less
accurate solution obtained
when keeping only the main
terms in μ (*dashed line*)

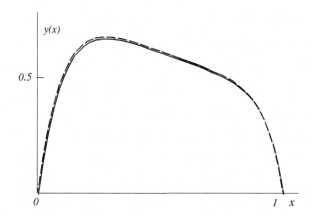

$$+\frac{1}{c}+\frac{2\mu^2}{c^4}+O\left(\mu^3\right), \quad c=c(x).$$

In Fig. 4.9, we plotted the obtained asymptotic solution (solid line) and the less accurate solution obtained when keeping only the main terms in μ (dashed line). The asymptotic solution differs from the exact one by less than 0.001.

4.3.2.

$$y(x,\mu)=\frac{\mu}{2}e^{x/\mu}, \quad x\le 0,$$

$$y(x,\mu)=x+\frac{\mu}{2}e^{-x/\mu}-\frac{1+\mu}{2}e^{(x-1)/\mu}, \quad 0\le x\le 1,$$

$$y(x,\mu)=\frac{1-\mu}{2}e^{(1-x)/\mu}, \quad x\ge 1.$$

The error of the obtained solution is not larger than $\varepsilon=1/2e^{-1/\mu}$.

In Fig. 4.10 the graphs of the function $q(x)$ (curve 1) and of the function $y(x,\mu)$ for $\mu=0.2$ (curve 2) and for $\mu=0.05$ (curve 3) are plotted. As μ decreases the curves $q(x)$ and $y(x,\mu)$ approach each other.

4.3.3. The approximate solution

$$y(x,\mu)=\frac{1}{2}e^{x/\mu}, \quad x\le 0,$$

$$y(x,\mu)=1-\frac{1}{2}e^{-x/\mu}-\frac{1}{2}e^{(x-1)/\mu}, \quad 0\le x\le 1,$$

$$y(x,\mu)=\frac{1}{2}e^{(1-x)/\mu}, \quad x\ge 1.$$

has a discontinuity of order $\varepsilon=1/2e^{-1/\mu}$ at the points $x=0$ and $x=1$.

Fig. 4.10 Function $q(x)$
(*curve 1*), function $y(x, \mu)$
for $\mu = 0.2$ (*curve 2*) and
$\mu = 0.05$ (*curve 3*)

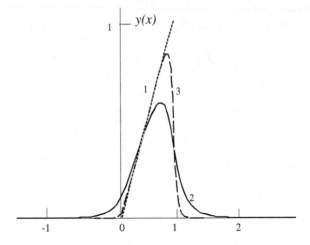

Fig. 4.11 Exact solution
(*curve 1*) and asymptotic
expansion (*curve 2*) for
$y(x, \mu)$ with $\mu = 0.5$

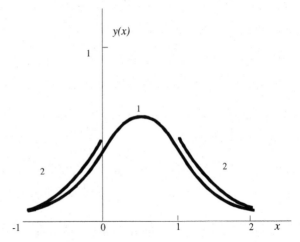

The exact solution has the form

$$y(x, \mu) = C\, e^{x/\mu}, \quad C = \frac{1}{2}\left(1 - e^{-1/\mu}\right), \quad x \le 0,$$

$$y(x, \mu) = 1 - \frac{1}{2}\, e^{-x/\mu} - \frac{1}{2}\, e^{(x-1)/\mu}, \quad 0 \le x \le 1,$$

$$y(x, \mu) = C\, e^{(1-x)/\mu}, \quad x \ge 1.$$

In Fig. 4.11 the graphs of the exact solution (curve 1) and of the asymptotic
expansion (curve 2) for the function $y(x, \mu)$ for $\mu = 0.5$ are plotted. For $0 \le x \le 1$
the discontinuous asymptotic solution coincides with the exact solution.

Fig. 4.12 Exact solution
(*solid line*) and asymptotic
solution (4.3.22) (*dashed
line*)

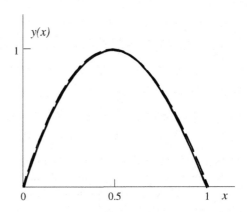

4.3.4. In the problem under consideration

$$q(x) = (1 + \alpha x)^{-3/4}, \quad u_0 = (1 + \alpha x)^{-15/8}, \quad a = -\frac{9\alpha}{2}(1 + \alpha x)^{-7/4}, \quad y_0^* = 1.$$

In Fig. 4.12 the exact solution (solid line) and the asymptotic solution (4.3.22) (dashed
line) are plotted.

4.3.5. The solution with an error of order μ^2 has the form (4.3.21), where

$$C_1 = \frac{r_1 y_0^*(0)}{u_0(0)\sqrt{2}} + \frac{\mu i}{q(0)u_0(0)\sqrt{2}}\left[y_0^{*\prime}(0) - \frac{y_0^*(0)u_0'(0)}{u_0(0)}\right] + O\left(\mu^2\right),$$

$$C_3 = -\frac{r_3 y_0^*(1)}{u_0(1)\sqrt{2}} + \frac{\mu i}{q(1)u_0(1)\sqrt{2}}\left[y_0^{*\prime}(1) - \frac{y_0^*(1)u_0'(1)}{u_0(1)}\right] + O\left(\mu^2\right).$$

The solution with an error of order μ has the form

$$y(x, \mu) = y_0^*(x) - \Re(y_0^*(0)(1 - i)\,e^{r_1 q_0 x/\mu} + y_0^*(1)(1 + i)\,e^{r_3 q_1 x/\mu}) + O(\mu),$$

where $q_0 = q(0)$ and $q_1 = q(1)$.

4.3.6.

$$y^{(2)}(x, \mu) = y^{(1)}(x, \mu) + \mu^2\left[y_2(x) + C_1\,e^{-x/\mu} + C_2\,e^{(x-1)/\mu}\right],$$

where the function $y^{(1)}(x, \mu)$ is found in Sect. 4.3.6.

Firstly, we find $C_1 = -C_2 = y_1' = y_0'(0) + y_0'(1)$. Then from the boundary value
problem

$$\frac{d^2 y_2}{dx^2} = \frac{d^4 y_0}{dx^4} = -g''(x), \quad y_2(0) = -C_1, \quad y_2(1) = -C_2$$

we get $y_2(x) = -g(x) + (g(0) - C_1)(1 - x) + (g(1) - C_2)x$.

4.3.7. The main terms of (4.3.42) in μ after substituting in system (4.3.41) yields (the superscript k is omitted)

$$\lambda^2 u_0 - \lambda c_1 w_0 = 0,$$

$$\lambda c_1 u_0 - \left[\lambda^4(1 - \nu^2) + \frac{1}{R_1^2} + \frac{2\nu}{R_1 R_2} + \frac{1}{R_2^2}\right] w_0 = 0.$$

Setting the determinant of the system to zero, we obtain equation (4.3.43) for the evaluation of λ. The first equation of the system permits to express u_0 through w_0. To determine w_0 we consider the next approximation in powers of μ:

$$\lambda^2 u_1 - \lambda c_1 w_1 + 2\lambda \frac{du_0}{ds} + \lambda' u_0 + \frac{B'c_2 w_0 - (Bc_1)'w_0}{B} - c_1 \frac{dw_0}{ds} = 0,$$

$$\lambda c_1 u_1 - c_1^2 w_1 + c_1 \frac{du_0}{ds} \frac{B'c_2 u_0}{B} - 4\lambda^3 \frac{dw_0}{ds} - 6\lambda^2 \lambda' w_0 - \frac{2B'\lambda^3 w_0}{B} = 0.$$

Multiply the first equation by c_1 and the second by λ and subtract the second from the first. Then after cancellation of the terms with w_1 and u_1 we get an equation for w_0:

$$-\lambda^4 \left[4 \frac{dw_0}{ds} + \left(\frac{6\lambda'}{\lambda} + \frac{2B'}{B}\right) w_0\right] = 0,$$

whence taking the equation for λ into account we obtain the required expression (4.3.43) for w_0.

4.3.8.

$$u = C + \nu q_n s + \mu \nu q_n \sqrt{2}\,(e^{-s_1}\cos s_1 - e^{s_2}\cos s_2),$$

$$s_1 = \frac{s}{\mu\sqrt{2}}, \qquad s_2 = \frac{s - l}{\mu\sqrt{2}},$$

$$w = q_n \left[1 - e^{-s_1}(\cos s_1 + \sin s_1) - e^{s_2}(\cos s_2 - \sin s_2)\right].$$

As in Sect. 4.3.8, the constant C remains undetermined and the conditions C_2 in (4.3.39) are satisfied with an error of order $\exp\left(-l/(\mu\sqrt{2})\right)$.

4.3.9. The membrane solution is

$$u_0 = 0, \quad w_0 = \tilde{q}_n, \quad \tilde{q}_n = (1 - \nu^2)q_n.$$

After adding the edge effect integrals similar to Exercise 4.3.8 we get

$$u^{(1)} = \mu\nu\tilde{q}_n\sqrt{2}\,(e^{-s_1}\cos s_1 - e^{s_2}\cos s_2), \quad s_1 = \frac{s}{\mu\sqrt{2}}, \quad s_2 = \frac{s-l}{\mu\sqrt{2}},$$

$$w^{(1)} = \tilde{q}_n\left[1 - e^{-s_1}(\cos s_1 + \sin s_1) - e^{s_2}(\cos s_2 - \sin s_2)\right].$$

However this solution does not satisfy conditions $u(0) = 0$ and $u(l) = 0$. To remove this residual we add the first approximation of the membrane solution evaluated from the boundary value problem

$$\frac{d^2 u_0^{(1)}}{ds^2} = 0, \quad u_0^{(1)}(0) = -\mu\nu\tilde{q}_n\sqrt{2}, \quad u_0^{(1)}(l) = \mu\nu\tilde{q}_n\sqrt{2},$$

from where we get

$$u_0^{(1)} = \frac{\mu\nu\tilde{q}_n\sqrt{2}}{l}(2s - l), \quad w_0^{(1)} = \frac{\mu\nu\tilde{q}_n\sqrt{8}}{l}.$$

Then we add again the edge effect integral to the obtained solution to meet the conditions $w(0) = w(l) = 0$ and with an error of order μ^2 we obtain

$$u = \frac{\mu\nu\tilde{q}_n\sqrt{2}}{l}(2s - l) + \mu\nu\tilde{q}_n\sqrt{2}\,\left(e^{-s_1}\cos s_1 - e^{s_2}\cos s_2\right) + O\left(\mu^2\right),$$

$$w = \tilde{q}_n\left[1 + \frac{\mu\nu\sqrt{8}}{l}\right]\left[1 - e^{-s_1}(\cos s_1 + \sin s_1) - e^{s_2}(\cos s_2 - \sin s_2)\right] + O\left(\mu^2\right).$$

4.3.10. In a neighborhood of the edge $s = 0$ we have

$$w = w_0\left[1 - e^{-\alpha_1}(\cos\beta_1 + b\sin\beta_1)\right], \quad b = \sqrt{\frac{1+\tau}{1-\tau}}, \quad |\tau| < 1,$$

$$w = w_0\left[1 - \frac{1}{\lambda_2 - \lambda_1}\left(\lambda_2\, e^{-\lambda_1 s/\mu} - \lambda_1\, e^{-\lambda_2 s/\mu}\right)\right], \quad \tau > 1,$$

where the notation of Sect. 4.3.9 is used.

4.4.1. Using the notation of Sect. 4.4.2 for the next iteration we obtain $z_2^{(1)} = 2n\pi e^{-\xi_1}$, $z_2^{(2)} = (-1)^{n+1}2n\pi\, e^{-\xi_1}$.

For Λ_2 and $y_2(x)$ we get the eigenvalue problem

$$\frac{d^2 y_2}{dx^2} + \Lambda_0 y_2 + \Lambda_1 y_1 + \Lambda_2 y_0 - \frac{d^4 y_0}{dx^4} = 0, \quad y_2(0) = -2n\pi, \quad y_2(1) = (-1)^n 2n\pi,$$

from which we obtain

$$\Lambda_2 = 12n^2\pi^2 + n^4\pi^4, \quad y_2(x) = 2n^2\pi^2(-1 + x - x^2)\sin n\pi x + 2n\pi(2x - 1)\cos n\pi x.$$

4.4.2. The degenerate boundary value problem has the same form as (4.4.15) and the same solution $\Lambda_0 = n^2\pi^2$, $y_0(x) = \sin n\pi x$ as in Sect. 4.4.2. However, the solution $y_0(x)$ satisfies the boundary conditions $y = y'' = 0$ at $x = 0, 1$. If we assume that $\Lambda = n^2\pi^2 + \mu^2 n^4\pi^4$ and $y(x) = y_0(x) = \sin n\pi x$, we obtain the exact solution of the problem. Here in contrast to the problem considered in Sect. 4.4.2 no edge effect integrals appear.

4.4.3. Differentiate the general solution (4.4.20) with respect to x and substitute into the boundary conditions. Then for evaluating μ_n we get the equation

$$\cot\frac{\varphi_1}{\mu} = \frac{U(0)G_1 - V(0)G_2}{U(0)G_2 + V(0)G_1} \simeq \sum_{k=0}^{\infty} b_k \mu^{2k+1}, \qquad (4.6.7)$$

where $G_1 = \varphi_1' V(1) - \mu U'(1)$, $G_2 = \varphi_1' U(1) + \mu V'(1)$, $\varphi_1' = \sqrt{\rho(1)}$, and the other notation is as in Sect. 4.4.3. Expanding the solution of equation (4.6.7) in series,

$$\mu_n \simeq \sum_{k=0}^{\infty} \frac{c_k^*}{[(n-1/2)\pi]^{2k+1}}, \qquad n \to \infty,$$

we find the first terms

$$c_0^* = \varphi_1, \quad c_1^* = b_0\varphi_1^2, \quad b_0 = \frac{v_1(1)}{u_0(1} - \frac{v_1'(1)}{\varphi_1' u_0(1)} - \frac{v_1(0)}{u_0(0)}.$$

4.4.4. For $\rho(x) = 1 + \alpha x$, the solution is found in Sect. 4.1.1. Assuming that $\alpha = 1$ with the notation (4.4.19) from Sect. 4.4.3, we obtain

$$U^{(K)} = \rho^{-1/4}\sum_{k=0}^{K}\frac{(-1)^k\mu^{2k}b_{2k}}{\rho^{3k}}, \quad V^{(K)} = \mu\rho^{-7/4}\sum_{k=0}^{K}\frac{(-1)^{k+1}\mu^{2k}b_{2k+1}}{\rho^{3k}},$$

$$b_0 = 1, \quad b_n = \frac{(6n-1)(6n-5)}{48n}b_{n-1}, \quad n = 1, 2, \ldots$$

Assuming that $K = 0, 1, 2, \ldots$, we get different asymptotic approximations for the functions U and V in Eq. (4.4.21) and, accordingly, different approximations for μ_n.

Table 4.1 lists the values of the non-dimensional frequencies Ω_n for the first five free vibrations frequencies ω_n of the vibrating string:

$$\omega_n = \sqrt{\frac{T}{\rho_0}}\,\Omega_n, \quad \Omega_n = \sqrt{\Lambda^{(n)}} = \frac{1}{\mu_n}, \quad n = 1, 2, \ldots, 5.$$

The first column contains the values calculated by a Runge–Kutta method, the next columns contain the values found by asymptotic formulas with different accuracy. Firstly, the less accurate approximation

Table 4.1 Exact and asymptotic non-dimensional frequencies Ω_n

n	Ω_n^{exact}	$\Omega_n = \frac{n\pi}{\varphi_1}$	$\Omega_n^{(0)}$	$\Omega_n^{(1)}$	$\Omega_n^{(2)}$	$\Omega_n^{(3)}$	$\Omega_n^{(4)}$
1	2.559	2.577	2.556	2.561	2.555	2.572	2.466
2	5.144	5.155	5.144	5.145	5.144	5.144	5.144
3	7.725	7.732	7.725	7.725	7.725	7.725	7.725
4	10.304	10.309	10.304	10.304	10.304	10.304	10.304
5	12.882	12.886	12.882	12.882	12.882	12.882	12.882

Fig. 4.13 Exact eigenfunction and asymptotic approximation

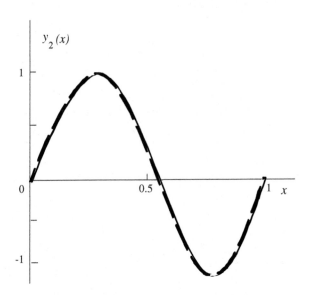

$$\Omega_n = \frac{n\pi}{\varphi_1}, \quad \varphi_1 = \int_0^1 \sqrt{\rho(x)}\,dx = \frac{2}{3}\left(\sqrt{8} - 1\right),$$

then $\Omega_n^{(K)} = 1/\mu_n^{(K)}$, where $\mu_n^{(K)}$ are the roots of equation (4.4.20) obtained when keeping the $K + 1$ first terms in series (4.4.19).

As it must be for asymptotic divergent series, the error firstly decreases with the number of terms and then starts to increase (see $n = 1$).

In Fig. 4.13 the exact eigenfunction $y_n(x)$ for $n = 2$ and the less accurate asymptotic approximation

$$y_n(x) \simeq C(1+x)^{-1/4} \sin\left(\frac{2}{3\mu_n}\left[(1+x)^{3/2} - 1\right]\right)$$

are plotted. For comparison, both eigenfunctions are normalized by the condition $\max y_n = 1$. In the graph, these functions are indiscernible since the maximal difference has order 10^{-3}.

4.4.5. The equation for the eigenvalue problem is a particular case of the equation from Exercise 4.1.1 for $p(x) = \rho(x) = S(x), r(x) = 0$. Its general solution is

$$y(x, \mu) = C_1 \left[U \cos\left(\frac{\varphi}{\mu}\right) - V \sin\left(\frac{\varphi}{\mu}\right) \right] + C_2 \left[U \sin\left(\frac{\varphi}{\mu}\right) + V \cos\left(\frac{\varphi}{\mu}\right) \right],$$

where C_1 and C_2 are arbitrary constants, $U = u_0 + \mu^2 u_2 + \cdots$, $V = \mu v_1 + \mu^3 v_3 + \cdots$, $v_{2n+1} = i u_{2n+1}$ and the functions $u_k(x)$ are obtained in Exercise 4.1.1.
The equation for μ is

$$\tan \frac{1}{\mu} = \frac{U(0) F_1 + V(0) F_2}{U(0) F_2 - V(0) F_1},$$

where

$$F_1 = S(1) \left[\mu U(1) + \mu^2 V'(1) \right] - a V(1),$$
$$F_2 = S(1) \left[\mu V(1) - \mu^2 U'(1) \right] + a U(1).$$

If $a \sim 1$, the solution has the same structure as in Sect. 4.4.3,

$$\mu_n \simeq \sum_{k=0}^{\infty} \frac{c_k}{(n\pi)^{2k+1}}, \quad \text{as } n \to \infty,$$

where

$$c_0 = 1, \quad c_1 = \frac{v_1(1)}{u_0(1)} - \frac{v_1(0)}{u_0(0)} - \frac{S(1)}{a}.$$

For $S(x) = 1 + \alpha x$, the approximate expressions for U and V are

$$U^{(K)} = \sum_{k=0}^{K} \frac{(-1)^k (\mu\alpha)^{2k} b_{2k}}{S^{2k+1/2}}, \quad V^{(K)} = \sum_{k=0}^{K} \frac{(-1)^{k+1} (\mu\alpha)^{2k+1} b_{2k+1}}{S^{2k+3/2}},$$

where $b_0 = 1$, $b_n = \frac{(2n-1)^2}{8n} b_{n-1}$, $n = 1, 2, \dots$.
The exact values of $\Omega_n = 1/\mu_n$ and the successive asymptotic approximations in the sense of Table 4.1 are given in Table 4.2 for the first eight roots of μ_n for $\alpha = a = 1$.

The difference between this problem and Exercise 4.4.4 is that the root Ω_0 cannot be found by asymptotic formulas.

4.4.6. For the problem at hand with the notation of Sect. 4.4.4 and by Exercise 4.1.3

$$p(x) = \rho(x) = 1 + \alpha x, \quad b_1 = \frac{3\alpha}{8}\left(1 - \frac{1}{1+\alpha}\right), \quad q \equiv 1, \quad f_1 = 1, \quad u_0 = p^{-1/2},$$

Table 4.2 Exact $\Omega_n = 1/\mu_n$ and successive asymptotic approximations

n	Ω_n^{exact}	$\Omega_n = n\pi$	$\Omega_n^{(0)}$	$\Omega_n^{(1)}$	$\Omega_n^{(2)}$	$\Omega_n^{(3)}$	$\Omega_n^{(4)}$
0	0.917	–	–	–	–	–	–
1	3.616	3.142	3.618	3.619	3.619	3.619	3.619
2	6.567	6.283	6.568	6.568	6.568	6.568	6.568
3	9.622	9.425	9.623	9.623	9.623	9.623	9.623
4	12.717	12.566	12.718	12.718	12.718	12.718	12.718
5	15.829	15.708	15.830	15.830	15.830	15.830	15.830
6	18.951	18.850	18.952	18.952	18.952	18.952	18.952
7	22.079	21.991	22.079	22.079	22.079	22.079	22.079
8	25.209	25.133	25.210	25.210	25.210	25.210	25.210

Table 4.3 First six values of Ω_n

n	Ω_n^{exact}	$\Omega_n^{(1)}$	$\Omega_n^{(2)}$
1	22.182	22.207	21.833
2	61.407	61.685	61.311
3	120.607	120.903	120.528
4	199.548	199.859	199.485
5	298.233	298.556	298.181
6	416.660	416.991	416.616

and the free vibrations frequencies, we have

$$\omega_n = \sqrt{\frac{E J_0}{\rho_0 S_0 l^4}}\, \Omega_n, \quad \Omega_n = c_n^2 + O(n^{-1}), \quad c_n = \frac{\pi}{2}(2n+1).$$

The approximate expression for the vibrations mode is

$$y_n(x) = \frac{\sin(c_n x - \pi/4)}{(1+\alpha x)^{1/2}} + \frac{1}{\sqrt{2}}e^{-c_n x} - \frac{(-1)^n}{\sqrt{2(1+\alpha)}}e^{c_n(x-1)} + O\left(c_n^{-1}\right).$$

Assume that $\alpha = 1$. Table 4.3 includes the first six values of Ω_n: firstly, the exact value, then the first and the second asymptotic approximations $\Omega_n^{(1)} = c_n^2$ and $\Omega_n^{(2)} = \left(c_n - \frac{b_1}{c_n}\right)^2$.

In Fig. 4.14 the vibrations mode for $n = 3$ is plotted. The difference between the exact mode and its asymptotic approximation is less than 0.01.

4.4.7. The comparison result for the vibrations frequencies of a non-membrane and a membrane type for $n \le 15$ are given in Table 4.4.

Fig. 4.14 Exact mode and its asymptotic approximation

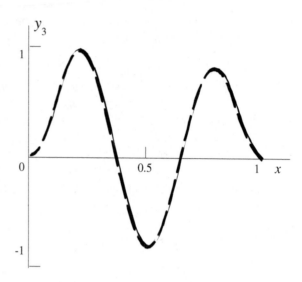

Table 4.4 Comparison of vibrations frequencies of non-membrane and membrane types

n	Λ_{n1}	Λ_{n1}^0	Λ_{n2}	Λ_{n2}^0
0	1.0989	1.0989	0.0000	0.0000
1	1.5013	1.5013	0.8027	0.8027
2	4.9443	4.9443	0.9751	0.9749
3	10.9546	10.9546	0.9910	0.9901
4	19.3856	19.3856	0.9974	0.9946
5	30.2293	30.2293	1.0035	0.9966
6	43.4841	43.4841	1.0119	0.9977
7	59.1496	59.1495	1.0247	0.9983
8	77.2254	77.2253	1.0438	0.9987
9	97.7115	97.7114	1.0712	0.9999
10	120.6078	120.6077	1.1092	0.9992
11	145.9144	145.9143	1.1604	0.9993
12	173.6311	173.6310	1.2277	0.9994
13	203.7581	203.7579	1.3139	0.9995
14	236.2952	236.2950	1.4225	0.9996
15	271.2425	271.2423	1.5569	0.9996

4.4.8. For the zeroth approximation we consider the solution (4.4.30) for the degenerate problem (4.4.29). Since the solution does not satisfy the conditions

$$w = \frac{dw}{ds} = 0 \quad \text{for} \quad s = 0, \ s = l, \tag{4.6.8}$$

we add to it edge effect integrals of accuracy of order μ,

$$u^e = \sum_{k=3,4} \nu p_k^{-1} C_k e^{p_k s} + \sum_{k=5,6} \nu p_k^{-1} C_k e^{p_k(s-l)},$$

$$w^e = \sum_{k=3,4} C_k e^{p_k s} + \sum_{k=5,6} C_k e^{p_k(s-l)},$$

$$p_{3,4} = \frac{(1-\Lambda^0)^{1/4}}{\mu} \left(-\frac{1}{\sqrt{2}} \pm \frac{i}{\sqrt{2}}\right), \quad p_{5,6} = \frac{(1-\Lambda^0)^{1/4}}{\mu} \left(\frac{1}{\sqrt{2}} \pm \frac{i}{\sqrt{2}}\right).$$

The constants C_3 and C_4 are found from conditions (4.6.8) for $s = 0$:

$$C_3 + C_4 + \frac{n\pi}{1-(1-\nu^2)\Lambda^0} = 0, \quad \frac{C_3}{p_3} + \frac{C_4}{p_4} = 0.$$

By now, the sum $u^0 + u^e$ does not satisfy condition $u(0) = 0$. Assume

$$\Lambda = \Lambda^0 + \mu\Lambda^1, \quad u = u^0 + u^1. \tag{4.6.9}$$

Then the function u^1 is a solution of the eigenvalue problem

$$\frac{d^2 u^1}{ds^2} + a(\Lambda^0)u^1 + \Lambda^1 a_\Lambda' u^0 = 0, \quad a_\Lambda' = \frac{da}{d\Lambda^0}, \tag{4.6.10}$$

with solution

$$u^1(0) = -\frac{\nu^2 n\pi\sqrt{2}}{l(1-(1-\nu^2)\Lambda^0)(1-\Lambda^0)^{1/4}}, \quad u^1(l) = (-1)^n u^1(0),$$

where $a(\Lambda^0)$ is given by (4.4.29) and $u^1(l)$ is found as for $u^1(0)$.

The value of Λ^1 is found from the compatibility conditions for the non-homogeneous problem (4.6.10) for the spectrum. To obtain Λ^1 we multiply equation (4.6.10) by u^0 and integrate with respect to s from 0 to l. Thus, we get

$$\Lambda^1 = \frac{4\nu^2 n^2 \pi^2 \sqrt{2}}{l^3(1-(1-\nu^2)\Lambda^0)a_\Lambda'(1-\Lambda^0)^{1/4}}.$$

Only the first few roots Λ can be evaluated by the approximate formula (4.6.9) since by the estimate (4.4.27) Λ_{n2}^0 approaches 1 rapidly when n increases and $\Lambda_1 \sim (1-\Lambda)^{-9/4}$ as $\Lambda \to 1$. Therefore the correction term is small only if $1 - \Lambda \gg \mu^{4/9}$.

4.4.9. The 26 lower eigenvalues ($\Lambda < 5$) are listed in Table 4.5. The table includes the exact values Λ_n^{exact} and the asymptotic approximations applicable in different variation ranges of Λ. Values out of the applicability domain for the corresponding

Table 4.5 The asymptotic and exact lower eigenvalues

n	Λ_n^{exact}	$\Lambda_n^{(4.38)}$	$\Lambda_n^{(4.35)}$	$\Lambda_n^{membrane}$	$\Lambda_n^{(6.9)}$
1	0.82157	(0.93403)	(1.00043)	0.80270	0.82039
2	0.98213	0.98593	(1.00165)	0.97493	0.98126
3	0.99603	0.99681	(1.00452)	0.99006	0.99329
4	1.00418	1.00467	(1.01008)	(0.99462)	(0.99667)
5	1.01465	1.01524	1.01966	(0.99662)	(0.99807)
6	1.02986	1.03096	1.03484	(0.99767)	(0.99877)
7	1.05275	1.05394	1.05749		
8	1.07678	1.08638	1.08970		
9	1.1028			1.0989	
10	1.12840	1.13071	1.13386		
11	1.19342	1.18958	1.19261		
12	1.26062	1.26592	1.26886		
13	1.36556	1.36291	1.36579		
14	1.44440	1.48399	1.48681		
15	1.52334			1.50128	
16	1.63506	1.63287	1.63565		
17	1.82056	1.81351	1.81626		
18	2.03198	2.03015	2.03287		
19	2.29137	2.28727	2.28997		
20	2.59107	2.58964	2.59232		
21	2.94535	2.94228	2.94494		
22	3.35119	3.35048	3.35313		
23	3.82226	3.81978	3.82242		
24	4.35368	4.35600	4.35864		
25	4.93833			4.94430	
26	4.96726	4.96523	4.96786		

formula are bracketed in parentheses, and the blank spaces indicate that the application of the formula is senseless.

The values $\Lambda_n^{(4.38)}$ are obtained from relation (4.4.38) derived under the assumption that $\Lambda \approx 1$. It occurs that relation (4.4.38) approximate well most values $\Lambda > 1$ except for $n = 9, n = 15$ and $n = 25$.

The values $\Lambda_n^{(4.35)}$ are found by formula (4.4.35), which is applicable only for $\Lambda > 1$. Values of $\Lambda > 1$ close to 1 are badly approximated and this formula is not applicable for $n = 9, n = 15$ and $n = 25$.

The values $\Lambda_n^{membrane}$ are membrane eigenvalues for $\mu = 0$. For $\Lambda < 1$ similar to Exercise 4.4.9 there is an infinite number of such values with accumulation point $\Lambda = 1$. However, only the first three membrane values are close to the exact values. For $\Lambda > 1$ the membrane values are widely spaced (see $n = 9, n = 15$ and $n = 25$).

To refine the membrane values, $\Lambda_n^{(6.9)}$ are obtained by formula (4.6.9) for $\Lambda < 1$, i.e. in the domain of regular degeneracy.

From Table 4.5 it follows that in a neighborhood of the point $\Lambda = 1$ the eigenvalues are the most dense.

4.5.1. Write the general solution of equation (4.5.1),

$$w(x_1) = C_1 e^{-sx_1} + C_2 e^{-rx_1} + C_3 e^{sx_1} + C_4 e^{rx_1}, \qquad (4.6.11)$$

where C_k are arbitrary constants obtained as a result of the substitution (4.5.3) in the boundary conditions and $w = w' = 0$ for $x_1 = \pi a/b$. Then instead of evaluating Λ by (4.5.4), we use the equation

$$s(r^2 - \nu)^2 - r(\nu - s^2)^2 + \frac{r+s}{r-s}\left[s(r^2 - \nu)^2 + r(\nu - s^2)^2\right]\exp\left(-\frac{2\pi as}{b}\right) = 0. \qquad (4.6.12)$$

We solve this equation for $a \gg b$. Thus, in its derivation the terms of order $\exp(-2\pi a/b)$ are neglected. Expressing r as a function of s by the formula $r = \sqrt{2 - s^2}$ we find that, for sufficiently large a/b, this equation has only a root $s > 0$ through which $\Lambda = (1 - s^2)^2$ may be expressed.

For $\nu = 0.3$, Eq. (4.6.12) has a real root if $\pi a/b \geq 23.42$. For different a/b the values of s and Λ are the following:

$\pi a/b$	25	30	35	40	∞
s	0.0186	0.0324	0.0375	0.0400	0.0436
Λ	0.9993	0.9979	0.9972	0.9968	0.9962

For $\pi a/b < 23.42$, the solution should be searched in a form different from (4.6.11). This case is not considered here.

4.5.2. After separation of variables (4.5.1) we come to the same Eq. (4.5.2), for which $\Lambda = Tb^2/(D\pi^2)$. Since the boundary conditions also coincides with those considered in Sect. 4.5.1 for $\nu = 0.3$, the critical value is $T = 0.9962\pi^2 D/b^2$. This decrease of the critical load was first established by Yu. Ishlinsky [16].

4.5.3. The results are given in Table. 4.6. The dash line means the absence of roots.

From Table 4.6, it follows that six types of boundary conditions are weak: S_4, C_6, S_6, S_7, C_8, S_8 and the equation has two roots only for the last type (S_8).

Since the listed equations have the root $\theta = 0$, one cannot conclude that the other 10 types of boundary conditions are not weak. This question is discussed in Exercise 4.5.4, where it is shown that the type S_5 is also weak.

4.5.4. The roots $\Lambda(q)$ of Eq. (4.5.15) for seven types of boundary conditions are plotted in Fig. 4.15. The curves are numbered in the order of increasing $\Lambda(q)$: 1 for the boundary conditions S_8, 2 for C_8, 3 for S_6, 4 for C_6, 5 for S_4, 6 for S_7, and 7 for S_5.

For the types S_4, C_6, S_6, S_7, C_8, S_8, the roots $\Lambda(q)$ converge to the values obtained in Exercise 4.5.3 as $q \to 0$, and for the boundary conditions S_5 $\Lambda(0) = 1$.

Table 4.6 Eigenvalue equations for different types of boundary conditions

Type of boundary conditions	Equation	Roots of equation
C_1	$\sin 2\theta\,(\cos\theta - \sin\theta)^2 = 0$	–
S_1	$\cos\theta\,(\cos\theta - \sin\theta) = 0$	–
C_2	$\sin 4\theta\,(\cos 3\theta + \sin 3\theta) = 0$	–
S_2	$\sin^2 4\theta = 0$	–
C_3	$\sin 4\theta\,(\cos\theta - \sin\theta) = 0$	–
S_3	$\sin^2 2\theta + \sin 6\theta = 0$	–
C_4	$\sin^2 4\theta = 0$	–
S_4	$\sin 4\theta\,(\cos 5\theta - \sin 5\theta) = 0$	$\Lambda = \cos\pi/5 = 0.809$
C_5	$\sin 4\theta\,(\cos 3\theta + \sin 3\theta) = 0$	–
S_5	$\sin^2 4\theta = 0$	–
C_6	$\sin 10\theta - \sin^2 2\theta = 0$	$\Lambda = 0.419$
S_6	$\sin 4\theta\,(\cos 7\theta + \sin 7\theta) = 0$	$\Lambda = \cos 3\pi/7 = 0.223$
C_7	$\sin^2 4\theta = 0$	–
S_7	$\sin 4\theta\,(\cos 5\theta - \sin 5\theta) = 0$	$\Lambda = \cos\pi/5 = 0.809$
C_8	$\sin 4\theta\,(\cos 7\theta + \sin 7\theta) = 0$	$\Lambda = \cos 3\pi/7 = 0.223$
S_8	$\sin 2\theta - \sin^2 6\theta = 0$	$\Lambda^{(1)} = 0.113,\ \ \Lambda^{(2)} = 0.973$

Fig. 4.15 Roots $\Lambda(q)$ for seven types of boundary conditions

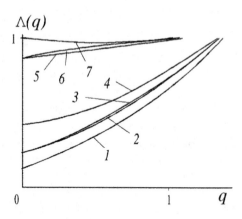

For $q > 1$, for weak support the inequality $\Lambda(q) < (q^2 + q^{-2})/2$ must be satisfied since only under this condition equation (4.5.16) has four roots with negative real parts. Those are required for construction of localized solution (4.5.11).

4.5.5. In the problem at hand with the notation introduced in Sect. 4.5.2 we have

$$T_2^{(0)} = -2Eh\mu^2\Lambda, \quad k_1 = 1,\ k_2 = 0,\ t_1 = 0,\ t_2 = 1$$

Fig. 4.16 $\Lambda(q)$ for eight types of boundary conditions

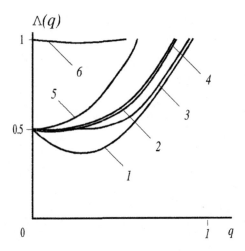

and Eq. (4.5.13) coincides with (4.5.16). Thus all results obtained for vibrations can be used for buckling analysis. The difference is that, in studying buckling, only the lowest eigenvalue is of interest. Thus in (4.5.16) one should assume $n = 1$.

4.5.6. In Fig. 4.16 $\Lambda(q)$ is plotted for eight types of boundary conditions (4.5.10) for which $\Lambda(q) < 1$.

The curves are numbered in the order of increasing $\Lambda(q)$: 1 for boundary conditions S_8, 2 for S_7, 3 for S_6, 4 for S_5, 5 for S_4, S_3, C_8 (the values of $\Lambda(q)$ coincide for three types of boundary conditions), 6 for C_6. The functions $\Lambda(q)$ are obtained numerically by Eq. (4.5.15) and the use of the exact solutions of the roots (4.5.20). Here to the seven types of boundary conditions (S_3, S_4, S_5, S_6, S_7, C_8, S_8) producing $\Lambda(0) = 1/2$ (see Sect. 4.5.4) one more type, C_6, is added for which $\Lambda(0) = 1$ and which gives $\Lambda_{min} = 0.97$. The curves 1, 2, 6 in Fig. 4.16 attain their minima at $q = 0.31, q = 0.20, q = 0.36$ respectively. The other curves attain their minima at $q = 0$. This fact is important in the evaluation of n in (4.5.20), which is taken from the condition that $\Lambda(q)$ is minimum.

Chapter 5
Singularly Perturbed Linear Ordinary Differential Equations with Turning Points

In this chapter, we consider systems of linear ordinary differential equations with variable coefficients and a small parameter μ in the derivative terms. Asymptotic expansions for solutions as $\mu \to 0$ are obtained under the assumption that there exists a turning point (or points) in the integration interval. These expansions are used in solving boundary value problems.

5.1 Airy Functions

Airy's functions play a significant role in the construction of asymptotic expansions. For the reader's convenience, we briefly discuss the properties of the Airy functions that will be used in this book (see also Sect. 2.5.2).

Airy's functions are entire functions of a complex variable, η. They satisfy the differential equation

$$\frac{d^2 v}{d\eta^2} - \eta v = 0. \tag{5.1.1}$$

Two standard Airy functions, $\mathrm{Ai}(\eta)$ and $\mathrm{Bi}(\eta)$, which are real for real η are introduced. These functions have the following Maclaurin series expansion:

$$\mathrm{Ai}(\eta) = a_1 f_1(\eta) - a_2 f_2(\eta), \quad \mathrm{Bi}(\eta) = \sqrt{3}\,[a_1 f_1(\eta) + a_2 f_2(\eta)], \tag{5.1.2}$$

where

$$f_1 = \sum_{k=0}^{\infty} b_k \eta^{3k}, \quad b_0 = 1, \quad b_k = \frac{b_{k-1}}{(3k-1)3k}, \quad a_1 = \frac{3^{-2/3}}{\Gamma(2/3)},$$

$$f_2 = \sum_{k=0}^{\infty} d_k \eta^{3k+1}, \quad d_0 = 1, \quad d_k = \frac{d_{k-1}}{3k(3k+1)}, \quad a_2 = \frac{3^{-1/3}}{\Gamma(1/3)},$$

© Springer International Publishing Switzerland 2015
S.M. Bauer et al., *Asymptotic Methods in Mechanics of Solids*,
International Series of Numerical Mathematics 167,
DOI 10.1007/978-3-319-18311-4_5

and $\Gamma(z)$ is the gamma function.

Note the relation

$$Bi(0) = Ai(0)\sqrt{3}. \tag{5.1.3}$$

The asymptotic expansions of the Airy functions as $\eta \to \infty$ are

$$Ai(\eta) \simeq \frac{1}{2\sqrt{\pi}} \eta^{-1/4} e^{-\zeta} \sum_{k=0}^{\infty} \frac{(-1)^k c_k}{\zeta^k}, \quad |\arg \eta| < \pi;$$

$$Bi(\eta) \simeq \frac{1}{\sqrt{\pi}} \eta^{-1/4} e^{\zeta} \sum_{k=0}^{\infty} \frac{c_k}{\zeta^k}, \quad |\arg \eta| < \frac{\pi}{3};$$

$$Ai(-\eta) \simeq \frac{1}{\sqrt{\pi}} \eta^{-1/4} \left[D_1(\zeta) \sin\left(\zeta + \frac{\pi}{4}\right) - D_2(\zeta) \cos\left(\zeta + \frac{\pi}{4}\right) \right],$$

$$|\arg \eta| < \frac{2\pi}{3}; \tag{5.1.4}$$

$$Bi(-\eta) \simeq \frac{1}{\sqrt{\pi}} \eta^{-1/4} \left[D_1(\zeta) \cos\left(\zeta + \frac{\pi}{4}\right) + D_2(\zeta) \sin\left(\zeta + \frac{\pi}{4}\right) \right],$$

$$|\arg \eta| < \frac{2\pi}{3};$$

$$D_1(\zeta) \simeq \sum_{k=0}^{\infty} \frac{(-1)^k c_{2k}}{\zeta^{2k}}, \quad D_2(\zeta) \simeq \sum_{k=0}^{\infty} \frac{(-1)^k c_{2k+1}}{\zeta^{2k+1}},$$

$$\zeta = \frac{2}{3} \eta^{3/2}, \quad c_k = \frac{\Gamma(3k + 1/2)}{54^k k! \Gamma(k + 1/2)}.$$

Series (5.1.4) are applicable not only for real η but also for complex η. We show sectors in a neighborhood of the point $\eta = \infty$ where expansions (5.1.4) are asymptotic. Series (5.1.4) are divergent.

In Fig. 5.1, the graphs of the functions $Ai(\eta)$ and $Bi(\eta)$ are plotted for real η.

5.1.1 Exercises

5.1.1. Compare the values of $Ai(\eta)$ and $Bi(\eta)$ for real η obtained with the convergent series (5.1.2) and the asymptotic series (5.1.4). Explore the possibility of using these series to get the values of the above functions accurate to five decimal places with computation to twelve decimal places.

Fig. 5.1 Ai(η) and Bi(η) for
real η

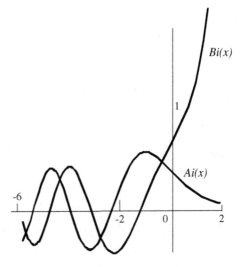

5.1.2. Find the first six zeros η_n and η'_n of the functions Ai(η) and $v_0(\eta) = \sqrt{3}$ Ai(η) − Bi(η) and obtain the asymptotic formulas for η_n and η'_n as $n \to \infty$. Compare the exact and the asymptotic values of the zeros of these functions.

5.2 Solutions of Second-Order Ordinary Differential Equations with Turning Points

Consider the linear second-order differential equation

$$\mu^2 \frac{d^2y}{dx^2} - q(x, \mu)y = 0, \quad q(x, \mu) = \sum_{k=0}^{\infty} q_k(x)\mu^k, \tag{5.2.1}$$

where $\mu > 0$ is a small parameter and the functions $q_k(x)$ are real analytic for $x \in S = [x_1, x_2] \subset \mathbb{R}$. Many studies are devoted to analysis of Eq. (5.2.1) (see, for example, [14, 20, 27, 32, 53, 65]).

The more general equation

$$\mu^2 a_0 \frac{d^2y_1}{dx^2} + \mu a_1 \frac{dy_1}{dx} + a_2 y_1 = 0, \quad a_n(x, \mu) = \sum_{k=0}^{\infty} a_{nk}(x)\mu^k, \ n = 0, 1, 2,$$

can be transformed into Eq. (5.2.1) by the substitution

$$y_1(x, \mu) = y(x, \mu) \exp\left(-\frac{1}{2\mu} \int \frac{a_1 dx}{a_0}\right),$$

provided $a_{00}(x) \neq 0$. Then in (5.2.1)

$$q = \frac{1}{4}\left(\frac{a_1}{a_0}\right)^2 - \frac{a_2}{a_0} + \frac{\mu}{2}\left(\frac{a_1}{a_0}\right)' \qquad ()' = \frac{d}{dx}.$$

If $q_0(x) \neq 0$ for $x \in S$, then the asymptotic expansion of the solutions of Eq. (5.2.1) has the form (4.1.2) for $\lambda(x) = \pm\sqrt{q_0(x)}$.

Now let $q_0(x_*) = 0$, $x_1 \leq x_* \leq x_2$. Then solution (4.1.2) becomes inapplicable since $u_k(x) \to \infty$ as $x \to x_*$. In particular, $u_0(x) = q_0(x)^{-1/4}$ [see (4.1.11))] and $u_0(x_*) = \infty$. The points $x = x_*$, where $q_0(x_*) = 0$ are called *turning points*. If $q_0'(x_*) \neq 0$, the turning point $x = x_*$ is called *a simple turning point*.

5.2.1 Asymptotic Expansion of Solutions

The asymptotic expansions of solutions of Eq. (5.2.1) in neighborhoods of a simple turning point $x = x_*$ for $q_0'(x_*) > 0$ have the form [39]:

$$y(x, \mu) = a^{(0)}(x, \mu)v\left[\eta(x, \mu)\right] + \mu^{1/3}a^{(1)}(x, \mu)\frac{dv}{d\eta}, \qquad (5.2.2)$$

where $v(\eta)$ is one of Airy's functions (see Sect. 5.1):

$$\eta(x, \mu) = \mu^{-2/3}\xi(x), \qquad \xi(x) = \left[\frac{3}{2}\int_{x_*}^x \sqrt{q_0(x)}\,dx\right]^{2/3},$$

$$a^{(j)}(x, \mu) \simeq \sum_{k=0}^{\infty} a_k^{(j)}(x)\mu^k, \quad j = 0, 1. \qquad (5.2.3)$$

The function $\xi(x)$ and the coefficients $a_k^{(j)}(x)$ are evaluated as a result of the substitution of solution (5.2.2) into (5.2.1) and equating the coefficients of $\mu^k v$ and $\mu^k(dv/d\eta)$. If there are no other turning points in S, the functions $\xi(x)$ and $a_k^{(j)}(x)$ will be analytic in S, including the point $x = x_*$.

Calculating the first coefficients in (5.2.3) and assuming that solution $y(x, \mu)$ is analytic at the turning point, where $\xi = 0$, we get

$$a_0^{(0)} = \frac{1}{\sqrt{\xi'}}\cosh \nu, \quad a_0^{(1)} = \frac{1}{\sqrt{\xi\xi'}}\sinh \nu, \quad \nu = \int_{x_*}^x \frac{q_1\,dx}{2\xi'\sqrt{\xi}}. \qquad (5.2.4)$$

In particular, for $q_1 = 0$ we have $\nu = 0$ and

$$a_0^{(0)} = \frac{1}{\sqrt{\xi'}}, \quad a_0^{(1)} = 0. \qquad (5.2.5)$$

All the coefficients $a_k^{(j)}(x)$ are expressed in terms of $q_i(x)$ by quadratures (see also Exercise 5.2.1).

Expansion (5.2.2) is uniformly applicable in the entire interval S including a neighborhood of the point $x = x_*$. If $|\eta| \gg 1$, we may use asymptotic expansions for Airy's functions (5.1.4) and express (5.2.2) in the form of a linear combination of functions (5.1.2). Let the function $q(x, \mu)$ be real for real x and the conditions

$$q_0(x_*) = 0, \quad q_0'(x_*) > 0, \quad q_1(x) \equiv 0,$$
$$q_0(x) < 0 \quad \text{for} \quad x < x_*, \quad q_0(x) > 0 \quad \text{for} \quad x > x_*, \tag{5.2.6}$$

be satisfied. Then, for two solutions, $y_1(x, \mu)$ and $y_2(x, \mu)$, the following asymptotic expansions hold:

$$y_1(x, \mu) = \frac{1}{\sqrt{\xi'}} \mathrm{Ai}(\eta) \, [1 + O(\mu)] + C \mathrm{Ai}'(\eta), \quad C = O\left(\mu^{4/3}\right),$$

$$y_1(x, \mu) = \frac{a}{2} \exp\left(-\frac{1}{\mu} \int_{x_*}^{x} \sqrt{q_0(x)} \, dx\right) [1 + O(\mu)], \quad x > x_*,$$
$$y_1(x, \mu) = a \left[\sin\left(\frac{1}{\mu} \int_{x}^{x_*} \sqrt{-q_0(x)} \, dx + \frac{\pi}{4}\right) + O(\mu)\right], \quad x < x_*. \tag{5.2.7}$$

$$y_2(x, \mu) = \frac{1}{\sqrt{\xi'}} \mathrm{Bi}(\eta) \, [1 + O(\mu)] + C \mathrm{Bi}'(\eta), \quad C = O\left(\mu^{4/3}\right),$$

$$y_2(x, \mu) = a \exp\left(\frac{1}{\mu} \int_{x_*}^{x} \sqrt{q_0(x)} \, dx\right) [1 + O(\mu)], \qquad \qquad x > x_*,$$
$$y_2(x, \mu) = a \left[\cos\left(\frac{1}{\mu} \int_{x}^{x_*} \sqrt{-q_0(x)} \, dx + \frac{\pi}{4}\right) + O(\mu)\right], \qquad x < x_*. \tag{5.2.8}$$

where $a = \mu^{1/6} \Big/ \left(|q_0|^{1/4} \sqrt{\pi}\right)$.

The first formulas of (5.2.7) and (5.2.8) are uniformly applicable for $x_1 \leqslant x \leqslant x_2$, but the next two are uniformly applicable only for $x > x_*$ and for $x < x_*$, respectively. For $x > x_*$ the function y_1 decreases exponentially and the function y_2 increases. For $x < x_*$ both functions oscillate.

5.2.2 Turning Points at the Ends of Integration Intervals

Find the asymptotic expansions for the eigenvalues Ω_n and the eigenfunctions $y_n(x)$ of the Sturm–Liouville problem as $n \to \infty$:

$$y'' + \Omega^2 f(x) y = 0, \quad y(0) = y(l) = 0 \tag{5.2.9}$$

under the assumption that $f(0) = 0$, $f'(0) > 0$, and $f(x) > 0$ for $0 < x \le l$. The problem of the free vibrations of a string with variable density is reduced to this problem (see Sect. 4.4.3).

Assume that $\Omega = \mu^{-1}$. Then Eq. (5.2.9) is transformed into (5.2.1) with solution (5.2.2). Limiting ourselves to the main terms of the asymptotic expansions we find

$$y(x, \mu) = \frac{1}{\sqrt{\xi'}} v_0(\eta) \left[1 + O\left(\mu^2\right) \right] + v_0'(\eta) O\left(\mu^{4/3}\right), \quad \eta = -\mu^{-2/3} \xi,$$

$$v_0(\eta) = \sqrt{3}\, \mathrm{Ai}(\eta) - \mathrm{Bi}(\eta), \ \xi = \left[\frac{3}{2} \varphi(x) \right]^{2/3}, \ \varphi(x) = \int_0^x \sqrt{f(x)}\, dx.$$

$$\tag{5.2.10}$$

Note that $v_0(0) = 0$. The required value $\mu = \mu_n$ is calculated from the condition $v_0[\eta(l)] = 0$. We get

$$\mu_n = \frac{3}{2} |\eta_n'|^{-3/2} \varphi^0 + O\left(n^{-3}\right), \quad \Omega_n = \frac{2|\eta_n'|^{3/2}}{3\varphi^0} + O\left(n^{-1}\right),$$

$$\varphi^0 = \varphi(l) = \int_0^l \sqrt{f(x)}\, dx, \tag{5.2.11}$$

where η_n' is the nth zero of the function $v_0(\eta)$ (see Exercise 5.1.2 and Table 5.5). Replacing η_n' with their asymptotic representations by formula (5.5.1) (see the solution of Exercise 5.1.2) we obtain

$$\Omega_n = \frac{\pi(n - 1/12)}{\varphi^0} + O\left(n^{-1}\right), \quad \text{as} \ \ n \to \infty. \tag{5.2.12}$$

Now find the asymptotic expansions of the eigenvalues Ω_n under the assumption that $f(0) = 0$, $f'(0) > 0$, $f(l) = 0$, $f'(l) < 0$, and $f(x) > 0$ for $0 < x < l$.

There are two turning points in this problem, namely $x = 0$ and $x = l$. Thus, one cannot construct the asymptotic expansions of the solutions uniformly applicable to the entire interval $0 \le x \le l$ with the help of the standard Airy functions. We find two different solutions, $y^{(1)}(x)$ and $y^{(2)}(x)$. The solution $y^{(1)}(x)$ is applicable for $0 \le x \le l - \varepsilon$ ($\varepsilon > 0$) and satisfies the condition $y^{(1)}(0) = 0$ and the solution $y^{(2)}(x)$ is applicable for $\varepsilon \le x \le l$ and satisfies the condition $y^{(2)}(l) = 0$. We find the eigenvalues Ω_n from the condition

$$y^{(1)}(x) = C y^{(2)}(x), \quad C = \text{const}, \ \ \varepsilon \le x \le l - \varepsilon. \tag{5.2.13}$$

Taking (5.2.8) and (5.2.10) into account and omitting the residual terms for $\varepsilon \leq x \leq l - \varepsilon$, we have

$$y^{(1)}(x) \simeq \frac{1}{\sqrt{\xi_1'}} v_0\left(\eta^1\right) \simeq \frac{1}{\sqrt{\pi f(x)}} \sin\left(\mu^{-1}\varphi_1(x) + \pi/12\right),$$

$$\eta^1 = -\mu^{-2/3}\xi_1, \quad \xi_1 = \left[\frac{2}{3}\varphi_1(x)\right]^{3/2}, \quad \varphi_1(x) = \int_0^x \sqrt{f(x)}\,dx;$$

$$y^{(2)}(x) \simeq \frac{1}{\sqrt{\xi_2'}} v_0(\eta_2) \simeq \frac{1}{\sqrt{\pi f(x)}} \sin\left(\mu^{-1}\varphi_2(x) + \pi/12\right),$$

$$\eta_2 = -\mu^{-2/3}\xi_2, \quad \xi_2 = \left[\frac{2}{3}\varphi_2(x)\right]^{3/2}, \quad \varphi_2(x) = \int_x^l \sqrt{f(x)}\,dx.$$

Identity (5.2.13) holds only for

$$\mu^{-1}\varphi_1(x) + \pi/12 + \mu^{-1}\varphi_2(x) + \pi/12 = n\pi,$$

whence we get the required asymptotic formula

$$\Omega_n = \mu_n^{-1} = \frac{\pi(n - 1/6)}{\varphi^0} + O\left(n^{-1}\right), \quad \text{as } n \to \infty, \tag{5.2.14}$$

where

$$\varphi^0 = \varphi_1 + \varphi_2 = \int_0^l \sqrt{f(x)}\,dx.$$

5.2.3 Interior Turning Points

Find asymptotic expansions of eigenvalues Ω_n and eigenfunctions of the Sturm–Liouville problem

$$y'' + \Omega^2 f(x)y = 0, \quad y(x_1) = y(x_2) = 0, \quad \text{as } n \to \infty, \tag{5.2.15}$$

under assumption that

$$f(x_*) = 0, \quad f'(x_*) > 0, \quad x_1 < x_* < x_2,$$
$$f(x) < 0 \text{ for } x < x_*, \quad f(x) > 0 \text{ for } x > x_*. \tag{5.2.16}$$

We seek only eigenvalues Ω_n such that $\Omega_n^2 > 0$. Problem (5.2.15) also has a countable set of eigenvalues such that $\Omega_n^2 < 0$. Those may be obtained in a similar manner by the substitution $x' = -x$, $\Omega' = i\Omega$, $i = \sqrt{-1}$.

As in Sect. 5.2.2, we assume that $\Omega = \mu^{-1}$ and transform the problem into Eq. (5.2.1). We represent solution (5.2.2) in the form

$$y(x) = \frac{1}{\sqrt{\xi'}} \mathrm{Ai}(\eta)[(1 + O(\mu)] + C\mathrm{Ai}'(\eta), \quad C = O(\mu^{4/3}), \qquad (5.2.17)$$

where

$$\eta = -\mu^{-2/3}\xi, \quad \xi = \left(\frac{3}{2}\varphi(x)\right)^{2/3}, \quad \varphi(x) = \int_{x_*}^x \sqrt{f(x)}\,dx.$$

The function $y(x)$ approximately satisfies the condition $y(x_1) = 0$ since $\mathrm{Ai}(\eta) \to 0$ as $\eta \to \infty$. Satisfying condition $y(x_2) = 0$, we get

$$\Omega_n = \frac{2|\eta_n|^{3/2}}{3\varphi^0} + O\left(n^{-1}\right), \quad \varphi^0 = \varphi(x_2) = \int_{x_*}^{x_2} \sqrt{f(x)}\,dx, \qquad (5.2.18)$$

where η_n is the nth zero of the function $\mathrm{Ai}(\eta)$ (see Exercise 5.1.2 and Table 5.5). Replacing η_n with the asymptotic expansions by formula (5.5.1) (see solution of Exercise 5.1.2) we obtain

$$\Omega_n = \frac{\pi(n - 1/4)}{\varphi^0} + O\left(n^{-1}\right), \quad \text{as } n \to \infty. \qquad (5.2.19)$$

It should be noted that formulas (5.2.18) and (5.2.19) do not depend on the boundary condition at $x = x_1$ and also on the behavior of the function $f(x)$ for $x < x_*$, if only $f(x) < 0$. The eigenfunctions exponentially decreases with x for $x < x_*$ (see Fig. 5.7).

5.2.4 Vibrations of Strings on Elastic Foundations

The free vibrations of a string on an elastic foundation are described by the equation (see Sect. 4.3.3)

$$T\frac{d^2y}{dx_1^2} - c_1(x_1)y + \omega^2\rho_1(x_1)y = 0, \quad y(0) = y(l) = 0, \qquad (5.2.20)$$

where l is the string length, T is the tension, ω is the frequency. The linear density of the string, $\rho_1(x_1)$, and the foundation stiffness, $c_1(x_1)$, are assumed to be variable.

The non-dimensional variables $x = x_1/l$, $c_1 = c_0 c(x)$, $\rho_1 = \rho_0 \rho(x)$, and $c(x), \rho(x) \sim 1$ transform problem (5.2.20) into the standard form (5.2.1)

$$\mu^2 y'' - q(x)y = 0, \quad y(0) = y(1) = 0, \qquad (5.2.21)$$

where $q(x) = q(x, \Lambda) = c(x) - \Lambda \rho(x)$, $\mu^2 = T/(c_0 l^2)$ and $\Lambda = \omega^2 \rho_0/c_0$.

Under the assumption that $\mu > 0$ is a small parameter, we study the frequency spectrum of the free vibrations of the string. We assume also that $c(x) > 0$ and $\rho(x) > 0$ for $0 \le x \le 1$.

We introduce the auxiliary function $z(x)$ and the variables Λ^- and Λ^+ by the formulas:

$$z(x) = \frac{c(x)}{\rho(x)}, \quad \Lambda^- = \min_x z(x), \quad \Lambda^+ = \max_x z(x). \qquad (5.2.22)$$

For $\Lambda < \Lambda^-$, the function $q(x)$ is positive for all x and problem (5.2.21) does not have nontrivial solutions.

For $\Lambda > \Lambda^+$, the function $q(x)$ is negative for all x and for $\Lambda > \Lambda^+ + \varepsilon$, $\varepsilon > 0$, one may use solution (4.4.21) obtained in Sect. 4.4.3. Limiting ourselves to the zeroth approximation, we write the equation for Λ in the form

$$\varphi^0(\Lambda) = \mu n\pi + O\left(\mu^2\right), \quad \varphi^0 = \int_0^1 \sqrt{-q(x)}\, dx. \qquad (5.2.23)$$

For $\Lambda^- < \Lambda < \Lambda^+$, the interval of integration $[0, 1]$ contains the turning points which, in contrast to 5.2.2–5.2.8, move along the x-axis when $\Lambda \in [\Lambda^-, \Lambda^+]$ changes.

Here we consider only the case $z'(x) > 0$, $x \in [0, 1]$. Then the interval of integration for all $\Lambda \in [\Lambda^-, \Lambda^+]$ has one turning point, $x_*(\Lambda)$, and $\Lambda = z(x_*)$ and $q'(x_*) > 0$.

In the construction of the asymptotic expansions of Eq. (5.2.21) we find only the main term in (5.2.2). Then

$$y(x) = \frac{1}{\sqrt{\xi'}} \left[C_1 \mathrm{Ai}(\eta) + C_2 \mathrm{Bi}(\eta)\right], \quad \varphi = \int_{x_*}^x \sqrt{-q(x)}\, dx, \qquad (5.2.24)$$

where C_1 and C_2 are arbitrary constants, $\eta = \mu^{2/3}\xi$, $\xi = (3\varphi/2)^{2/3}$. Moreover, $\eta < 0$ for $x < x_*$ and $\eta > 0$ for $x > x_*$.

Substituting (5.2.24) into the boundary conditions (5.2.21), we obtain an approximate equation for Λ in the form

$$\mathrm{Ai}\left(\eta^0\right) - \gamma \mathrm{Bi}\left(\eta^0\right) = 0, \quad \gamma = \frac{\mathrm{Ai}\left(\eta^1\right)}{\mathrm{Bi}\left(\eta^1\right)}, \qquad (5.2.25)$$

where

$$\eta^0 = -\left(\frac{3\varphi^0}{2\mu}\right)^{2/3} < 0, \quad \varphi^0 = \int_0^{x_*} \sqrt{-q(x)}\,dx,$$

$$\eta^1 = \left(\frac{3\varphi^1}{2\mu}\right)^{2/3} > 0, \quad \varphi^1 = \int_{x_*}^1 \sqrt{q(x)}\,dx.$$

We consider two particular cases.

(1) If the turning point x_* is situated far from the edge $x = 1$, then $\eta^1 \gg 1$ and $\gamma \ll 1$, and Eq. (5.2.25) reduces to the form $\mathrm{Ai}(\eta^0) = 0$ (see Sect. 5.2.3 and Exercise 5.2.5) or

$$\varphi^0 = \frac{2\mu|\eta_n|^{3/2}}{3} \simeq \mu\pi\left(n - \frac{1}{4}\right), \quad n = 1, 2, \ldots, \tag{5.2.26}$$

where η_n is the nth zero of the function $\mathrm{Ai}(\eta)$. The eigenfunction oscillates to the left of the turning point x_* and decreases exponentially to the right of x_*.

(2) If the turning point x_* is situated close to the edge $x = 1$, then Eq. (5.2.26) becomes inapplicable since we cannot approximately assume $\gamma = 0$. In particular, if $x_* = 1$, i.e. the turning point coincides with the edge, then $\gamma = 1/\sqrt{3}$ and Eq. (5.2.26) is replaced with Eq. (5.2.12) $\varphi^0 \simeq \mu\pi(n - 1/12)$.

If we assume that $\tan(\alpha\pi) = \gamma$, then it follows from Eq. (5.2.25) that

$$\varphi^0 \simeq \mu\pi(n + \alpha - 1/4). \tag{5.2.27}$$

For $\Lambda > \Lambda^+$, the formula for calculating η^1 becomes inapplicable. If we assume that the functions $c(x)$ and $\rho(x)$ can be analytically extended to the right of the point $x = 1$, then $x_* > 1$ and we use the formula

$$\eta^1 = -\left(\frac{3\varphi^1}{2\mu}\right)^{2/3}, \quad \varphi^1 = \int_1^{x_*} \sqrt{-q(x)}\,dx,$$

to evaluate η^1 in (5.2.25)

The function $\alpha(\Phi)$,

$$\Phi = -\frac{1}{\mu}\int_{x_*}^1 \sqrt{q(x)}\,dx, \ x_* < 1; \quad \Phi = \frac{1}{\mu}\int_1^{x_*} \sqrt{-q(x)}\,dx, \ x_* > 1,$$

is plotted in Fig. 5.2.

By formulas (5.1.4),

$$\alpha(\Phi) = \frac{\Phi}{\pi} + \frac{1}{4} + O(\Phi^{-1}), \quad \text{as} \quad \Phi \to \infty,$$

and formula (5.2.27) transforms into (5.2.23).

Fig. 5.2 The function $\alpha(\Phi)$

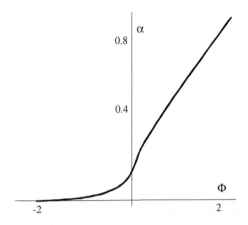

5.2.5 Asymptotic Expansions of Bessel Functions

Find the main terms of the asymptotic expansions of the Bessel functions $I_\nu(z)$, $K_\nu(z)$, $J_\nu(z)$ and $N_\nu(z)$ under the assumption that the order ν and the argument z of these functions are simultaneously large positive numbers.

Assume that $z = \nu x$ and consider $\nu \to \infty$ for fixed x. As $z \to 0$ (see [1])

$$\{I_\nu(z),\ J_\nu(z)\} \simeq \frac{z^\nu}{2^\nu \Gamma(\nu+1)}, \quad K_\nu(z) \simeq \frac{2^{\nu-1}\Gamma(\nu)}{z^\nu}, \quad N_\nu(z) \simeq -\frac{2^\nu \Gamma(\nu)}{\pi z^\nu}.$$

$$(5.2.28)$$

On substituting $z = \nu x$ and applying Stirling's approximation [1],

$$\Gamma(\nu) \simeq \sqrt{2\pi\nu}\,\nu^{\nu-1}\,e^{-\nu}, \quad \text{as}\ \nu \to \infty,$$

we write relations (5.2.28) in the form

$$\{I_\nu(z),\ J_\nu(z)\} \simeq \frac{1}{\sqrt{2\pi\nu}}\left(\frac{ex}{2}\right)^\nu, \quad K_\nu(z) \simeq \sqrt{\frac{\pi}{2\nu}}\left(\frac{ex}{2}\right)^{-\nu},$$

$$N_\nu(z) \simeq -\sqrt{\frac{2}{\pi\nu}}\left(\frac{ex}{2}\right)^{-\nu}, \quad (5.2.29)$$

as $x \to 0$ and $\nu \to \infty$. The functions $I_\nu(\nu x)$ and $K_\nu(\nu x)$ satisfy Bessel's equation

$$x^2 y'' + xy' - \nu^2(x^2 + 1)y = 0, \quad ()' = \frac{d}{dx}.$$

If we assume that $y = y_1 x^{-1/2}$ and $\mu = \nu^{-1}$, then this equation becomes

$$\mu^2 y_1'' - q(x, \mu) y_1 = 0, \quad \text{as} \quad \mu \to 0, \tag{5.2.30}$$

where

$$q(x, \mu) = q_0(x) + \mu^2 q_2(x), \quad q_0(x) = 1 + \frac{1}{x^2}, \quad q_2(x) = -\frac{1}{4x^2}.$$

For $x > 0$, Eq. (5.2.30) does not contain turning points. The main terms of the asymptotic expansions of its two solutions are

$$y_1(x) \simeq (q_0(x))^{-1/4} \exp\left[\pm\frac{1}{\mu} \int \sqrt{q_0(x)}\, dx\right]. \tag{5.2.31}$$

Evaluating the integral and returning to the variable y we get

$$I_\nu(\nu x) \simeq C_1 (x^2 + 1)^{-1/4} \left[\frac{x}{1 + \sqrt{x^2 + 1}}\right]^\nu e^{\nu\sqrt{x^2+1}},$$

$$K_\nu(\nu x) \simeq C_2 (x^2 + 1)^{-1/4} \left[\frac{x}{1 + \sqrt{x^2 + 1}}\right]^{-\nu} e^{-\nu\sqrt{x^2+1}}, \quad \nu \to \infty. \tag{5.2.32}$$

The constants C_1 and C_2 are evaluated by formula (5.2.32), as $x \to 0$, and (5.2.29):

$$C_1 = \frac{1}{\sqrt{2\pi\nu}}, \quad C_2 = \sqrt{\frac{\pi}{2\nu}}.$$

The construction of the asymptotic expansions for the functions $J_\nu(\nu x)$ and $N_\nu(\nu x)$ is reduced to the same Eq. (5.2.30), in which, now, $q_0(x) = x^{-2} - 1$. Equation (5.2.30) has the turning point $x_* = 1$ and, by (5.2.2) and (5.2.5), the asymptotic expansions of its solution are expressed in terms of Airy's functions.

Returning to the function y, we find as $\nu \to \infty$

$$J_\nu(\nu x) \simeq C_3 \left(\frac{\xi}{1 - x^2}\right)^{1/4} \text{Ai}(\eta), \quad N_\nu(\nu x) \simeq C_4 \left(\frac{\xi}{1 - x^2}\right)^{1/4} \text{Bi}(\eta), \tag{5.2.33}$$

where

$$\eta = \nu^{2/3} \xi(x), \quad \xi(x) = -\left(\frac{3}{2} \Phi(x)\right)^{2/3} \text{sgn}(x - 1),$$

and

$$\Phi(x) = \int_1^x \sqrt{-q(x)}\, dx = \sqrt{x^2 - 1} - \arctan\sqrt{x^2 - 1}, \quad x \geq 1,$$

$$\Phi(x) = \int_x^1 \sqrt{q(x)}\, dx = -\sqrt{1 - x^2} + \ln\frac{x}{1 + \sqrt{1 - x^2}}, \quad x \leq 1.$$

If $|x - 1| \gg \nu^{-2/3}$, then $|\eta| \gg 1$ and one can use formulas (5.2.7) and (5.2.8) from which it follows that

$$J_\nu(\nu x) \simeq C_3 \frac{\nu^{-1/6}(1 - x^2)^{-1/4}}{2\sqrt{\pi}} \left[\frac{x}{1 + \sqrt{1 - x^2}}\right]^\nu e^{\nu\sqrt{1-x^2}}, \quad x < 1,$$

$$N_\nu(\nu x) \simeq C_4 \frac{\nu^{-1/6}(1 - x^2)^{-1/4}}{\sqrt{\pi}} \left[\frac{x}{1 + \sqrt{1 - x^2}}\right]^{-\nu} e^{-\nu\sqrt{1-x^2}}, \quad x < 1,$$

$$J_\nu(\nu x) \simeq C_3 \frac{\nu^{-1/6}(x^2 - 1)^{-1/4}}{\sqrt{\pi}} \sin(\nu\Phi + \pi/4), \quad x > 1,$$

$$N_\nu(\nu x) \simeq C_4 \frac{\nu^{-1/6}(x^2 - 1)^{-1/4}}{\sqrt{\pi}} \cos(\nu\Phi + \pi/4), \quad x > 1. \tag{5.2.34}$$

Comparing formulas (5.2.34), as $x \to 0$, and (5.2.29) we obtain

$$C_3 = \sqrt{2}\nu^{-1/3}, \quad C_4 = -\sqrt{2}\nu^{-1/3}.$$

Obviously, $\Phi(x) \simeq x - \pi/2$ as $x \gg 1$. Substituting this approximate expression into formula for $J_\nu(x)$ we obtain the asymptotic formula (2.4.9).

5.2.6 Exercises

5.2.1. Develop recurrent formulas for the coefficients $a_k^{(j)}(x)$, $j = 0, 1$, of the expansion (5.2.3) if $q(x, \mu) = q_0(x)$.

5.2.2. Under the conditions of Sect. 5.2.2, find the asymptotic expansions of the eigenvalues Ω_n for $\rho(x) = x$ for $0 \le x \le l$.

5.2.3. Under the conditions of Sect. 5.2.2 find the asymptotic expansions of the eigenvalues Ω_n for $\rho(x) = x + x^2$ over the interval $0 \le x \le 1$. Compare the exact and asymptotic results.

5.2.4. Under the conditions of Sect. 5.2.2 compare the exact eigenvalues Ω_n and their asymptotic expansions (5.2.14) for $\rho(x) = x(1 - x)$, $l = 1$.

5.2.5. Under the conditions of Sect. 5.2.3 find the asymptotic expansions of the eigenvalues Ω_n for $\rho(x) = x + x^2$ over the interval $-1 \le x \le 1$. Compare the exact eigenvalues Ω_n and their asymptotic approximations. Plot the eigenfunction for $n = 3$. Study the effect of the boundary condition at $x = -1$ assuming that $y'(-1) = 0$ instead of $y(-1) = 0$.

5.2.6. Find the asymptotic expansions of the eigenvalues Ω_N and eigenfunctions $Y_N(x)$ for problem (5.2.15) as $N \to \infty$ if

$$\rho(x_*^{(k)}) = 0, \quad \rho'(x_*^{(k)}) \neq 0, \quad k = 1, 2, \quad x_1 < x_*^{(1)} < x_*^{(2)} < x_2;$$

$$\rho(x) > 0 \text{ for } x < x_*^{(1)} \text{ and for } x > x_*^{(2)};$$

$$\rho(x) < 0 \text{ for } x_*^{(1)} < x < x_*^{(2)}.$$

5.2.7. The function $\rho(x) = x^2 - a^2$ satisfies the conditions of Exercise 5.2.6 if $x_1 < -a$ and $x_2 > a$. Study the spectrum of the eigenvalues for $a = 0.5$, $x_1 = -1$, $x_2 = 1$ and compare the exact and approximate eigenvalues.

5.2.8. The function $\rho(x) = x^2 - a^2$ satisfies the conditions of Exercise 5.2.6 if $a = 0.5$, $x_1 = -0.9$, $x_2 = 1$, $x_1 < -a$ and $x_2 > a$. Find the exact first ten eigenvalues and compare them with those defined by formula (5.5.4). Plot some of the eigenfunctions.

5.2.9. For problem (5.2.21) compare the first ten exact eigenvalues Λ_n with the values obtained by the asymptotic formulas (5.2.23), (5.2.26) and (5.2.27) if $c(x) = 1 + x$, $\rho(x) = 1$ and $\mu = 0.03$.

5.2.10. The free vibrations of a circular membrane are described by the equation

$$T\left[\frac{1}{r}\frac{\partial}{\partial r}\left(r\frac{\partial w}{\partial r}\right) + \frac{1}{r^2}\frac{\partial^2 w}{\partial \varphi^2}\right] + \rho\omega^2 w = 0, \quad w(R, \varphi) = 0,$$

where $w(r, \varphi)$ is the membrane deflection, T is the tension, ρ is the area density, ω is the vibrations frequency and R is the membrane radius. Separating the variables $w(r, \varphi) = y(r)\cos m\varphi$ and scaling $r = ax$, $a = \sqrt{T/(\rho\omega^2)}$ we come to Bessel's equation for the function $y(x)$, $x^2 y'' + xy' + (x^2 - m^2) = 0$, with the boundary conditions $y(R/a) = 0$, $y(0) < \infty$.

Develop the approximate asymptotic formulas for the free vibrations frequencies of a circular membrane with a large number of waves in the circumferential direction m.

5.2.11. For $m = 8$, compare the exact values of α_{mn} which are the roots of the Bessel functions $J_m(x)$ with their asymptotic approximations (5.5.10)–(5.5.15) (see the solution of Exercise 5.2.10).

5.3 Solutions of Systems of Linear Ordinary Differential Equations with Turning Points

Consider the system of Eq. (4.2.1),

$$\mu\frac{dy}{dx} = A(x, \mu)y, \quad \mu > 0, \quad x_1 \leq x \leq x_2. \tag{5.3.1}$$

under the same assumptions on the matrix $A(x, \mu)$ as in Sect. 4.2.

We consider cases where the multiplicity of the roots of the characteristic equation (4.2.4),

$$\det(A_0(x) - \lambda E_n) = 0, \tag{5.3.2}$$

changes at some points $x = x_*$ ($x_1 \leq x_* \leq x_2$), called turning points.

5.3.1 Splitting Theorem

In the general case, the problem of the construction of asymptotic expansions of solutions of system (5.3.1) as $\mu \to 0$ has not been solved yet. Here, we limit ourselves to the simple case, where the multiplicity of the roots of equation (5.3.2) at $x = x_*$ is not larger than 2. Additionally, we assume that these roots, denoted $\lambda_1(x)$ and $\lambda_2(x)$, coincide at $x = x_*$ and can be written in the form

$$\lambda_{1,2} = p(x) \pm \sqrt{q(x)}, \quad q(x_*) = 0, \quad q'(x_*) \neq 0, \tag{5.3.3}$$

where the functions $p(x)$ and $q(x)$ are real analytic in a neighborhood of the point $x = x_*$.

Under the above assumptions, $\lambda_1(x)$ and $\lambda_2(x)$ provide integrals of system (5.3.1) and their asymptotic expansions as $\mu \to 0$ (similar to Sect. 5.2) are expressed by means of Airy's functions:

$$y(x, \mu) = \left[a^{(0)}(x, \mu)v(\eta) + \mu^{1/3} a^{(1)}(x, \mu)v'(\eta) \right] \exp\left(\frac{1}{\mu} \int p(x)\,dx \right), \tag{5.3.4}$$

where

$$\eta(x, \mu) = \mu^{-2/3}\xi(x), \quad \xi(x) = \left[\frac{3}{2} \int_{x_*}^{x} \sqrt{q(x)}\,dx \right]^{2/3},$$

$$a^{(j)}(x, \mu) \simeq \sum_{k=0}^{\infty} a_k^{(j)}(x)\mu^k, \quad j = 0, 1,$$

and $v(\eta)$ is one of the solutions of Airy's equation (5.1.1). Here, in contrast to solution (5.2.3), y, $a^{(j)}$ and $a_k^{(j)}$ are vector-functions. As in (5.2.3), the function $\xi(x)$ and the coefficients $a_k^{(j)}(x)$ are real analytic at $x = x_*$. Series (5.3.4) diverge.

Further, we mostly pay attention to the evaluation of the functions $a_k^{(j)}$, in particular, $a_0^{(j)}$. The following *splitting theorem* may be useful for this purpose [25, 65]. The theorem states

Splitting theorem *Suppose the roots of Eq. (5.3.2) can be split into two groups,*
$\lambda_1(x), \ldots, \lambda_p(x)$ *and* $\lambda_{p+1}(x), \ldots, \lambda_n(x)$, *such that*

$$\lambda_j(x_*) \neq \lambda_k(x_*), \quad j = 1, \ldots, p, \quad k = p+1, \ldots, n. \tag{5.3.5}$$

Then there exist a formal transformation

$$y = Pz, \quad P = P(x, \mu) \simeq \sum_{k=0}^{\infty} P_k(x)\mu^k, \quad \det P_0(x_*) \neq 0. \tag{5.3.6}$$

with real analytic coefficients $P_k(x)$ *at* $x = x_*$ *which transform system (5.3.1) into the form*

$$\mu \frac{dz}{dx} = B(x, \mu)z, \quad B = P^{-1}(AP - \mu P'), \tag{5.3.7}$$

where the blockdiagonal matrix B *has real analytic entries at* $x = x_*$

$$B = \begin{bmatrix} B_{11} & 0 \\ 0 & B_{22} \end{bmatrix} \tag{5.3.8}$$

with square matrices B_{11} *and* B_{22} *of sizes* p *and* $n - p$, *respectively, and the eigenvalues of the matrices* $B_{11}(x, 0)$ *and* $B_{22}(x, 0)$ *are equal to* $\lambda_1(x), \ldots, \lambda_p(x)$ *and* $\lambda_{p+1}(x), \ldots, \lambda_n(x)$, *respectively.*

As a result of repeated applications of the splitting theorem under the above assumptions on the roots of Eq. (5.3.2), system (5.3.1) can be split into the separated first-order equations

$$\mu \frac{dz_j}{dx} = b^{(j)}(x, \mu)z_j, \quad b^{(j)}(x, \mu) \simeq \sum_{k=0}^{\infty} b_k^{(j)}(x)\mu^k, \quad b_0^{(j)}(x) = \lambda_j(x), \tag{5.3.9}$$

for the simple roots $\lambda_j(x_*)$, or into systems of two equations

$$\mu \frac{dz_j}{dx} = b^{(j,j)}z_j + b^{(j,j+1)}z_{j+1},$$

$$\mu \frac{dz_{j+1}}{dx} = b^{(j+1,j)}z_j + b^{(j+1,j+1)}z_{j+1}, \tag{5.3.10}$$

if $\lambda_j(x_*) = \lambda_{j+1}(x_*)$.

Let $\lambda_1(x_*) = \lambda_2(x_*)$ at $x = x_*$ and relations (5.3.3) be satisfied. Find expressions for the vectors $a_0^{(0)}(x)$ and $a_0^{(1)}(x)$ in (5.3.4). For $x \neq x_*$, to construct the integrals $y^{(j)}(x, \mu)$ corresponding to the roots $\lambda_j(x)$, $j = 1, 2$, one may use formulas (4.2.3)–(4.2.9). Then, for $j = 1, 2$,

$$y^{(j)}(x, \mu) = U_0^{(j)}(x) \exp\left(\frac{1}{\mu} \int_{x_0}^{x} \lambda_j(x)\, dx\right)[1 + O(\mu)],$$

$$U_0^{(j)} = \varphi_0^{(j)} V^{(j)}, \quad (A_0 - \lambda_j E)V^{(j)} = 0, \quad \varphi_0^{(j)'} = b_1^{(j)} \varphi_0^{(j)},$$

$$V^{(j)} = V_1(x) - (-1)^j \sqrt{q}\, V_2(x), \quad b_1^{(j)} = \frac{c_1(x)}{q} - (-1)^j \frac{c_2(x)}{\sqrt{q}},$$

$$(5.3.11)$$

where the vector-functions $V_1(x)$ and $V_2(x)$ and the functions $c_1(x)$ and $c_2(x)$ are real analytic at $x = x_*$. However, $\varphi_0^{(j)}(x) \to \infty$ as $x \to x_*$.

In a neighborhood of the point $x = x_*$, the main term of the asymptotic expansions (5.3.4) as $\mu \to \infty$ have the form

$$y(x, \mu) = d_0(x)\left[\left(\frac{V_1 \cosh \nu}{\sqrt{\xi'}} + V_2\sqrt{\xi\xi'} \sin \nu + O(\mu)\right) v(\eta) + \right.$$
$$\left. \mu^{1/3}\left(\frac{V_1 \sinh \nu}{\sqrt{\xi\xi'}} + V_2\sqrt{\xi'} \cos \nu + O(\mu)\right) v'(\eta)\right] \exp\left(\frac{1}{\mu}\int_{x_*}^{x} p(x)\, dx\right),$$

$$(5.3.12)$$

where $\eta = \mu^{-2/3}\xi$,

$$d_0(x) = \exp\left(\int_{x_*}^{x} \frac{4c_1 + q'}{4q}\, dx\right), \quad \nu = \int_{x_*}^{x} \frac{c_2}{\sqrt{q}}\, dx, \quad \xi = \left(\frac{3}{2}\int_{x_*}^{x} \sqrt{q}\, dx\right)^{2/3}.$$

5.3.2 Vibrations of Circular Plates

The equation of free vibrations of a circular plate with m waves in the circumferential direction has the form

$$D \Delta\Delta w - \rho h \omega^2 w = 0, \quad 0 \le r \le R, \tag{5.3.13}$$

where

$$\Delta w = \frac{1}{r}\frac{d}{dr}\left(r\frac{dw}{dr}\right) - \frac{m^2}{r^2}w, \quad D = \frac{Eh^3}{12(1 - \nu^2)},$$

$w(r, \varphi) = w(r) \cos m\varphi$ is the plate deflection, R is the plate radius, E, ν, ρ, h are Young's modulus, Poisson's ratio, plate density and plate thickness, respectively.

The general solution of equation (5.3.13) bounded at $r = 0$ is expressed by means of the Bessel functions $J_m(x)$ and $I_m(x)$:

$$w(r) = C_1 J_m(\alpha r) + C_2 I_m(\alpha r), \quad \alpha^4 = \frac{\rho h \omega^2}{D}. \tag{5.3.14}$$

For a clamped edge $r = R$, the boundary conditions have the form

$$w = 0, \quad \frac{dw}{dr} = 0 \quad \text{for} \quad r = R.$$

Denote $\beta = \alpha R$. Then the frequency equation reduces to

$$f_1(\beta) = J_m(\beta) I'_m(\beta) - I_m(\beta) J'_m(\beta) = 0, \quad ()' = \frac{d}{d\beta}. \tag{5.3.15$_1$}$$

For a freely supported edge $r = R$, the boundary conditions and the equation are

$$w = 0, \quad \frac{d^2 w}{dr^2} + \nu \left(\frac{1}{r} \frac{dw}{dr} - \frac{m^2}{r^2} w \right) = 0 \quad \text{for} \quad r = R,$$

$$f_2(\beta) = J_m(\beta) I'_m(\beta) - I_m(\beta) J'_m(\beta) - \frac{2\beta}{1-\nu} J_m(\beta) I_m(\beta) = 0. \tag{5.3.15$_2$}$$

Finally for a free edge $r = R$, the boundary conditions and the equation are

$$\frac{d^2 w}{dr^2} + \nu \left(\frac{1}{r} \frac{dw}{dr} - \frac{m^2}{r^2} w \right) = 0,$$

$$\frac{d}{dr} \left[\frac{d^2 w}{dr^2} + \nu \left(\frac{1}{r} \frac{dw}{dr} - \frac{m^2}{r^2} w \right) \right]$$
$$+ (1 - \nu) \left(\frac{1}{r} \frac{d^2 w}{dr^2} - \frac{2m^2 + 1}{r^2} \frac{dw}{dr} + \frac{3m^2}{r^2} w \right) = 0, \quad \text{for} \quad r = R,$$

$$f_3(\beta) = \left(a_1 J_m(\beta) - \frac{1}{\beta} J'_m(\beta) \right) \left(\frac{m^2}{\beta^3} I_m(\beta) - a_1 I'_m(\beta) \right)$$

$$- \left(\frac{m^2}{\beta^3} J_m(\beta) - a_2 J'_m(\beta) \right) \left(a_2 I_m(\beta) - \frac{1}{\beta} I'_m(\beta) \right) = 0, \tag{5.3.15$_3$}$$

where

$$a_1 = \frac{m^2}{\beta^2} - \frac{1}{1-\nu}, \quad a_2 = \frac{m^2}{\beta^2} + \frac{1}{1-\nu}.$$

In developing Eqs. (5.3.2) and (5.3.2) we used equations satisfied by the functions $J_m(\beta)$ and $I_m(\beta)$ (see Sect. 5.2.5).

The frequencies ω_{mn} are expressed by means of the roots β_{mn} of Eqs. (5.3.15) and the formula

$$\omega_{mn} = \sqrt{\frac{D}{\rho h} \frac{\beta_{mn}^2}{R^2}}. \qquad (5.3.16)$$

As in Sect. 5.2.5, for the asymptotic analysis as $m \to \infty$, we make the substitution $r = mx/\alpha$ and assume that $\mu = 1/m$. Then, Eq. (5.3.13) takes the shape of (4.1.1):

$$\Delta_1^2 w - w = 0, \quad \Delta_1 w = \mu^2 \left(\frac{d^2 w}{dx^2} + \frac{1}{x}\frac{dw}{dx}\right) - \frac{w}{x^2}. \qquad (5.3.17)$$

Equation (5.3.17) has a turning point at $x_* = 1$. The above splitting theorem permits to consider two second order equations instead of the single fourth-order equation (5.3.17)

(a) $\Delta_1 w + w = 0, \quad w = J_m(mx),$

(b) $\Delta_1 w - w = 0, \quad w = I_m(mx),$ \qquad (5.3.18)

The first of equations contains the turning point $x_* = 1$.

We use the asymptotic formulas obtained in Sect. 5.2.5 to write formula (5.2.31) as

$$I_m(\beta) \simeq (m^2 + \beta^2)^{1/4} \exp\left(\int \frac{\sqrt{m^2 + \beta^2}}{\beta} d\beta\right), \quad (m, \beta) \to \infty, \qquad (5.3.19)$$

from where we have

$$I_m'(\beta) \simeq \frac{\sqrt{m^2 + \beta^2}}{\beta} I_m(\beta), \quad (m, \beta) \to \infty. \qquad (5.3.20)$$

For the roots of equation (5.3.15), $\beta \to \infty$ as $m \to \infty$. Thus, the applicability conditions for formula (5.3.20) hold and with an accuracy of order $1/m$, Eq. (5.3.15) transform to

(1) $\sqrt{m^2 + \beta^2} J_m(\beta) - \beta J_m'(\beta) = 0,$

(2) $J_m(\beta) = 0,$

(3) $a_1^2 \sqrt{m^2 + \beta^2} J_m(\beta) - a_2^2 \beta J_m'(\beta) = 0,$ \qquad (5.3.21)

where a_1 and a_2 are as in (5.3.15).

For a further simplification, replace $J_m(\beta)$ with its asymptotic expansion (5.2.33). Then Eq. (5.3.21) become

$$\mathrm{Ai}(\eta) + b_k \mathrm{Ai}'(\eta) = 0, \quad k = 1, 2, 3, \qquad (5.3.22)$$

where

$$\eta = -\left[\frac{3m}{2}\Phi\left(\frac{\beta}{m}\right)\right]^{2/3} \mathrm{sgn}(\beta - m),$$

$$b_1 = \left[\frac{m^2 - \beta^2}{\eta(m^2 + \beta^2)}\right]^{1/2}, \quad b_2 = 0, \quad b_3 = \frac{a_2^2}{a_1^2}\left[\frac{m^2 - \beta^2}{\eta(m^2 + \beta^2)}\right]^{1/2},$$

and the function $\Phi(x)$ is the same as in (5.2.33).

5.3.3 Vibrations of Shells of Revolution

Find the main terms of the asymptotic expansions of the integrals of system (4.2.35) which describes the free vibrations of a shell of revolution with large wave number, m, in the circumferential direction if the characteristic equation (4.2.38) has two roots with changing multiplicity

$$\lambda_{1,2} = \pm\sqrt{q(s)}, \quad q(s_*) = 0, \quad q'(s_*) < 0, \tag{5.3.23}$$

and the other roots of the equation are simple. The integrals corresponding to the simple roots are found in Sect. 4.2.6. Consider the integrals, corresponding to the roots (5.3.23).

To reduce system (4.2.35) to the standard form (5.3.1) we introduce the vector-function $y(s, \mu)$ by means of formulas (4.2.36). Then, for $s \neq s_*$, by formulas (4.2.3)–(4.2.9) and (4.2.39) we have

$$y(s, \mu) \simeq U_0(s)\exp\left(\frac{1}{\mu}\int\lambda(s)\,ds\right), \quad U_0 = \varphi_0(s)V(s), \tag{5.3.24}$$

where

$$\varphi_0 = (bf_\lambda)^{-1/2}, \quad f_\lambda = \frac{\partial f}{\partial \lambda}, \quad V = (v_1, v_2, \ldots, v_8)^T,$$

$$v_1 = \lambda^2 - \frac{r^2}{b^2}, \quad v_5 = -\frac{k_2\lambda^2 - k_1 r^2/b^2}{\lambda^2 - r^2/b^2}, \quad v_{k+1} = \lambda v_k, \quad k = 2, 3, \ldots, 8.$$

Here the function $f = f(\lambda, s)$ is given by formula (4.2.38) and the symbol ^{T} denotes transposition.

The multiple roots (5.3.23) are of the form (5.3.3) for $p(x) \equiv 0$. From formulas (5.3.9), (4.2.9) and (5.3.11),

$$b_1^{(j)}(s) = \frac{\varphi_0'}{\varphi_0} = \frac{c_1}{q} - (-1)^j\frac{c_2}{\sqrt{q}}, \quad j = 1, 2. \tag{5.3.25}$$

Since $f(\lambda, s)$ is a polynomial in even powers of λ, we get $c_2(s) \equiv 0$ and, therefore, $\nu(s) \equiv 0$ in (5.3.12).

The components $v_k, k = 1, \ldots, 8$, of the vector V are either even or odd functions of λ. Therefore, in representating the vector (5.3.11), each component v_k of $V^{(j)}$ contains only one of the two summands. Accordingly, formulas (5.3.12) can be simplified to

$$y_k(s, \mu) \simeq \frac{d_0 v_k}{\sqrt{\xi'}} v(\eta) = \mu^{1/6} \frac{d_0 v_k \eta^{1/4}}{\lambda^{1/2}} v(\eta), \quad k = 1, 3, 5, 7,$$

$$y_k(s, \mu) \simeq \mu^{1/3} \frac{d_0 \sqrt{\xi'} v_k}{\lambda} v'(\eta) = \mu^{1/6} \frac{d_0 v_k}{\eta^{1/4} \lambda^{1/2}} v'(\eta), \quad k = 2, 4, 6, 8, \quad (5.3.26)$$

where v_k come from formulas (5.3.24), $v(\eta)$ is Airy's function,

$$d_0 = \frac{\lambda}{\sqrt{bf_\lambda}}, \quad \eta = \mu^{-2/3}\xi, \quad \xi = \left(\frac{3}{2} \int_{s_*}^{s} \sqrt{q(s)}\, ds\right)^{2/3},$$

and $\xi > 0$ for $q > 0$ and $\xi < 0$ for $q < 0$. The right sides in (5.3.26) are real analytic for $s = s_*$.

We develop an approximate equation for evaluating the free vibrations frequencies of a shell of revolution with a large wave number, m, in the circumferential direction if in the interval of integration there exists one simple turning point of type (5.3.23).

Under the above assumptions, the turning point $s = s_*$ divides the interval of integration into two parts. For $s_1 \le s < s_*$, the characteristic equation (4.2.38) of degree eight has four roots with positive real parts and four with negative real parts. For $s_1 \le s < s_*$, the equation has three roots with positive real parts, three with negative parts and two pure imaginary roots.

We seek the vibrations mode which decays to the left of the turning point in the form

$$z(s, \mu) \simeq \sum_{j=1,3,4,5} C_j z^{(j)}(s, \mu), \quad (5.3.27)$$

where C_j are arbitrary constants, $z^{(1)}(s, \mu)$ are solutions of the form (5.5.22) or (5.5.23), in which $v(\eta) = \text{Ai}(\eta)$ is the Airy function which decays for $\eta > 0$, and $z^{(j)}(s, \mu)$, $j = 3, 4, 5$, are solutions of the form (5.5.20) for which $\Re(\lambda_j) > 0$. A substitution of (5.3.27) into the boundary conditions at $s = s_2$ leads to the frequency equation

$$\text{Ai}(-\eta_2) + d\left[-\frac{q(s_2)}{\eta_2}\right]^{1/2} \text{Ai}'(-\eta_2) = 0, \quad \eta_2 = \left(\frac{3}{2\mu} \int_{s_*}^{s_2} \sqrt{-q(s)}\, ds\right)^{2/3} > 0,$$
$$(5.3.28)$$

where $q(s)$ is as in (5.3.23) and the coefficient d depends on the boundary conditions at hand and it is a zero of a fourth-order determinant.

For example, for the clamped boundary conditions C_1 (see formulas (4.5.9)) $u = v = w = \gamma_1 = 0$, and equation for d is

$$
\begin{vmatrix}
\lambda_1^{-1} u_0^{(1)} & u_0^{(3)} & u_0^{(4)} & u_0^{(5)} \\
d v_0^{(1)} & v_0^{(3)} & v_0^{(4)} & v_0^{(5)} \\
d w_0^{(1)} & w_0^{(3)} & w_0^{(4)} & w_0^{(5)} \\
\lambda_1^{-1} \gamma_{10}^{(1)} & \gamma_{10}^{(3)} & \gamma_{10}^{(4)} & \gamma_{10}^{(5)}
\end{vmatrix} = 0,
\tag{5.3.29}
$$

where the functions $u_0^{(j)}$, $v_0^{(j)}$, $w_0^{(j)}$ and $\gamma_{10}^{(j)}$ are calculated by formulas (5.5.21) for $\lambda = \lambda_j(s)$. In the first column of the determinant (5.3.29) the functions (5.5.22) have multiplier d, and the functions (5.5.23) have multiplier λ^{-1}.

In particular, for boundary conditions of free support type S_2 ($T_1 = v = w = \gamma_1 = 0$) in the first column of determinant (5.3.29), all functions $z_0^{(1)}$ have multiplier d. Thus, $d = 0$ in (5.3.28).

For $\eta_2 \gg 1$, if one uses formulas (5.1.4), then Eq. (5.3.28) simplifies to

$$
\tan\left(\frac{1}{\mu} \int_{s_*}^{s_2} \sqrt{-q(s)}\, ds + \frac{\pi}{4}\right) = d\sqrt{-q(s_2)},
\tag{5.3.30}
$$

from where

$$
\int_{s_*}^{s_2} \sqrt{-q(s)} = \mu(\pi(n - 1/4) + \arctan\left(d\sqrt{-q(s_2)}\right), \quad n = 1, 2, \ldots
\tag{5.3.31}
$$

The free vibrations mode is localized near the edge $s = s_2$ (see Fig. 5.9). If the other edge, $s = s_1$, is weakly supported or free (see Sect. 4.5.2), then free vibrations modes localized near this edge may also appear.

5.3.4 Exercises

5.3.1. Develop formulas (5.3.12).

5.3.2. For the three types of boundary conditions considered in Sect. 5.3.2, compare the first six values of β_{mn}, obtained from the exact Eq. (5.3.15) and the approximate Eqs. (5.3.21)–(5.3.22), for $m = 8$ and $\nu = 0.3$.

5.3.3. Using the expansions (4.2.5), (5.3.26) and formulas (4.5.12), obtain the main terms of the asymptotic expansions of the functions u, v, w, $\gamma_1 = w'$, T_1, S, Q_1^* and M_1, which describe the stress-strain state for the free vibrations of a shell of revolution and which enter into the boundary conditions (4.5.12).

5.3.4. Find the integrals of system (4.2.53) which describe the stability of an axisymmetric stress-strain state of a momentless shell in the presence of a turning point of type (5.3.23).

5.3.5. Using Eq. (5.3.28) find the free vibrations frequencies for a paraboloidal shell of revolution in a neighborhood of the lowest frequency. The parameters of the paraboloidal shell entering formulas (4.2.38) and (5.5.21) are

$$k_2 = \frac{1}{\sqrt{1+b^2}}, \quad k_1 = k_2^3, \quad \frac{ds}{db} = \sqrt{1+b^2}, \tag{5.3.32}$$

where $b(s)$ is the dimensionless distance from the axis of revolution.

Consider the four types of boundary conditions, C_1, S_1, C_2, S_2 [see formulas (4.5.9)]. Assume the following values of the parameters $R/h = 250$, $\nu = 0.3$, $s_2 = 2$. As a characteristic size for R, take the radius of curvature at the top of the cupola.

5.3.6. Find the critical load for the paraboloidal shell of revolution considered in Exercise 5.3.5 under external normal pressure p. The initial axisymmetric stress state is determined by the non-dimensional stresses

$$t_1 = \frac{1}{2k_2}, \quad t_2 = \frac{1 - t_1 k_1}{k_2}, \tag{5.3.33}$$

where k_1 and k_2 are the same as in (5.3.32). The loading parameter Λ entering (4.2.53) is related to the pressure p by the formula

$$\Lambda = \frac{pR}{\mu^2 Eh}. \tag{5.3.34}$$

5.4 Localized Eigenfunctions

In Sects. 5.2 and 5.3, we considered eigenfunctions which were exponentially decreasing when approaching one of the edges of the interval $[x_1, x_2]$ and equations for evaluating the corresponding eigenvalues Λ within an accuracy of order $e^{-c/\mu}$ independent of the boundary conditions on that edge (see Exercises 5.2.5–5.2.9, 5.3.5–5.3.6). Here we study cases where system (5.3.1) or Eq. (4.1.1) has an oscillating solution on the interval $x_*^{(1)} < x < x_*^{(2)}$ ($x_1 < x_*^{(1)} < x_*^{(2)} < x_2$) and an exponentially decreasing solution when approaching the edges x_1 and x_2. With an error of order $e^{-c/\mu}$ one may assume that the solution satisfies any homogeneous boundary conditions at $x = x_1$ and $x = x_2$. Below we call this solution *localized*. We assume that the coefficients of Eq. (4.1.1) depend linearly on the parameter Λ and seek eigenvalues Λ for which the localized solution exists.

5.4.1 Existence Conditions for Localized Solutions

Firstly, consider the second-order equation

$$\mu^2 \frac{d^2 y}{dx^2} - q(x, \Lambda)y = 0, \quad q(x, \Lambda) = q_1(x) + \Lambda q_2(x). \tag{5.4.1}$$

Let $q(x_*^{(k)}, \Lambda) = 0, q_x'(x_*^{(k)}, \Lambda) \neq 0, k = 1, 2$, and $q < 0$ for $x_*^{(1)} < x < x_*^{(2)}$. Then an existence condition for localized solution is [22, 30]:

$$\frac{1}{\mu} \int_{x_*^{(1)}}^{x_*^{(2)}} g(x, \Lambda) \, dx = \frac{\pi}{2}(2n+1) + O(\mu), \quad n = 0, 1, 2, \ldots, \quad g = \sqrt{-q}. \tag{5.4.2}$$

Now, we assume that the order of system (5.3.1) or Eq. (4.1.1) is larger than two and for $x_*^{(1)} < x < x_*^{(2)}$ the characteristic equation (5.3.2) or (4.1.7) has two pure imaginary roots

$$\lambda_{1,2} = \pm i g(x), \quad g(x) \sim \left(x_*^{(2)} - x\right)^{1/2} \left(x - x_*^{(1)}\right)^{1/2}, \quad \text{as} \quad x \to x_*^{(k)}, \tag{5.4.3}$$

which, for $x \bar{\in} \left[x_*^{(1)}, x_*^{(2)}\right]$, go over into real roots of opposite signs. The other roots of the characteristic equation have nonzero real parts. Then, if condition (5.4.2) is satisfied, system (5.3.1) has a *formal* localized solution.

We note one more case where localized solutions appear for a system of at least fourth order. For $x_*^{(1)} < x < x_*^{(2)}$, let the characteristic equation (5.3.2) have four pure imaginary roots,

$$\lambda_{1,2,3,4} = \pm i q(x) \pm i g(x), \quad g(x) \sim \left(x_*^{(2)} - x\right)^{1/2} \left(x - x_*^{(1)}\right)^{1/2}, \quad x \to x_*^{(k)}, \tag{5.4.4}$$

where the function $q(x)$ is real analytic. For $x \bar{\in} \left[x_*^{(1)}, x_*^{(2)}\right]$ the roots of equation (5.3.2) have nontrivial real part. Under the same assumptions on the other roots of equation (5.3.2) the existence condition for localized solution is again relation (5.4.2).

5.4.2 Construction of Localized Solutions

In vibrational and buckling analysis, the evaluation of the lower eigenvalues Λ and the construction of the corresponding eigenfunctions is of special interest. It often occurs that an eigenfunction for such eigenvalue is localized and the turning points $x_*^{(1)}$ and $x_*^{(2)}$ are close apart ($x_*^{(2)} - x_*^{(1)} \sim \mu^{1/2}$). In this case, the algorithm described below can be applied to the example of the following self-adjoint equation of order $2m$:

$$\sum_{k=0}^{m}(-i\mu)^{2k}\frac{d^k}{dx^k}\left(a_k(x)\frac{d^k y}{dx^k}\right)=0,\quad x_1\le x\le x_2.\tag{5.4.5}$$

The coefficients a_k are assumed to be real analytic and depending linearly on the parameter Λ:

$$a_k = a_{1k} + \Lambda a_{2k},\quad \Lambda > 0.\tag{5.4.6}$$

We seek the eigenvalues Λ for which localized solutions of equation (5.4.5) exist in a neighborhood of some point x_0 ($x_1 < x_0 < x_2$). The point x_0 is called *the weakest point*. We construct a solution in the form of an asymptotic series

$$y(x,\mu)=\sum_{k=0}^{\infty}\mu^{k/2}y_k(\xi)\exp\left[i\left(\mu^{-1/2}p_0\xi + 1/2a\xi^2\right)\right],$$

$$\xi = \mu^{-1/2}(x - x_0),\quad \Lambda = \Lambda_0 + \mu\Lambda_1 + \cdots,\tag{5.4.7}$$

where $y_k(\xi)$ are polynomial in ξ, p_0, x_0 and Λ_k are real, and $\Im(a) > 0$. The last condition guarantees that solution (5.4.7) decreases as $|x - x_0|$ increases.

Substituting $\lambda = ip$ into Eq. (4.1.7) and solving it, we find Λ:

$$\Lambda = \left(\sum_{k=0}^{m}a_{1k}(x)p^{2k}\right)\left(\sum_{k=0}^{m}a_{2k}(x)p^{2k}\right)^{-1}\equiv f(p,x).\tag{5.4.8}$$

Assume that there exists a unique point (p_0, x_0) such that

$$\Lambda_0 = \min_{p,x}^{(+)}\{f\} = f(p_0, x_0)\tag{5.4.9}$$

and

$$d^2 f = f_{pp}^0\, dp^2 + 2f_{px}^0\, dp\, dx + f_{xx}^0\, dx^2 > 0,\quad f_{px}^0 = \frac{\partial^2 f(p_0, x_0)}{\partial p_0\partial x_0},\ldots\tag{5.4.10}$$

The minimum (5.4.9) is searched for all $x \in [x_1, x_2]$, $p \ge 0$, for which $f > 0$. Then a solution of the form (5.4.7) exists and

$$y_0^{(n)}(\xi) = H_n(z),\quad z = c^{1/2}\xi,\quad \Lambda^{(n)} = \Lambda_0 + \mu\Lambda_1^{(n)} + O\left(\mu^2\right),$$

$$\Lambda_1^{(n)} = r(n + 1/2),\quad n = 0, 1, 2, \ldots,\tag{5.4.11}$$

where

$$a = \frac{ir - f_{px}^0}{f_{pp}^0},\quad c = \frac{r}{f_{pp}^0},\quad r = \left[f_{pp}^0 f_{xx}^0 - (f_{px}^0)^2\right]^{1/2} > 0,$$

and $H_n(z)$ is the Hermite polynomial of degree n. In particular, $H_0 = 1$, $H_1 = z$, $H_2 = z^2 - 1/2$. Formulas (5.4.7) and (5.4.11) determine the series of eigenvalues $\Lambda^{(m)}$ the lowest of which is obtained for $n = 0$.

Concerning the eigenfunctions, we should distinguish two cases:

$$\text{(A)} \quad p_0 = 0 \quad \text{and} \quad \text{(B)} \quad p_0 > 0. \tag{5.4.12}$$

In the case $p_0 = 0$, the eigenfunctions have the form

$$y^{(n)}(x, \mu) = \left[H_n(z) + O(\mu)^{1/2} \right] e^{-z^2/2}, \quad z = \left(\frac{c}{\mu} \right)^{1/2} (x - x_0), \tag{5.4.13}$$

and the corresponding eigenvalues $\Lambda^{(n)}$ are simple.

In the case $p_0 > 0$, the function (5.4.7) is complex. Since the coefficients a_k in (5.4.5) are real, the real and imaginary parts of function (5.4.7) are solutions of equation (5.4.5). But it would be a mistake to consider eigenvalues $\Lambda^{(n)}$ to be double and assume that any arbitrary combination of the real and imaginary parts of (5.4.7) provides an eigenfunction. The point is that (5.4.7) is not convergent but asymptotic and two fixed real functions correspond to the parameter $\Lambda^{(n)}$:

$$y^{(n,j)}(x, \mu) = \left[H_n(z) \cos \Psi_j + O(\mu^{1/2}) \right] \exp \left(-\frac{\Im(a)(x - x_0)^2}{2\mu} \right),$$

$$\Psi_j = \frac{p_0 x}{\mu} + \frac{\Re(a)(x - x_0)^2}{2\mu} + \Theta_j, \quad j = 1, 2, \quad z = \left(\frac{c}{\mu} \right)^{1/2} (x - x_0),$$

$$\tag{5.4.14}$$

where the phases Θ_j are fixed ($0 \le \Theta_1, \Theta_2 < 2\pi$). The corresponding exact eigenvalues $\Lambda^{(n,1)}$ and $\Lambda^{(n,2)}$ are different, but

$$\Delta\Lambda = \Lambda^{(n,2)} - \Lambda^{(n,1)} = O\left(\mu^N \right) \tag{5.4.15}$$

for any N. Such eigenvalues are called *asymptotically doubled*. When the characteristic equation (5.3.2) has two pairs of pure imaginary roots (5.4.3) between the turning points, the eigenvalues are also asymptotically double.

In Fig. 5.3a: the eigenfunction for $p_0 = 0$, and Fig. 5.3b, c: two eigenfunctions for $p_0 > 0$ are plotted for $n = 0$.

5.4.3 Vibrations of Prolate Ellipsoidal Shells of Revolution

We use formulas (5.4.7)–(5.4.11) to find approximate vibrations frequencies of a thin shell in the shape of a prolate *ellipsoidal* shell of revolution ($a_0 < b_0$, see Fig. 5.4).

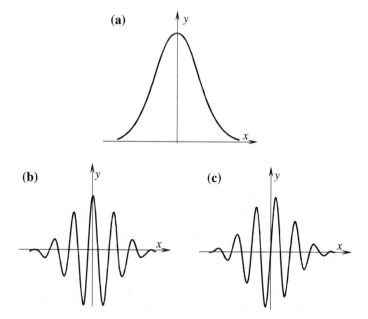

Fig. 5.3 **a** Eigenfunction for $p_0 = 0$, and **b** and **c** two eigenfunctions for $p_0 > 0$ for $n = 0$

Fig. 5.4 A prolate
ellipsoidal shell of revolution

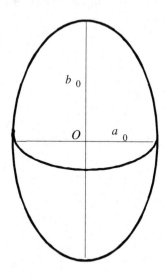

The vibrations of an ellipsoidal shell with m waves in the circumferential direction
are given by the system of Eq. (4.2.35):

$$\Delta \Delta w - \Delta_k \Phi - \Lambda w = 0, \quad \Delta \Delta \Phi + \Delta_k w = 0, \quad (5.4.16)$$

where

$$\Delta w = \mu^2 \frac{1}{b} (bw')' - \frac{r^2}{b^2} w, \quad \Delta_k w = \mu^2 \frac{1}{b} (bk_2 w')' - \frac{k_1 r^2}{b^2} w, \quad ()' = \frac{d()}{ds}.$$

Here we use the notation of Sect. 4.2.5.

The numbers a_0 and b_0 are the ellipse semi-axes ($a_0 < b_0$) and $R = a_0$ is the characteristic size in formulas (4.2.34). Then

$$k_2 = \sqrt{\sin^2 \theta + \delta^2 \cos^2 \theta}, \quad k_1 = \frac{k_2^3}{\delta^2}, \quad b = \frac{\sin \theta}{k_2},$$

$$\delta = \frac{b_0}{a_0} > 0, \quad \frac{d\theta}{ds} = k_1, \tag{5.4.17}$$

where θ is the angle between the axis of rotation and the normal to the shell.

The function (5.4.8),

$$f(p, s) = \left(p^2 + \frac{r^2}{b^2} \right)^2 + \frac{(k_2 p^2 + k_1 r^2/b^2)^2}{(p^2 + r^2/b^2)^2}, \quad r = \mu m, \tag{5.4.18}$$

attains its minimal value (5.4.9)

$$\Lambda_0 = \left(\frac{r^4}{b^4} + k_1^2 \right)_{\theta=\theta_0} = r^4 + \frac{1}{\delta^4} \tag{5.4.19}$$

at $p_0 = 0$, $\theta_0 = \pi/2$.

The derivatives

$$f_{pp}^0 = \left[\frac{4r^2}{b^2} + \frac{4k_1(k_2 - k_1)r^2}{b^2} \right]_{\theta=\theta_0} = 4r^2 + \frac{4(\delta^2 - 1)}{r^2 \delta^4},$$

$$f_{ss}^0 = \frac{d^2}{ds^2} \left[\frac{r^4}{b^4} + k_1^2 \right]_{\theta=\theta_0} = \frac{6(\delta^2 - 1)}{\delta^6} + \frac{4r^4}{\delta^2}, \quad f_{ps}^0 = 0, \tag{5.4.20}$$

we substitute in (5.4.11) and for $n = 0$, we get

$$\Lambda \simeq r^4 + \frac{1}{\delta^4} + \frac{\mu}{r\delta^5} \sqrt{(r^4 \delta^4 + \delta^2 - 1)(4r^4 \delta^4 + 6(\delta^2 - 1))}, \quad r = \mu m. \tag{5.4.21}$$

The minimum wavenumber, m, in the circumferential direction is attained for $r \ll 1$. Neglecting terms with the factor r^4 under the radical sign in (5.4.21) we approximately find

$$\Lambda_{\min} = \frac{1}{\delta^4} + \frac{5}{\delta}\left(\frac{\sqrt{6(\delta^2-1)}\mu}{4}\right)^{4/5}, \quad m_{\min} = \frac{1}{\mu\delta}\left(\frac{\sqrt{6(\delta^2-1)}\mu}{4}\right)^{1/5}.$$

$$(5.4.22)$$

Now the lowest frequency may be found by formula (4.2.34).

5.4.4 Buckling of Cylindrical Shells Under Non-uniform Compression

The buckling of a circular cylindrical thin shell of radius R with freely supported edges under non-uniform compression is described by the system of equations of type (4.2.53):

$$\Delta\Delta w - 2\Lambda r^2 t(\varphi)w + r^2\Phi = 0, \quad \Delta\Delta\Phi - r^2 w = 0, \qquad (5.4.23)$$

where

$$\Delta w = \mu^2 \frac{\partial^2 w}{\partial\varphi^2} - r^2 w, \quad r = \frac{\mu m \pi}{l}, \quad m = 1, 2, \ldots, \quad l = \frac{L}{R}, \quad t(\varphi) = -\frac{T_1^0(\varphi)}{2\Lambda E h \mu^2}.$$

Here $T_1^0(\varphi)$ is the membrane initial stress-resultant, L is the shell length, m is the number of semi-waves in the longitudinal direction when the variables are separated as

$$w(s, \varphi) = w(\varphi)\sin\frac{m\pi s}{l}, \quad 0 \le \varphi \le 2\pi, \qquad (5.4.24)$$

where s and φ are the longitudinal and circumferential coordinates of the shell surface.

System (5.4.23) has variable coefficient $t(\varphi)$. The loading parameter $\Lambda \ge 0$ is introduced in such a way that

$$\max_{\varphi} t(\varphi) = t(\varphi_0) = 1. \qquad (5.4.25)$$

Without loss of generality, we assume that $\varphi_0 = 0$.

For fixed value of r we seek a loading parameter Λ for which system (5.4.24) has a nontrivial solution satisfying the periodicity conditions of φ. We also assume that $t''(0) < 0$.

When using the algorithm given by formulas (5.4.5)–(5.4.15), the periodicity condition is replaced by a damping condition for the solution away from the weakest generatrix $\varphi_0 = 0$.

The function (5.4.8) can be written in the form

$$f(p, \varphi, r) = \frac{1}{2t(\varphi)}\left(z + \frac{1}{z}\right), \quad z = \frac{(p^2 + r^2)^2}{r^2}. \qquad (5.4.26)$$

As in the general case (see (5.4.12)) when one seeks the minimum (5.4.9),

$$\Lambda_0 = \min_{p,\varphi} \ f(p,\varphi,r) = f(p_0,0,r), \tag{5.4.27}$$

two cases are possible depending on the value of r:

(A) $p_0 = 0, \quad \Lambda_0 = \dfrac{1}{2}\left(r^2 + \dfrac{1}{r^2}\right)$ for $r > 1$,   (5.4.28)

(B) $p_0 = \sqrt{r - r^2}, \quad \Lambda_0 = 1$ for $r < 1$.   (5.4.29)

In case (A), the inequality $r > 1$ holds. Therefore, this case can happen only for buckling of rather short shells ($m = 1, l < \pi\mu$) and we obtain the loading parameter by the formulas (5.4.7) and (5.4.11),

$$\Lambda^{(n)} = \frac{1}{2}\left(r^2 + \frac{1}{r^2}\right) + \mu\left(n + \frac{1}{2}\right)\sqrt{\frac{2(r^8 - 1)t''(0)}{r^6}} + O\left(\mu^2\right),\ n = 0, 1, \dots \tag{5.4.30}$$

In case (B), from the same formulas (5.4.7) and (5.4.11) we find

$$\Lambda^{(n)} = 1 + \mu\left(n + \frac{1}{2}\right)\sqrt{-\frac{16(r - 1)t''(0)}{r}} + O\left(\mu^2\right),\ n = 0, 1, \dots \tag{5.4.31}$$

In contrast to case (A), the eigenvalues (5.4.31) are asymptotically double (see also Exercise 5.4.10).

Let $l \sim 1$. Then for $\mu \ll 1$, we have $r = \mu m\pi/l < 1$ for several values of m for which formula (5.4.31) gives approximately equal values for the load $\Lambda \simeq 1$. The difference is only in the term of order μ. Thus, it follows that the lowest value of Λ corresponds to $r = 1$.

Both formulas (5.4.30) and (5.4.31) give the same value of $\Lambda = 1$ for $r = 1$. However, for $r = 1$ the expansion (5.4.7) together with formulas (5.4.31) is not applicable since $f_{pp} = 0$ violates assumption (5.5.29). For $r \simeq 1$ the asymptotic expansion for the solution is obtained in Exercise 5.4.11.

Remark The problem considered here is a good illustration of asymptotic analysis. However, for estimate of a real critical loading under compression for real shells, formula (5.4.31) is not applicable because of the significant influence of the initial imperfections of the shape of the neutral surface which can reduce the critical load by a factor of two of three (see [31, 56]). Moreover, buckling of long cylindrical shells ($l \gg 1$) is different from the above scheme. Indeed, long shells behave like a beam that is compressed with an axial force.

5.4.5 Exercises

5.4.1. Find the coefficients p_0, Λ_0, Λ_1 in formulas (5.4.7) for Eq. (5.4.1).

5.4.2. The free vibrations of a string on an elastic foundation are described by Eq. (5.4.1), in which (see also Sects. 4.3.3 and 5.2.4)

$$\mu^2 = \frac{T}{c_0 l^2}, \quad \Lambda = \frac{\rho_0 \omega^2}{c_0},$$

and $q_1(x) = c(x)$ and $q_2(x) = \rho(x)$ are the non-dimensional foundation stiffness and the string density, respectively (see notation in Sect. 5.2.4).

Find the solution of equation (5.4.1) satisfying the damping conditions:

$$y(-\infty) = y(\infty) = 0, \tag{5.4.32}$$

for $q_1(x) = x^2$ and $q_2(x) = 1$.

5.4.3. Consider the eigenvalue problem

$$\mu^2 \frac{d^2 y}{dx^2} - \left(x^2 - \Lambda\right) y = 0, \quad y(-1) = y(1) = 0. \tag{5.4.33}$$

Compare the exact eigenvalues $\Lambda^{(n)}$ and their asymptotic approximations found with formulas (5.4.2) and (5.5.28) with $\mu = 0.1$ and $\Lambda < 1$.

5.4.4. Consider the eigenvalue problem

$$\mu^2 \frac{d^2 y}{dx^2} - \left(x^2 + \frac{1}{3}x^3 - \Lambda\right) y = 0, \quad y(-2) = y(1) = 0. \tag{5.4.34}$$

Compare the exact eigenvalues $\Lambda^{(n)}$ and their asymptotic approximations found by formulas (5.4.2) and (5.5.30) with $\mu = 0.1$ and $\Lambda < 1$.

Consider also the boundary conditions $y(-2) = y'(1) = 0$ instead of (5.4.34).

5.4.5. Compute the values of the parameter Λ by formulas (5.4.2) and (5.4.21) for an ellipsoidal shell of revolution with $R/h = 500$, $\nu = 0.3$ and $\delta = \sqrt{2}$.

5.4.6. Find the critical value of the external pressure q for the buckling of a thin shell of the shape of a prolate ellipsoidal shell of revolution ($a_0 > b_0$, see Fig. 5.5).

5.4.7. Compare the numerical results obtained by formulas (5.4.2) and (5.5.34) applied to an ellipsoidal shell of revolution with $R/h = 500$, $\nu = 0.3$ and $\delta = \sqrt{2}$.

5.4.8. Analyze the buckling of a thin ellipsoidal shell of revolution under the *internal* pressure $q > 0$ for different ratios of the semi-axes δ (see formula (5.4.17)).

5.4.9. Find a formula similar to (5.5.41) for an oblate ellipsoidal shell of revolution ($\delta < 1/2$) under an internal pressure $q > 0$.

5.4.10. Consider the buckling of a circular cylindrical shell with freely supported edges under axial force P and bending moment M applied to the shell edges (see Fig. 5.6). Then the initial stress-resultant $T_1^{(0)}$ and the function $t(\varphi)$ are [see (5.4.23)]

Fig. 5.5 Thin shell in the
shape of a prolate *ellipsoidal
shell* of revolution

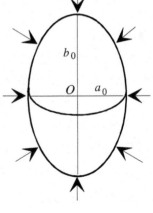

Fig. 5.6 Buckling of a
circular cylindrical shell
with freely supported edges

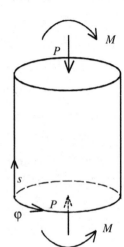

$$T_1^{(0)} = \frac{P}{2\pi R} + \frac{M \cos \varphi}{\pi R^2}, \quad t(\varphi) = \frac{\alpha + \cos \varphi}{\alpha + 1},$$

$$M = -\frac{2\pi R^2 \Lambda E h \mu^2}{\alpha + 1}, \quad \alpha = \frac{PR}{2M} > -1. \tag{5.4.35}$$

The generatrix $\varphi_0 = 0$ is the weakest one.

For $R/h = 100$, $\nu = 0.3$, $\alpha = 1$ (see notation in Sect. 5.4.4) and different q, compare the exact values of Λ obtained by means of the numerical integration of system (5.4.23) and their asymptotic approximations obtained by formulas (5.4.30) and (5.4.31).

5.4.11. By formulas (5.4.30) and (5.4.31), the lowest value of $\Lambda(r)$ is attained at $r = 1$, where these formulas are not applicable. Under the same assumption as in Sect. 5.4.4, study the asymptotics of the first two eigenvalues of $\Lambda(r)$ for problem (5.4.23) as $\mu \to 0$ and $r \simeq 1$.

5.4.12. Consider the free vibrations of a thin non-circular cylindrical shell with freely supported edges. A part of the frequency spectrum close to the lowest frequency can be found by the following system of equations:

$$\mu^4 \frac{d^4 w}{dx^4} - \Lambda w + k(y)\Phi = 0, \quad \mu^4 \frac{d^4 \Phi}{dx^4} - k(y)w = 0, \tag{5.4.36}$$

$$\mu^8 = \frac{h^2 L^4}{12\pi(1-\nu)r^6}, \quad k(y) = \frac{R}{R_2(y)}, \quad \Lambda = \frac{\rho\omega^2 r^2}{E\mu^4},$$

where y is the non-dimensional coordinate in the transverse direction, L is the shell length, $k(y)$ is the non-dimensional curvature of a directrix, the other notation being the same as in Sect. 4.2.5.

Obtain the asymptotic formula for the lowest vibrations frequency, i.e. the minimal value of Λ as $\mu \to 0$.

Consider the particular case of an elliptic cylindrical shell with ellipse semi-axes a_0 and b_0, for which

$$R = a_0, \quad \frac{d\theta}{dy} = k(y) = \delta^{-2}\left(\sin^2\theta + \delta^2\cos^2\theta\right)^{3/2}, \quad \delta = \frac{b_0}{a_0} > 1. \tag{5.4.37}$$

Here θ is the angle between the major axis of the ellipse and the normal to the shell.

Compare the asymptotic and the numerical results for $\delta = \sqrt{2}$ and two values of μ ($\mu = 0.1$ and $\mu = 0.2$).

5.5 Answers and Solutions

5.1.1. Results on the evaluation of the function $\text{Ai}(\eta)$ for $\eta > 0$ and $\eta < 0$ are listed in Tables 5.1 and 5.3, respectively. Similar results are presented in Tables 5.2 and 5.4 for the function $\text{Bi}(\eta)$. The values of $\text{Ai}^{(1.2)}$ and $\text{Bi}^{(1.2)}$ were obtained by the convergent series (5.1.2). The values of $\text{Ai}^{(N)}$ and $\text{Bi}^{(N)}$ were obtained by the first terms of the divergent series (5.1.4) with $N = 1, 4, 7, 10$. The decimal order is shown inside parentheses.

Generally the convergent series (5.1.2) can be used for small $|\eta|$ and the asymptotic series (5.1.4) for large $|\eta|$. For the intermediate values of $|\eta|$, the results, according to (5.1.2) and (5.1.4), are expected to coincide. From the above tables, it follows that, for $\eta > 0$, the domain where these results coincide (the domain of overlap) is narrower ($4 \le \eta \le 5$) for the function $\text{Ai}(\eta)$ and wider ($\eta \ge 4$) for the function $\text{Bi}(\eta)$. For $\eta < 0$, the domain of overlap is $4 \le |\eta| \le 10$ for both functions.

The observed difference in the domains of overlap is explained by the fact that a loss of accuracy occurs for calculation by series (5.1.2) for large $|\eta|$ because small differences of large values are computed. The highest loss of accuracy occurs with $\text{Ai}(\eta)$ when $\eta > 0$. Actually, for $\eta \ge 7$ even the decimal order of $\text{Ai}(\eta)$ calculated

Table 5.1 Value of Ai(η) for $\eta > 0$

η	Ai$^{(1.2)}$	Ai$^{(1)}$	Ai$^{(4)}$	Ai$^{(7)}$	Ai$^{(10)}$
1	0.1353(−0)	0.1448(−0)	0.1233(−0)	0.5189(−0)	−0.5894(+2)
2	0.3492(−1)	0.3599(−1)	0.3484(−1)	0.3506(−1)	0.3390(−1)
3	0.6591(−2)	0.6709(−2)	0.6589(−2)	0.6592(−2)	0.6590(−2)
4	0.9516(−3)	0.9630(−3)	0.9515(−3)	0.9516(−3)	0.9516(−3)
5	0.1084(−3)	0.1093(−3)	0.1083(−3)	0.1083(−3)	0.1083(−3)
6	0.9979(−5)	0.1002(−4)	0.9948(−5)	0.9948(−5)	0.9948(−5)
7	0.1134(−5)	0.7533(−6)	0.7492(−6)	0.7492(−6)	0.7492(−6)
8	0.5797(−5)	0.4713(−7)	0.4692(−7)	0.4692(−7)	0.4692(−7)
9	0.1029(−3)	0.2481(−8)	0.2471(−8)	0.2471(−8)	0.2471(−8)
10	0.2184(−2)	0.1108(−9)	0.1105(−9)	0.1105(−9)	0.1105(−9)

Table 5.2 Value of Bi(η) for $\eta > 0$

η	Bi$^{(1.2)}$	Bi$^{(1)}$	Bi$^{(4)}$	Bi$^{(7)}$	Bi$^{(10)}$
1	0.1208(1)	0.1099(1)	0.1446(1)	0.6386(1)	0.6310(3)
2	0.3298(1)	0.3127(1)	0.3292(1)	0.3342(1)	0.3563(1)
3	0.1404(2)	0.1370(2)	0.1403(2)	0.1404(2)	0.1404(2)
4	0.8385(2)	0.8263(2)	0.8384(2)	0.8385(2)	0.8385(2)
5	0.6578(3)	0.6512(3)	0.6578(3)	0.6578(3)	0.6578(3)
6	0.6536(4)	0.6488(4)	0.6536(4)	0.6536(4)	0.6536(4)
7	0.8033(5)	0.7986(5)	0.8033(5)	0.8033(5)	0.8033(5)
8	0.1200(7)	0.1194(7)	0.1200(7)	0.1200(7)	0.1200(7)
9	0.2147(8)	0.2139(8)	0.2147(8)	0.2147(8)	0.2147(8)
10	0.4556(9)	0.4541(9)	0.4556(9)	0.4556(9)	0.4556(9)

by series (5.1.2) appears to be incorrect. For Bi(η) no loss of accuracy occurs for $\eta > 0$ since all terms in (5.1.2) are positive.

It should be noted that, for small $|\eta|$, increasing the number of terms in the asymptotic series (5.1.4) makes the result worse (as it must be for divergent series).

5.1.2. Using the asymptotic formulas (5.1.4) as $n \to \infty$ we find

$$\eta_n = \eta_{na}\left[1 + O\left(n^{-2}\right)\right], \qquad \eta_{na} = -[1.5\pi(n - 1/4)]^{2/3},$$

$$\eta_n' = \eta_{na}'\left[1 + O\left(n^{-2}\right)\right], \qquad \eta_{na}' = -[1.5\pi(n - 1/12)]^{2/3}. \qquad (5.5.1)$$

The values of $\eta_n, \eta_{na}, \eta_n', \eta_{na}'$, for $n = 1, 2, \ldots, 6$, are given in Table 5.5.

5.2.1. Substituting (5.2.2) into (5.2.1) and equating the coefficients of v and $dv/d\eta$, and taking (5.2.3) into account, we obtain the system of equations

Table 5.3 Value of Ai(η) for $\eta < 0$

η	Ai$^{(1.2)}$	Ai$^{(1)}$	Ai$^{(4)}$	Ai$^{(7)}$	Ai$^{(10)}$
−1	0.5356(+0)	0.5602(+0)	0.5150(+0)	−0.1241(+1)	0.1227(+2)
−2	0.2274(+0)	0.2151(+0)	0.2260(+0)	0.2277(+0)	0.2448(+0)
−3	−0.3788(+0)	−0.3836(+0)	−0.3787(+0)	−0.3788(+0)	−0.3788(+0)
−4	−0.7027(−1)	−0.6531(−1)	−0.7025(−1)	−0.7027(−1)	−0.7027(−1)
−5	0.3508(+0)	0.3497(+0)	0.3508(+0)	0.3508(+0)	0.3507(+0)
−6	−0.3292(+0)	−0.3303(+0)	−0.3291(+0)	−0.3292(+0)	−0.3292(+0)
−7	0.1843(+0)	0.1860(+0)	0.1843(+0)	0.1843(+0)	0.1843(+0)
−8	−0.5271(−1)	−0.5423(−1)	−0.5271(−1)	−0.5271(−1)	−0.5271(−1)
−9	−0.2213(−1)	−0.2088(−1)	−0.2214(−1)	−0.2213(−1)	−0.2213(−1)
−10	0.4024(−1)	0.3921(−1)	0.4024(−1)	0.4024(−1)	0.4024(−1)

Table 5.4 Value of Bi(η) for $\eta < 0$

η	Bi$^{(1.2)}$	Bi$^{(1)}$	Bi$^{(4)}$	Bi$^{(7)}$	Bi$^{(10)}$
−1	0.1040(+0)	0.6683(−1)	0.4777(−1)	0.3392(+0)	0.2630(+3)
−2	−0.4123(+0)	−0.4229(+0)	−0.4117(+0)	−0.4099(+0)	−0.4114(+0)
−3	−0.1983(+0)	−0.1914(+0)	−0.1982(+0)	−0.1983(+0)	−0.1983(+0)
−4	0.3922(+0)	0.3936(+0)	0.3922(+0)	0.3922(+0)	0.3922(+0)
−5	−0.1384(+0)	−0.1417(+0)	−0.1384(+0)	−0.1384(+0)	−0.1384(+0)
−6	−0.1467(+0)	−0.1444(+0)	−0.1467(+0)	−0.1467(+0)	−0.1467(+0)
−7	0.2938(+0)	0.2928(+0)	0.2938(+0)	0.2938(+0)	0.2938(+0)
−8	−0.3313(+0)	−0.3311(+0)	−0.3313(+0)	−0.3313(+0)	−0.3313(+0)
−9	0.3250(+0)	0.3251(+0)	0.3250(+0)	0.3250(+0)	0.3250(+0)
−10	−0.3147(+0)	−0.3148(+0)	−0.3147(+0)	−0.3147(+0)	−0.3147(+0)

Table 5.5 Values of η_n, η_{na}, η'_n, η'_{na}

n	η_n	η_{na}	η'_n	η'_{na}
1	−2.3381	−2.3203	−2.6664	−2.6524
2	−4.0880	−4.0818	−4.3425	−4.3370
3	−5.5206	−5.5172	−5.7410	−5.7379
4	−6.7867	−6.7845	−6.9861	−6.9840
5	−7.9441	−7.9425	−8.1288	−8.1272
6	−9.0227	−9.0214	−9.1961	−9.1949

$$\left(a^{(0)}\xi'\right)' + a^{(0)\prime}\xi' + \mu a^{(1)\prime\prime} = 0,$$

$$\left(a^{(1)}\xi\xi'\right)' + a^{(1)\prime}\xi\xi' + \mu a^{(0)\prime\prime} = 0,$$

where $\xi(x)$ is defined by (5.2.3). This system admits the asymptotic solution

$$a^{(0)}(x, \mu) \simeq \sum_{n=0}^{\infty} a_{2n}^{(0)}(x)\mu^{2n}, \quad a^{(1)}(x, \mu) \simeq \sum_{n=0}^{\infty} a_{2n+1}^{(1)}(x)\mu^{2n+1}, \qquad (5.5.2)$$

where $a_0^{(0)}(x) = 1/\sqrt{\xi'}$,

$$a_{2n}^{(0)}(x) = -\frac{1}{2\sqrt{\xi'}} \int_{x_*}^{x} \frac{a_{2n-1}^{(1)''}}{\sqrt{\xi'}} \, dx, \quad n = 1, 2, \dots,$$

$$a_{2n+1}^{(1)}(x) = -\frac{1}{2\sqrt{\xi\xi'}} \int_{x_*}^{x} \frac{a_{2n}^{(0)''}}{\sqrt{\xi\xi'}} \, dx, \quad n = 0,1,\dots, \qquad (5.5.3)$$

and all functions $a_k^{(j)}(x)$ are real analytic.

5.2.2. The exact solution is

$$y^{(n)}(x) = v_0 \left(-\Omega_n^{2/3} x \right), \quad \Omega_n = \frac{|\eta_n'|^{3/2}}{l^{3/2}},$$

where the function $v_0(\eta)$ is the same as in (5.2.10).

5.2.3. We have $\varphi^0 = 0.75\sqrt{2} - 0.125 \log \left(3 + \sqrt{8} \right)$. The exact values Ω_n^{exact} and the asymptotic approximations $\Omega_n^{2.11}$ and $\Omega_n^{2.12}$ found by formulas (5.2.11) and (5.2.12) are compared in Table 5.6.

5.2.4. Since $\varphi^0 = \pi/8$, from formula (5.2.14) we get $\Omega_n = 8(n - 1/6)$. The exact, Ω_n^{exact}, and approximate, $\Omega_n^{\text{approx}} = 8(n - 1/6)$, values of Ω_n are compared in Table 5.7. As expected, the accuracy of the asymptotic formula increases with n.

5.2.5. In this problem, $\varphi^0 = \int_0^1 \sqrt{x + x^2} \, dx = 0.75\sqrt{2} - 0.125 \log \left(3 + \sqrt{8} \right)$, $x_* = 0$. Values of Ω_n are compared in Table 5.8: (1) Ω_n^{exact} for problem (5.2.15), (2) $\Omega_n'^{\text{exact}}$ for the problem when the boundary condition at the left end is replaced with $y' = 0$, (3) asymptotic values $\Omega_n^{2.18}$, and (4) $\Omega_n^{2.19}$ obtained by formulas (5.2.18) and (5.2.19) and independent of the boundary condition at the left end. It should be noted that Ω_n^{exact} and $\Omega_n'^{\text{exact}}$ get closer as n increases.

The eigenfunction $y_3(x)$ is plotted in Fig. 5.7.

Table 5.6 Exact and asymptotic values

n	Ω_n^{exact}	$\Omega_n^{2.11}$	$\Omega_n^{2.12}$
1	3.4368	3.4270	3.4270
2	7.1679	7.1656	7.1656
3	10.9047	10.9042	10.9041
4	14.6427	14.6428	14.6428
5	18.3810	18.3814	18.3814
6	22.1195	22.1200	22.1199

Table 5.7 Exact and asymptotic values

n	Ω_n^{exact}	Ω_n^{approx}
1	6.7264	6.6667
2	14.6892	14.6667
3	22.6794	22.6667
4	30.6752	30.6667
5	38.6730	38.6667
6	46.6716	46.6667

Table 5.8 Exact and asymptotic values

n	Ω_n^{exact}	$\Omega_n'^{\text{exact}}$	$\Omega_n^{2.18}$	$\Omega_n^{2.19}$
1	2.8183	2.6891	2.8364	2.8039
2	6.5296	6.52310	6.5573	6.5425
3	10.2726	10.2723	10.2906	10.2811
4	14.0137	14.0137	14.0266	14.0197
5	17.7537	17.7537	17.7638	17.7583
6	21.4931	21.4931	21.5014	21.4968

Fig. 5.7 Eigenfunction $y_3(x)$,

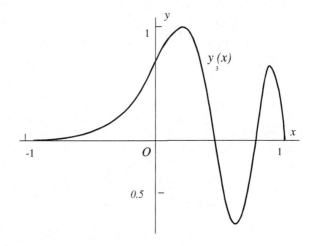

5.2.6. The interval of integration contains two turning points, $x_*^{(1)}$ and $x_*^{(2)}$, and two domains where the eigenfunctions oscillate, $x_1 \leq x \leq x_*^{(1)}$ and $x_*^{(2)} \leq x \leq x_2$. Each of these domains provides a specific set of eigenvalues $\Omega_n^{(1)}$ and $\Omega_n^{(2)}$. For $\Omega_n^{(2)}$, the eigenfunctions oscillate for $x_*^{(2)} \leq x \leq x_2$ and are exponentially small in the remaining part of the interval $[x_1, x_2]$ (see Fig. 5.8a). The eigenfunction corresponding to $\Omega_n^{(1)}$ oscillates for $x_1 \leq x \leq x_*^{(1)}$ and decreases exponentially in the remaining part of the interval (Fig. 5.8b). If, for some m and n, $\Omega_m^{(1)} \simeq \Omega_n^{(2)}$, then

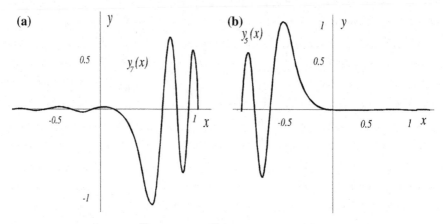

Fig. 5.8 Eigenfunctions of $\Omega_n^{(2)}$ *left* and of $\Omega_n^{(1)}$ *right*

the two eigenfunctions oscillate for both $x_1 \le x \le x_*^{(1)}$ and $x_*^{(2)} \le x \le x_2$ (see Exercise 5.2.7).

The asymptotic formulas for the evaluation of $\Omega_n^{(k)}$ are similar to formulas (5.2.18) and (5.2.19):

$$\Omega_n^{(k)} = \frac{2|\eta_n|^{3/2}}{3\varphi_k^0} + O\left(n^{-1}\right) = \frac{\pi(n - 1/4)}{\varphi_k^0} + O\left(n^{-1}\right), \quad n \to \infty,$$

$$\varphi_k^0 = \left|\int_{x_*^{(k)}}^{x_k} \sqrt{\rho(x)}\, dx\right|, \quad k = 1, 2. \quad (5.5.4)$$

5.2.7. The interval of integration contains the two turning points $x_*^{(1,2)} = \pm 0.5$. Because the function $\rho(x)$ is even and the problem is symmetric with respect to the point $x = 0$, the problem splits into two problems:

$$y^{(1)\prime\prime} + \Omega^{(1)\,2} y^{(1)} = 0, \quad y^{(1)}(-1) = 0, \qquad y^{(1)\prime}(0) = 0;$$
$$y^{(2)\prime\prime} + \Omega^{(2)\,2} y^{(2)} = 0, \quad y^{(2)}(-1) = 0, \qquad y^{(2)}(0) = 0. \quad (5.5.5)$$

For $x < 0$, the functions $y^{(1)}(x)$ should be continued as even and the functions $y^{(2)}(x)$ as odd.

From formula (5.5.4),

$$\Omega_n^{(1)} \simeq \Omega_n^{(2)} \simeq \frac{\pi(n - 1/4)}{\varphi^0}, \quad \varphi^0 = \frac{1}{4}\sqrt{3} - \frac{1}{8}\log\left(2 + \sqrt{3}\right) \quad (5.5.6)$$

as $n \to \infty$.

Table 5.9 Exact and approximate values

n	$\Omega_n^{(1)}$	$\Omega_n^{(2)}$	Ω_n^{approx}
1	8.7614	8.8779	8.7789
2	20.5058	20.5070	20.4841
3	32.2039	32.2039	32.1893
4	43.9053	43.9053	43.8945
5	55.6082	55.6082	55.5997
6	67.3120	67.3120	67.3049

Table 5.10 Exact and asymptotic values

N	Ω_N	n	$\Omega_n^{(1)}$	n	$\Omega_n^{(2)}$
1	8.8200			1	8.7789
2	12.6489	1	12.5573		
3	20.5064			2	20.4841
4	29.3440	2	29.3004		
5	32.2039			3	32.1893
6	43.9053			4	43.8945
7	46.0717	3	46.0435		
8	55.6082			5	55.5997
9	62.8074	4	62.7865		
10	67.3200			6	67.3049

A pair of close exact eigenvalues, and one odd and one even eigenfunctions correspond to each value of $\Omega_n^{(k)}$ obtained by formula (5.5.6). Such eigenvalues will be called *asymptotically double* (see also Sect. 5.4).

Table 5.9 lists the exact values of $\Omega_n^{(1)}$, $\Omega_n^{(2)}$ and the values obtained by formula (5.5.6). We see that, within the accepted accuracy, the values of $\Omega_n^{(1)}$ and $\Omega_n^{(2)}$ coincide for $n \geq 4$.

5.2.8. From formula (5.5.4) we get

$$\Omega_n^{(k)} = \frac{\pi(n - 1/4)}{\varphi_k^0}, \quad \varphi_k^0 = \int_a^{|x_k|} \sqrt{x^2 - a^2}\, dx, \quad k = 1, 2. \tag{5.5.7}$$

Table 5.10 lists the first ten exact eigenvalues Ω_N and their asymptotic approximations, $\Omega_n^{(1)}$ and $\Omega_n^{(2)}$, obtained from formula (5.5.7). It occurs that four actual eigenvalues are from the first series, $\Omega_n^{(1)}$, and six are from the second series, $\Omega_n^{(2)}$.

The eigenfunctions $y_7(x) = y_4^{(2)}(x)$ and $y_5(x) = y_3^{(1)}(x)$ are plotted in Fig. 5.8 (left) and (right), respectively. As expected, the function $y_5(x)$ differs significantly from zero and oscillates near the left edge of the interval. On the other hand, the function $y_7(x)$ oscillates near the right edge. Since Ω_7 is close to Ω_6, the function $y_7(x)$ slightly oscillates near the left edge.

Table 5.11 Exact and approximate values

n	Λ_n^{exact}	$\Lambda_n^{(1)}$	$\Lambda_n^{(2)}$	$\Lambda_n^{(3)}$	$\Lambda_n^{(3)}$
1	1.2257	1.2257	1.2240	1.2240	
2	1.3947	1.3947	1.3941	1.3941	
3	1.5330	1.5330	1.5327	1.5327	
4	1.6553	1.6553	1.6550	1.6550	
5	1.7671	1.7670	1.7668	1.7670	
6	1.8730	1.8711	1.8710	1.8729	
7	1.9819	1.9694	1.9693	1.9818	
8	2.1055			2.1054	2.1082
9	2.2487			2.2486	2.2498
10	2.4119			2.4118	2.4124

5.2.9. In this problem, $q(x) = 1 + x - \Lambda$. The turning point $x_* = \Lambda - 1$ belongs to the interval $[0, 1]$ for $\Lambda \in [1, 2]$.

The calculated results are listed in Table 5.11. The exact values, Λ_n^{exact}, are obtained from Eq. (5.2.25), where $\eta^0 = -(\Lambda - 1)\mu^{-2/3}$ and $\eta^1 = -(\Lambda - 2)\mu^{-2/3}$. In contrast to the general case, here Eq. (5.2.25) is exact since the function $q(x)$ depends linearly on x. The next two columns contain the approximate values of $\Lambda_n^{(1)}$ and $\Lambda_n^{(2)}$:

$$\Lambda_n^{(1)} = 1 + \mu^{2/3}|\eta_n|, \tag{5.5.8}$$

$$\Lambda_n^{(2)} = 1 + \left[\frac{2}{3}\mu\pi(n - 1/4)\right]^{2/3}, \tag{5.5.9}$$

obtained by formula (5.2.26) which holds for $\Lambda < 2$ only. The solutions $\Lambda_n^{(3)}$ of Eq. (5.2.27) and the solution $\Lambda_n^{(4)}$ of Eq. (5.2.23), which is valid for $\Lambda > 2$ only, are listed in the last two columns of the Table 5.11.

5.2.10. The exact solution is expressed through the zeros, α_{mn}, of the Bessel function $J_m(x)$:

$$\omega_{mn} = \frac{1}{R}\sqrt{\frac{T}{\rho}}\,\alpha_{mn}, \quad J_m(\alpha_{mn}) = 0, \quad n = 1, 2, \ldots.$$

Find the asymptotics for α_{mn} as $m \to \infty$. From formulas (5.2.33) and (5.2.34) it follows that $\alpha_{mn} > m$ for all n. Formulas (5.2.33) give

$$\alpha_{mn} = m\left(1 + \beta_{mn}^2\right)^{1/2} + O\left(m^{-1/3}\right), \tag{5.5.10}$$

where β_{mn} is a positive root of the equation

$$\beta - \arctan \beta = \gamma \tag{5.5.11}$$

for

$$\gamma = \gamma_{mn} = \frac{2}{3m} |\eta_n|^{3/2}. \tag{5.5.12}$$

For fixed n and $m \to \infty$, we have

$$\gamma_{mn} \ll 1, \quad \beta_{mn} \simeq (3\gamma_{mn})^{1/3}, \quad \alpha_{mn} \simeq m \left[1 + \left(\frac{2}{m} \right)^{2/3} |\eta_n| \right]^{1/2}, \tag{5.5.13}$$

where η_n it the nth root of the function $\mathrm{Ai}(\eta)$.

For $n \gg 1$, instead of (5.5.12) we may assume that

$$\gamma_{mn} \simeq \frac{\pi}{m}(n - 1/4). \tag{5.5.14}$$

For $\gamma_{mn} \gg 1$, Eq. (5.5.11) can also be solved approximately:

$$\beta_{mn} \simeq \gamma_{mn} + \pi/2 - (\gamma_{mn} + \pi/2)^{-1}. \tag{5.5.15}$$

5.2.11. The exact values $\alpha_{mn}^{\mathrm{exact}}$ and the approximate values $\alpha_{mn}^{(k)}$, $k = 1, 2, 3, 4$, obtained form (5.5.10) are given in Table 5.12 for $n \le 10$. To compute $\alpha_{mn}^{(1)}$, $\alpha_{mn}^{(2)}$, and $\alpha_{mn}^{(3)}$, the values β_{mn} in (5.5.10) are obtained from Eq. (5.5.11), in which $\gamma = \gamma_{mn}$ are found from (5.5.12)–(5.5.14) respectively. In the evaluation of $\alpha_{mn}^{(4)}$, formula (5.5.15) is used to get β_{mn}.

Table 5.12 Exact and approximate values

n	$\alpha_{mn}^{\mathrm{exact}}$	$\alpha_{mn}^{(1)}$	$\alpha_{mn}^{(2)}$	$\alpha_{mn}^{(3)}$	$\alpha_{mn}^{(4)}$
1	12.2251	12.2234	11.1079	12.1873	
2	16.0378	16.0362	12.9548	16.0219	
3	19.5545	19.5531		19.5443	
4	22.9452	22.9438		22.9376	
5	26.2668	26.2656		26.2607	26.4019
6	29.5457	29.5445		29.5405	29.6411
7	32.7958	32.7947		32.7913	32.8656
8	36.0256	36.0246		36.0217	36.0781
9	39.2405	39.2395		39.2369	39.2808
10	42.4439	42.4430		42.4407	42.4755

The breaks in Table 5.12 come from the inapplicability of the relevant approximate formula. Formula (5.5.13) provides acceptable accuracy only for $m \gg 8$.

5.3.1. Solutions (5.3.1) can be represented in the form $y^{(j)} \simeq V^{(j)}z_j$, where the scalar functions z_j satisfy Eqs. (5.3.9):

$$\mu \frac{dz_j}{dx} = \left[\lambda_j(x) + \mu b_1^{(j)}(x)\right] z_j, \quad j = 1, 2, \qquad (5.5.16)$$

and $b_1^{(j)}(x)$ are the same as b_0 in formulas (4.2.9) and (5.3.11). Here and further we write only those terms which affect the main terms in (5.3.12).

The functions $\lambda_j(x)$ and $b_1^{(j)}(x)$ have a singular point at $x = x_*$. In accordance with the splitting theorem there exists a linear transformation,

$$z_j = \left[\nu_{j1}^{(0)}(x) + \mu \nu_{j1}^{(1)}(x)\right]\theta_1 + \left[\nu_{j2}^{(0)}(x) + \mu \nu_{j2}^{(1)}(x)\right]\theta_2, \quad j = 1, 2, \quad (5.5.17)$$

which reduces Eq. (5.5.16) to a system of two equations without a turning point. The coefficients of this transform are $\nu_{11}^{(0)} = \nu_{21}^{(0)} = 1/2$, $\nu_{12}^{(0)} = -\nu_{22}^{(0)} = 1/(2\sqrt{q})$, $\nu_{11}^{(1)} = -b_1^{(2)}/(2\sqrt{q})$, $\nu_{12}^{(1)} = \nu_{21}^{(1)} = 0$, $\nu_{22}^{(1)} = b_1^{(1)}/(2\sqrt{q})$. Now, the obtained system of equations

$$\mu\theta_1' = p\theta_1 + \theta_2, \quad \mu\theta_2' = (q + 2\mu c_2)\theta_1 + \left(p + \mu\frac{4c_1 + q'}{2q}\right)\theta_2 \qquad (5.5.18)$$

has no singular point $x = x_*$.

The main terms of the asymptotic expansions for the solutions of system (5.5.18) are expressed by means of Airy's functions:

$$\begin{aligned}
\theta_1 &= d_0(x)\left[\frac{\cos\nu}{\sqrt{\xi'}}v(\eta) + \mu^{1/3}\frac{\sin\nu}{\sqrt{\xi\xi'}}v'(\eta)\right]\exp\left(\frac{1}{\mu}\int_{x_*}^x p(x)\,dx\right), \\
\theta_2 &= d_0(x)\left[\sqrt{\xi\xi'}\sin\nu\,v(\eta) + \mu^{1/3}\sqrt{\xi}\cos\nu\,v'(\eta)\right]\exp\left(\frac{1}{\mu}\int_{x_*}^x p(x)\,dx\right),
\end{aligned}$$
$$(5.5.19)$$

where the notation in formulas (5.3.11) and (5.3.12) is used. To get formulas (5.3.12) one must substitute (5.5.19) into (5.5.17) and then into the formulas $y^{(j)} = V^{(j)}z_j$.

5.3.2. For the three types of boundary conditions the results are compared in Table 5.13, where the first three columns are for clamped edges, the next three for freely supported edges and last three for free edges. The table includes the solutions $\beta_{mn}^{\text{exact}}$ of the exact equations (5.3.15$_1$), (5.3.15$_2$), and (5.3.15$_3$), the solutions $\beta_{mn}^{(1)}$ of the approximate equations (5.3.21$_1$), (5.3.21$_2$), and (5.3.21$_3$), and the solution $\beta_{mn}^{(2)}$ of the approximate equation (5.3.22). Note that for clamped edges, β_{mn} do not depend on Poisson's ratio ν. For freely supported edges, the roots $\beta_{mn}^{(1)}$ and $\beta_{mn}^{(2)}$ of the approximate equations (5.3.21$_2$) and (5.3.22) for $k = 2$ do not depend on ν.

Table 5.13 Results for clamped, freely supported and free edges, respectively

n	β_{mn}^{exact}	$\beta_{mn}^{(1)}$	$\beta_{mn}^{(2)}$	β_{mn}^{exact}	$\beta_{mn}^{(1)}$	$\beta_{mn}^{(2)}$	β_{mn}^{exact}	$\beta_{mn}^{(1)}$	$\beta_{mn}^{(2)}$
1	12.971	12.957	12.964	12.195	12.225	12.223	9.039	9.214	9.277
2	16.799	16.787	16.792	16.015	16.038	16.036	13.483	13.535	13.552
3	20.323	20.312	20.312	19.536	19.554	19.553	17.086	17.121	17.131
4	23.718	23.709	23.713	22.930	22.945	22.944	20.510	20.537	20.545
5	27.043	27.034	27.038	26.253	26.267	26.265	23.850	23.871	23.878
6	30.323	30.316	30.319	29.534	29.546	29.544	27.140	27.159	27.164

Therefore the roots of the exact equation (3.15) depend weakly on ν. For a free edge, one root of the equations under consideration is less than m. This root is not included in Table 5.13.

5.3.3. For the simple roots $\lambda(s)$ of the characteristics equation (4.2.38) we have

$$z(s, \mu) \simeq z_0(s) \exp\left(\frac{1}{\mu} \int \lambda(s)\, ds\right), \tag{5.5.20}$$

where z replaces any of the listed functions $(u, v, \ldots)$ and

$$u_0 = \frac{\mu}{\lambda}\left[k_1 w_0 - \left(\frac{r^2}{b^2} + \nu\lambda^2\right)\Phi_0\right], \quad T_{10} = -\frac{Ehr^2}{Rb^2}\Phi_0,$$

$$v_0 = \frac{\mu b}{r}\left[k_2 w_0 + \left(\lambda^2 + \frac{\nu r^2}{b^2}\right)\Phi_0\right], \quad S_0 = \frac{Ehr\lambda}{Rb^2}\Phi_0,$$

$$w_0 = \left(\lambda^2 - \frac{r^2}{b^2}\right)(bf_\lambda)^{-1/2},$$

$$\Phi_0 = -\left(k_2\lambda^2 - \frac{k_1 r^2}{b^2}\right)\left(\lambda^2 - \frac{r^2}{b^2}\right)^{-2} w_0,$$

$$Q_{10}^* = -\frac{Eh\mu}{R}\left(\lambda^3 - \frac{(2-\nu)\lambda r^2}{b^2}\right)w_0,$$

$$\gamma_{10} = \frac{\lambda}{\mu} w_0, \quad M_{10} = Eh\mu^2\left(\lambda^2 - \frac{\nu r^2}{b^2}\right)w_0, \tag{5.5.21}$$

and the notation in Sect. 4.2.5 is used.

For multiple roots, $s = s_*$, of Eq. (4.2.38) of type (5.3.28), the main terms of the asymptotic expansions for these functions are expressed either by means of Airy's functions $v(\eta)$ or their derivatives:

$$z(s, \mu) \simeq \frac{\mu^{1/6}\eta^{1/4}z_0}{\lambda^{1/2}} v(\eta), \quad z = \{v, w, T_1, M_1\}, \tag{5.5.22}$$

$$z(s, \mu) \simeq \frac{\mu^{1/6} z_0}{\lambda^{1/2} \eta^{1/4}} v'(\eta), \quad z = \{u, \ \gamma_1, \ S, \ Q_1^*\}, \tag{5.5.23}$$

where z replaces one of the functions inside braces. The expressions for z_0 in (5.5.22) and (5.5.23) given by formulas (5.5.21) are the same as in (5.5.20). The functions $z_0 \to \infty$ as $s \to s_*$. However, the right sides in (5.5.22) and (5.5.23) are regular at $s = s_*$.

The obtained integrals may be used for approximating the free vibrations frequencies.

5.3.4. We replace the function (4.2.38) with (4.6.5) and use the following formula for Q_1^*:

$$Q_{10}^* = -\frac{Eh\mu}{R} \left[\lambda^3 - \frac{(2-\nu)\lambda r^2}{b^2} + 2\Lambda t_1 \lambda \right] w_0, \tag{5.5.24}$$

The other formulas are the same as in Sect. 5.3.3 and Exercise 5.3.3.

5.3.5. The eigenvalues Λ related to the free vibrations frequencies ω obtained by means of formula (4.2.34),

$$\Lambda = \frac{\rho \omega^2 r^2}{E}, \tag{5.5.25}$$

are listed in Table 5.14.

The Table contains the lowest roots Λ_{m1} of Eq. (5.3.28) for the following boundary conditions at $s = s_2$:

$$
\begin{aligned}
S_2 &: \ T_1 = v = w = M_1 = 0, \\
C_2 &: \ u = v = w = M_1 = 0, \\
S_1 &: \ T_1 = v = w = \gamma_1 = 0, \\
C_1 &: \ u = v = w = \gamma_1 = 0.
\end{aligned}
\tag{5.5.26}
$$

For the boundary conditions S_2 we assume that $d = 0$ in (5.3.28), for conditions C_1 we find the value of d from Eq. (5.3.29), and for conditions C_2 and S_1 from the equations

Table 5.14 Eigenvalues Λ for listed boundary conditions

m	S_2	C_2	S_1	C_1
9	0.09844	0.09845	0.10001	0.10002
10	0.09495	0.09495	0.09675	0.09677
11	0.09311	0.09311	0.09516	0.09521
12	0.09277	0.09278	0.09511	0.09519
13	0.09387	0.09389	0.09654	0.09664
14	0.09641	0.09643	0.09945	0.09956
15	0.10041	0.10043	0.10387	0.10399

Fig. 5.9 Eigenfunction $w(s)$ for $m = 12$ and boundary conditions S_2

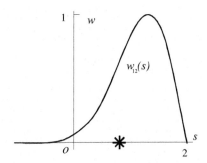

$$\begin{vmatrix} \lambda_1^{-1}u_0^{(1)} & u_0^{(3)} & u_0^{(4)} & u_0^{(5)} \\ dv_0^{(1)} & v_0^{(3)} & v_0^{(4)} & v_0^{(5)} \\ dw_0^{(1)} & w_0^{(3)} & w_0^{(4)} & w_0^{(5)} \\ dM_{10}^{(1)} & M_{10}^{(3)} & M_{10}^{(4)} & M_{10}^{(5)} \end{vmatrix} = 0, \qquad \begin{vmatrix} dT_{10}^{(1)} & T_{10}^{(3)} & T_{10}^{(4)} & T_{10}^{(5)} \\ dv_0^{(1)} & v_0^{(3)} & v_0^{(4)} & v_0^{(5)} \\ dw_0^{(1)} & w_0^{(3)} & w_0^{(4)} & w_0^{(5)} \\ \lambda_1^{-1}\gamma_{10}^{(1)} & \gamma_{10}^{(3)} & \gamma_{10}^{(4)} & \gamma_{10}^{(5)} \end{vmatrix} = 0,$$

respectively.

For the considered parameters and boundary conditions the lowest frequency is attained for $m = 12$. The vibrations frequencies depend rather slightly on the boundary conditions.

The eigenfunction $w(s)$ for $m = 12$ for the boundary conditions S_2 is plotted in Fig. 5.9. The turning point $s_* = 1.3914$ is marked with an asterisk.

5.3.6. The lowest eigenvalues Λ obtained for different m from Eq. (5.3.28) are given in Table 5.15. The boundary conditions (5.5.26) are introduced at the shell edge $s = s_2 = 2$.

Buckling occurs for $m = 20$. As in Exercise 5.3.5, the values of Λ depend slightly on the boundary conditions (5.5.26). Nevertheless, the higher critical load corresponds to the stiffer edge support.

5.4.1. Formula (5.4.8) takes the form

$$\Lambda = f(p, x) = \frac{p^2}{q_2(x)} + \gamma(x), \quad \gamma(x) = \frac{q_1(x)}{q_2(x)}. \qquad (5.5.27)$$

Table 5.15 Lowest eigenvalues Λ

m	S_2	C_2	S_1	C_1
17	0.36137	0.36160	0.37114	0.37172
18	0.35016	0.35037	0.36024	0.36077
19	0.34412	0.34431	0.35453	0.35501
20	0.34227	0.34248	0.35307	0.35350
21	0.34392	0.34407	0.35507	0.35547
22	0.34848	0.34863	0.36000	0.36035
23	0.35554	0.35568	0.36745	0.36777

Let the function $\gamma(x)$ attains its minimum at $x = x_0$ and $\gamma''(x_0) > 0$, in the interval $x_1 < x_0 < x_2$. Then, by (5.4.9)–(5.4.11) we obtain

$$p_0 = 0, \quad \Lambda^{(n)} = \Lambda_0 + \mu\Lambda_1^{(n)} + O(\mu^2), \quad n = 0, 1, \dots,$$

$$\Lambda_0 = \gamma(x_0), \quad \Lambda_1^{(n)} = \left(n + \frac{1}{2}\right)\sqrt{\frac{2\gamma''(x_0)}{q_2(x_0)}}. \tag{5.5.28}$$

The eigenfunction for $n = 0$ is shown in Fig. 5.3a.
5.4.2. From formulas (5.4.13) and (5.5.28) we have

$$\Lambda^{(n)} = \mu(2n + 1), \quad y^{(n)}(x) = H_n(z)\, e^{-z^2/2}, \quad n = 0, 1, \dots, \tag{5.5.29}$$

where $z = x/\sqrt{\mu}$ and $H_n(z)$ is the Hermite polynomial of degree n. In contrast to (5.4.13) the obtained solution is exact.
5.4.3. Formulas (5.4.2) and (5.5.28) provide the same result:

$$\Lambda^{(n)} = \mu(2n + 1), \quad n = 0, 1, \dots \tag{5.5.30}$$

The approximate values of $\Lambda^{(n)}_{\text{approx}}$ obtained from formula (5.5.30) and the exact values found by numerical integration are given in Table 5.16. As the turning points $s_*^{(1,2)} = \pm\sqrt{\Lambda}$ approach the ends of the segment $[-1, 1]$, the error of the approximate formula (5.5.30) increases.
5.4.4. Table 5.17 contains two approximate values, $\Lambda^{(n)}_{\text{approx}}$, which are independent of the boundary conditions and are obtained by formula (5.5.30) and $\Lambda^{(n)}_{\text{refined}}$ found by the refined formula (5.4.2) and also two exact values: $\Lambda^{(n)}_{\text{exact}}$ and $\Lambda'^{(n)}_{\text{exact}}$ for the boundary conditions at the right end $y(1) = 0$ and $y'(1) = 0$, respectively. The last two columns contain the coordinates of the turning points.

Formula (5.5.30) takes into account only the local properties of the function $q(x, \Lambda)$ near the point $x = 0$ and neglects the term $x^3/3$. Formula (5.4.2) takes into account the behavior of the function $q(x, \Lambda)$ between the turning points and provides a better approximation to the exact values than formula (5.5.30). As the turning points approach the ends of the segment $[-2, 1]$ the error of the approximate values $\Lambda^{(n)}_{\text{approx}}$ and $\Lambda^{(n)}_{\text{refined}}$ increases and the difference between the values $\Lambda^{(n)}_{\text{exact}}$ and $\Lambda'^{(n)}_{\text{exact}}$ also increases.

Table 5.16 Approximate and exact eigenvalues

n	0	1	2	3	4
$\Lambda^n_{\text{approx}}$	0.10000	0.30000	0.50000	0.70000	0.90000
Λ^n_{exact}	0.10003	0.30054	0.50413	0.71864	0.95714

Table 5.17 Approximate values and coordinates of turning points

n	$\Lambda_{\text{approx}}^{(n)}$	$\Lambda_{\text{refined}}^{(n)}$	$\Lambda_{\text{exact}}^{(n)}$	$\Lambda_{\text{exact}}^{\prime(n)}$	$s_*^{(1)}$	$s_*^{(2)}$
0	0.100	0.100	0.099	0.099	−0.335	0.301
1	0.300	0.295	0.295	0.295	−0.608	0.503
2	0.500	0.486	0.486	0.485	−0.817	0.633
3	0.700	0.671	0.672	0.668	−1.004	0.734
4	0.900	0.850	0.853	0.842	−1.185	0.817

Table 5.18 Computed $\Lambda^{(m)}$ for $(10 \leq m \leq 13)$

m	$\Lambda_{\text{refined}}^{(m)}$	$\Lambda_{\text{approx}}^{(m)}$
10	0.2851	0.2975
11	0.2841	0.2954
12	0.2842	0.2946
13	0.2853	0.2949

5.4.5. The computed values of $\Lambda^{(m)}$ for $(10 \leq m \leq 13)$ are given in Table 5.18, which contains the values of $\Lambda_{\text{refined}}^{(m)}$ and $\Lambda_{\text{approx}}^{(m)}$ obtained from the refined formula (5.4.2) and the approximate formula (5.4.21), respectively. The lowest value of $\Lambda^{(m)}$ we get for $m = 11$ [for $m = 12$ from formula (5.4.21)]. The approximate formulas (5.4.22) provide $m_{\min} = 12.4$, $\Lambda_{\min} = 0.2936$.

5.4.6. The buckling of an ellipsoidal shell of revolution is described by the system of Eq. (4.2.53), for which function (4.2.8) has the form

$$\Lambda = f(p, s, r) = \frac{(p^2 + r^2/b^2)^4 + (k_2 p^2 + k_1 r^2/b^2)^2}{(t_1 p^2 + t_2 r^2/b^2)(p^2 + r^2/b^2)^2}, \tag{5.5.31}$$

where, by (5.3.33) and (5.3.34),

$$t_1 = \frac{1}{2k_2}, \quad t_2 = \frac{1 - t_1 k_1}{k_2}, \quad \Lambda = \frac{qR}{Eh\mu^2}, \quad r = \mu m, \tag{5.5.32}$$

and the functions $k_1(s)$, $k_2(s)$ and $b(s)$ are defined by formulas (5.4.17). The minimum

$$\Lambda_0 = \min_{p,s,r} f(p, s, r) = f(p_0, s_0, r_0) = \frac{4}{2\delta^2 - 1} \tag{5.5.33}$$

is attained at $p = p_0 = 0$, $\theta = \theta_0 = \pi/2$, $r = r_0 = 1/\delta$. After calculating the derivatives (5.4.20), we use formula (5.4.11) for $n = 0$ to get

$$\Lambda = \frac{4}{2\delta^2 - 1} + \mu \frac{4(\delta^2 - 1)\sqrt{8\delta^2 - 2}}{(2\delta^2 - 1)^2} + O\left(\mu^2\right). \tag{5.5.34}$$

Table 5.19 Computed $\Lambda^{(m)}$ by formulas (5.4.2)

m	26	27	28	29	30	31	32
$\Lambda^{(m)}$	1.40523	1.38697	1.37691	1.37415	1.37793	1.38762	1.40268

In terms of physical (dimensional) variables, the critical values of the pressure q and the wave numbers m in the circumferential direction can be obtained by the formulas:

$$
q = \frac{2Eh^2}{\sqrt{3(1-\nu^2)}a_0^2(2\delta^2-1)}\left[1+\sqrt{\frac{h(4\delta^2-1)}{a_0\sqrt{3(1-\nu^2)}}\frac{(\delta^2-1)}{(2\delta^2-1)^2}}+O\left(\frac{h}{a_0}\right)\right],
$$

$$
m \simeq \sqrt{\frac{a_0\sqrt{12(1-\nu^2)}}{h\delta^2}}, \quad \delta = \frac{b_0}{a_0} > 0,
$$

$$
\tag{5.5.35}
$$

where a_0 and b_0 are the ellipse semi-axes, E, ν, and h are Young's modulus, Poisson's ratio and the shell thickness, respectively.

5.4.7. From formulas (5.5.34) we get $\Lambda = 1.37424$ and $m = 28.7$. The computed values of $\Lambda^{(m)}$ by formulas (5.4.2) for m close to $m = 29$ are listed in Table 5.19.

5.4.8. Introduce the loading parameter Λ by formula (5.5.32):

$$
\Lambda = \frac{qR}{Eh\mu^2}. \tag{5.5.36}
$$

Then, the non-dimensional stress-resultants t_1 and t_2 differ by their sign from those introduced by formulas (5.5.32),

$$
t_1 = -\frac{1}{2k_2}, \quad t_2 = -\frac{1+t_1k_1}{k_2}, \tag{5.5.37}
$$

where the functions $k_1(s)$ and $k_2(s)$ are defined by formulas (5.4.17). From relations (5.5.37), $t_1 < 0$ for all θ,

$$
t_2 = \frac{\sin^2\theta + \delta^2(\cos^2\theta - 2)}{2k_2\delta^2}, \quad \delta = \frac{b_0}{a_0}. \tag{5.5.38}
$$

Thus, $t_2 < 0$ for $2\delta^2 > 1$ and for all θ. By (4.2.53), the shell is only under tensile stresses and, for the semi-axes ratio $\sqrt{2}b_0 < a_0$, the ellipsoidal shell of revolution under internal pressure does not buckle.

Let $2\delta^2 < 1$. To find the minimum (5.5.33) we consider only such values of p, s, and r, for which $f(p, s, r) > 0$. Firstly, we find

$$
\gamma(\theta) = \min_{p,r} f(p, s, r) = \frac{2k_2}{t_1} = \frac{4(\sin^2\theta + \delta^2\cos^2\theta)^2}{\sin^2\theta + \delta^2(\cos^2\theta - 2)}, \quad p_0 = 0. \tag{5.5.39}
$$

Depending on δ, the function $\gamma(\theta)$ attains its minimum,

$$\Lambda_0 = \min_{\theta} \gamma(\theta) = \gamma(\theta_0), \qquad (5.5.40)$$

for different values of θ_0.

For $1/2 < \delta < 1/\sqrt{2}$, the weakest parallel is the equator $\theta = \pi/2$. For that, the loading parameter, Λ, and the wavenumber, m, in the circumferential direction are found by formulas similar to (5.5.34):

$$\Lambda = \frac{4}{1 - 2\delta^2} + \mu \frac{4(1 - \delta^2)\sqrt{8\delta^2 - 2}}{(2\delta^2 - 1)^2} + O\left(\mu^2\right), \quad m = \frac{1}{\mu\delta}. \qquad (5.5.41)$$

For $\delta < 1/2$, the function $\gamma(\theta)$ has a local maximum at the equator $\theta = \pi/2$. The global minimum, which is equal to $\Lambda_0 = 32\delta^2$, is attained at

$$\theta_0^{(1)} = \arcsin\sqrt{\frac{3\delta^2}{1 - \delta^2}}, \quad \theta_0^{(2)} = \pi - \theta_0^{(1)}. \qquad (5.5.42)$$

5.4.9. Applying formulas (5.4.7)–(5.4.11) and (5.5.31) we find

$$\Lambda = 32\delta^2 \left[1 + \mu \sqrt{\frac{193(1 - 4\delta^2)}{16\delta^2}} + O\left(\mu^2\right) \right], \quad m = \frac{1}{\mu}\sqrt{\frac{6\delta}{1 - \delta^2}}.$$

5.4.10. In formulas (5.4.30) and (5.4.31) for $\alpha = 1$ we have $t''(0) = -0.5$. We separately seek the even and odd buckling modes $w(\varphi)$ by numerical integration. The boundary conditions,

$$w' = w''' = \Phi' = \Phi''' = 0 \quad \text{for} \quad \varphi = 0, \ \varphi = \pi, \qquad (5.5.43)$$

and

$$w = w'' = \Phi = \Phi'' = 0 \quad \text{for} \quad \varphi = 0, \ \varphi = \pi, \qquad (5.5.44)$$

correspond to the even and odd buckling modes, respectively.

The first and second eigenvalues, $\Lambda^{(1)}$ and $\Lambda^{(2)}$, are listed in Table 5.20 for different values of r. The even and odd eigenfunctions $w(\varphi)$ correspond to $\Lambda^{(1)}$ and $\Lambda^{(2)}$, respectively. The table includes the exact values $\Lambda_{\text{exact}}^{(1)}$ and $\Lambda_{\text{exact}}^{(2)}$ obtained by numerical integration, and the approximate values $\Lambda_{\text{approx}}^{(1)}$ and $\Lambda_{\text{approx}}^{(2)}$ obtained by the asymptotic formulas (5.4.30) and (5.4.31) and the rounded values $\Lambda_{\text{rounded}}^{(1)}$ and $\Lambda_{\text{rounded}}^{(2)}$ calculated by formulas (5.5.50).

For $r < 1$, the asymptotic values are asymptotically double. This is why the values of $\Lambda_{\text{exact}}^{(1)}$ and $\Lambda_{\text{exact}}^{(2)}$ converge as the parameter $r < 1$ decreases and the same asymptotic value, $\Lambda_{\text{approx}}^{(1)} = \Lambda_{\text{approx}}^{(2)}$, for $r < 1$ corresponds to both of them.

Table 5.20 Exact, approximate and rounded values of Λ

r	$\Lambda^{(1)}_{exact}$	$\Lambda^{(1)}_{approx}$	$\Lambda^{(1)}_{rounded}$	$\Lambda^{(2)}_{exact}$	$\Lambda^{(2)}_{approx}$	$\Lambda^{(2)}_{rounded}$
0.50	1.0792	1.0778		1.0792	1.0778	
0.55	1.0713	1.0704		1.0713	1.0704	
0.60	1.0639	1.0635		1.0640	1.0635	
0.65	1.0571	1.0571		1.0571	1.0571	
0.70	1.0505	1.0509		1.0505	1.0509	
0.75	1.0438	1.0449		1.0440	1.0449	
0.80	1.0366	1.0389	1.0321	1.0376	1.0389	1.0330
0.85	1.0275	1.0327	1.0249	1.0321	1.0327	1.0291
0.90	1.0169	1.0259	1.0159	1.0292	1.0259	1.0275
0.95	1.0101	1.0178	1.0100	1.0310	1.0178	1.0304
1.00	1.0108	1.0000	1.0111	1.0388	1.0000	1.0398
1.05	1.0194	1.0212	1.0208	1.0531	1.0540	1.0568
1.10	1.0356	1.0403	1.0396	1.0740	1.0845	1.0822
1.15	1.0589	1.0653	1.0679	1.1011	1.1172	1.1164

For $r > 1$, the asymptotic approximations $\Lambda^{(1)}_{approx}$ and $\Lambda^{(2)}_{approx}$ for the even and odd modes $w(\varphi)$ are obtained form formula (5.4.30) for $n = 0$ and $n = 1$, respectively.

In a neighborhood of the point $r = 1$, formulas (5.4.30) and (5.4.31) are unreliable (see Exercise 5.4.11, where the method for calculating the values $\Lambda^{(1)}_{rounded}$ and $\Lambda^{(2)}_{rounded}$ is given).

5.4.11. To study the neighborhood of the point $r = 1$, we assume that

$$r = 1 + \varepsilon r_1, \quad \Lambda = 1 + \varepsilon^2 \Lambda_1, \quad \varphi = \varphi_0 + \varepsilon\eta, \quad \varepsilon = \mu^{2/3} \qquad (5.5.45)$$

and seek a solution of system (5.4.23) in the form

$$w \simeq \sum_{k=0}^{\infty} \varepsilon^k w_k(\eta), \quad \Phi \simeq \sum_{k=0}^{\infty} \varepsilon^k \Phi_k(\eta). \qquad (5.5.46)$$

Substituting formulas (5.5.46) in system (5.4.23) in the zeroth approximation we get the fourth-order equation

$$4\left(\frac{d^2}{d\eta^2} - r_1\right)^2 w_0 + (a\eta^2 - 2\Lambda_1)w_0 = 0, \quad a = -t_1''(0) > 0, \qquad (5.5.47)$$

and the boundary conditions $w_0 \to 0$ as $\eta \to \pm\infty$.

Equation (5.5.47) cannot be integrated in terms of known functions. To reduce the number of parameters in the numerical integration, we substitute

$$x = \frac{\eta}{b}, \quad b = \left(\frac{4}{a}\right)^{1/6}, \quad k = r_1 b^2, \quad \lambda = 2\Lambda_1 b^4 \qquad (5.5.48)$$

Fig. 5.10 Curves $\lambda_0(k)$ and $\lambda_1(k)$

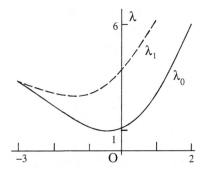

and transform equation (5.5.47) to

$$\left(\frac{d^2}{dx^2} - k\right)^2 w_0 + (x^2 - \lambda)w_0 = 0, \quad w(\pm\infty) = 0. \tag{5.5.49}$$

For each k there exists a countable set of values $\lambda_i(k)$ for the parameter λ for which the eigenvalue problem (5.5.49) has a nontrivial solution. The curves $\lambda_0(k)$ and $\lambda_1(k)$ are plotted in Fig. 5.10.

From formulas (5.5.45) and (5.5.48) we have

$$\Lambda^{(i)} = 1 + \frac{2\varepsilon^2}{b^4} \lambda_i(k) + O\left(\varepsilon^3\right), \quad k = \frac{(r-1)b^2}{\varepsilon}. \tag{5.5.50}$$

The function $\lambda_0(k)$ attains its minimum, $\lambda_0 = 0.905$, for $k = -0.44$. Therefore, as r changes, the parameter λ is minimum for $r = 1 - 0.44\varepsilon b^2$ and it is equal to

$$\Lambda_{\min} = 1 + 0.181\varepsilon^2 b^{-4} + O\left(\mu^2\right).$$

Table 5.20 contains the values $\Lambda_{\text{rounded}}^{(1)}(r)$ and $\Lambda_{\text{rounded}}^{(2)}(r)$ obtained by formula (5.5.50) for $r \simeq 1$ for the parameters from Exercise 5.4.10.

5.4.12. The function (5.4.8) has the form

$$f(p, y) = p^4 + \frac{k^2(y)}{p^4}. \tag{5.5.51}$$

After minimizing $f(p, y)$, we obtain

$$\Lambda_0 = 2k_0, \quad k_0 = k(y_0) = \min_y k(y), \quad p_0 = k_0^{1/4}, \tag{5.5.52}$$

where y_0 is the weakest generatrix.

Table 5.21 Results for the first four eigenvales

μ	$\Lambda_{\text{asympt.}}$	Λ_{oo}	Λ_{oe}	Λ_{eo}	Λ_{ee}
0.1	1.205976	1.210790	1.212372	1.212372	1.210790
0.2	1.411953	1.450047	1.387287	1.387290	1.450068

If $k''(y_0) = k_0'' > 0$, then, by formula (5.4.11), we have

$$\Lambda^{(n)} = 2k_0 + \mu \left(\frac{1}{2} + n \right) \sqrt{\frac{32k_0''}{p_0^2}} + O\left(\mu^2 \right), \quad n = 0, 1, \ldots, \qquad (5.5.53)$$

where the eigenvalues $\Lambda^{(n)}$ are asymptotically double.

For an elliptic cylindrical shell, we have

$$k_0 = \frac{1}{\delta^2}, \quad k_0'' = \frac{3\left(\delta^2 - 1\right)}{\delta^6}, \quad \theta_0^{(1)} = \frac{\pi}{2}, \quad \theta_0^{(2)} = \frac{3\pi}{2}.$$

Since the shell has two weakest generatrices, the eigenvalues

$$\Lambda^{(n)} = \frac{2}{\delta^2} \left[1 + \mu \left(\frac{1}{2} + n \right) \sqrt{\frac{24(\delta^2 - 1)}{\delta}} + O\left(\mu^2 \right) \right], \quad n = 0, 1, \ldots,$$
$$(5.5.54)$$

are asymptotically quadruple.

For the numerical evaluation of the parameters Λ, which are close to each other, we integrate system (5.4.23) making use of the evenness (5.5.43) or oddness (5.5.44) conditions for $\theta = 0$ or $\theta = \pi/2$.

The numerical results for $\delta = \sqrt{2}$ and $\mu = 0.1$, $\mu = 0.2$ are given in Table 5.21 for the first four eigenvalues.

The table contains the approximate asymptotic values $\Lambda_{\text{asympt.}}$, obtained from the asymptotic formula (5.5.54) for $n = 0$, which are equal for all four vibrations modes, and also four close to each other exact eigenvalues Λ_{oo}, Λ_{oe}, Λ_{eo}, and Λ_{ee} obtained by numerical integration. These eigenvalues correspond to four types of vibrations modes: odd in θ and $\theta - \pi/2$, odd in θ and even in $\theta - \pi/2$, even in θ and odd in $\theta - \pi/2$, even in θ and $\theta - \pi/2$, respectively.

The values of Λ_{oo} and Λ_{ee}, and also Λ_{oe} and Λ_{eo}, for $\mu = 0.1$, within the accepted accuracy, coincide with each other and for $\mu = 0.2$ are close to each other.

Chapter 6
Asymptotic Integration of Nonlinear Differential Equations

There are several types of asymptotic expansions for the solutions of nonlinear differential equations. Regularly perturbed nonlinear equations were considered in Chap. 3.

Some types of singular perturbed nonlinear equations are considered in monographs [9, 38, 57, 58, 60, 61].

Generally for solution of singular perturbed equations the methods of matched asymptotic expansions and multiscale methods are used. However, for example, in [15] nonlinear boundary value problems, for which these methods are inapplicable, are analyzed. In [52] one of the chapters deals with nonlinear boundary value problems. The author consider some of the second order equations those have exact solutions to illustrate various phenomena occur as a small parameter at the higher derivative converges to zero.

In this chapter we also consider a limited number of problems connected with singular perturbation and ramification of solutions of nonlinear equations.

6.1 Cauchy Problems for Ordinary Differential Equations with a Small Parameter

6.1.1 Problem Statement

Consider the Cauchy problem for the following system of $m+n$ nonlinear differential equations:

$$\frac{dy}{dt} = f(y, z, t), \quad y^T = (y_1, \ldots, y_m),$$

$$\mu \frac{dz}{dt} = F(y, z, t), \quad z^T = (z_1, \ldots, z_n),$$

$$y(0) = y^0, \quad z(0) = z^0, \quad 0 \le t \le T. \tag{6.1.1}$$

© Springer International Publishing Switzerland 2015
S.M. Bauer et al., *Asymptotic Methods in Mechanics of Solids*,
International Series of Numerical Mathematics 167,
DOI 10.1007/978-3-319-18311-4_6

Here y and f are m-vectors, and z and F are n-vectors. As a rule one cannot obtain an exact solution $(y(t, \mu), z(t, \mu))$ and the problem is to find an asymptotic solution taking the smallness of parameter μ into account.

As $\mu \to 0$, system (6.1.1) degenerates into a system of m differential and n algebraic equations:

$$\frac{dy_0}{dt} = f(y_0, z_0, t), \quad F(y_0, z_0, t) = 0. \tag{6.1.2}$$

The order of this system is lower than the order of the given system since the second equation is not differential but algebraic. One should not introduce for this system an initial condition for the function z, but only an initial condition for y,

$$y_0(0) = y^0.$$

Suppose that the second system of equations in (6.1.2),

$$F(y_0, z_0, t) = 0, \tag{6.1.3}$$

has an isolated solution

$$z_0 = z_0(y_0, t), \tag{6.1.4}$$

i.e. there exists $\eta > 0$ such that $F(y_0, z, t) \neq 0$ for $0 < |z - z_0(y_0, t)| < \eta$. Substituting this solution into system (6.1.2) we obtain the degenerate (unperturbed) problem

$$\frac{dy_0}{dt} = f(y_0, z_0(y_0, t), t), \quad y_0(0) = y^0. \tag{6.1.5}$$

Two questions arise as problem (6.1.1) degenerates into (6.1.5): firstly, how the solutions for (6.1.1) and (6.1.5) relate as $\mu \to 0$, and, secondly, what is the analytic structure of the solution of problem (6.1.1) for $\mu > 0$. The Tikhonov theorem answers the first question [60]:

Let

(1) the right sides $F(y, z, t)$ and $f(y, z, t)$ be real analytic functions of their arguments;

(2) the function (6.1.4) be an isolated solution of system (6.1.3);

(3) the solution (6.1.4) be an asymptotically stable equilibrium point for the adjoint system,

$$\frac{d\tilde{z}}{d\tau} = F(y_0, \tilde{z}, t), \tag{6.1.6}$$

where y_0 and t are considered as parameters, and all the roots of the complementary characteristic equation

$$\det(A - \lambda E) = 0, \quad A = \left.\frac{\partial F}{\partial \tilde{z}}\right|_{\tilde{z}=z_0(y_0,t)} \tag{6.1.7}$$

have negative real parts;

(4) the initial point z^0 belongs to the domain of attraction of the stability condition (6.1.4).

If conditions (1)–(4) are satisfied, then, for sufficiently small μ, problem (6.1.1) has a unique solution $y(t, \mu)$, $z(t, \mu)$ for which the following limit equalities are valid:

$$y(t, \mu) \to y_0(t) \quad \text{as} \quad \mu \to 0, \quad 0 \le t \le T,$$
$$z(t, \mu) \to z_0(t) = z_0(y_0(t), t) \quad \text{as} \quad \mu \to 0, \quad \varepsilon \le t \le T. \tag{6.1.8}$$

It is natural that the function $z_0(t)$ does not satisfy the boundary condition (6.1.1) in the general case.

We discuss conditions (3) and (4) of the Tikhonov theorem.

In Eq. (6.1.6) y and t are considered as parameters and by condition (2) $\tilde{z} = z_0(y_0, t)$ is a solution of equation (6.1.6). Since this solution does not depend on τ, then $dz_0(y_0, t)/d\tau = 0$, i.e. solution (6.1.4) is an equilibrium point for system (6.1.6).

Condition (4) means that the solution $\tilde{z}(\tau)$ for the problem

$$\frac{d\tilde{z}}{d\tau} = F(y_0, \tilde{z}, 0), \quad \tilde{z}(0) = z^0$$

converges to the equilibrium point $z_0(y_0, 0)$ as $\tau \to \infty$.

We note that system (6.1.3) may have a non-unique solution and conditions (3) and (4) help us select the required root (6.1.4). If one wishes to obtain only relations (6.1.8), then conditions (1)–(4) may be relaxed.

6.1.2 Construction of a Formal Asymptotic Solution

The solution of problem (6.1.1) in the form of an asymptotic series of powers of μ is obtained in [60] and is shortly discussed below.

Let x denote any of the symbols y and/or z. We seek a solution $x(t, \mu)$ of problem (6.1.1) in the form

$$x(t, \mu) \simeq \sum_{k=0}^{\infty} x_k(t)\mu^k + \sum_{k=0}^{\infty} X_k(\tau)\mu^k, \tag{6.1.9}$$

where $\tau = t/\mu$ is the fast variable, and we apply the restriction

$$X_k(\tau) \to 0 \quad \text{as} \quad \tau \to \infty, \tag{6.1.10}$$

to the functions $X_k(\tau)$, i.e. $X_k(\tau)$ are the boundary layers functions.

The construction of successive terms in series (6.1.9) is notably similar to the Vishik–Lyusternik algorithm (see Chap. 4). The zeroth approximation for $y_0(t)$ is evaluated from (6.1.5), and $Y_0(\tau) \equiv 0$. The function $z_0(t)$ is determined from formula (6.1.4). To obtain $Z_0(\tau)$, one should solve the problem

$$\frac{dZ_0}{d\tau} = F\left(y^0, z_0(0) + Z_0(\tau), 0\right) - F\left(y^0, z_0(0), 0\right),$$
$$Z_0(0) = z^0 - z_0(0), \tag{6.1.11}$$

which, by conditions (3) and (4), has a solution that satisfies the relation (6.1.10).

The next terms of series (6.1.9) are obtained from linear equations in the following order: $Y_1, y_1, z_1, Z_1, Y_2, \ldots$. The functions Y_1 are found by quadratures:

$$Y_1(\tau) = \int_\infty^\tau \left[f\left(y^0, z_0(0) + Z_0(\tau), 0\right) - f\left(y^0, z_0(0), 0\right) \right] d\tau. \tag{6.1.12}$$

To obtain y_1, and z_1, we solve the linear Cauchy problem:

$$\frac{dy_1}{dt} = \left(\frac{\partial f}{\partial y}\right)_0 y_1 + \left(\frac{\partial f}{\partial z}\right)_0 z_1, \quad y_1(0) = -Y_1(0), \tag{6.1.13}$$

$$\frac{dz_0}{dt} = \left(\frac{\partial F}{\partial y}\right)_0 y_1 + \left(\frac{\partial F}{\partial z}\right)_0 z_1, \tag{6.1.14}$$

where the subscript "0" means that the corresponding derivatives are evaluate at the solution z_0 defined in (6.1.4). By condition (4), the matrix $A = (\partial F/\partial z)_0$ has a nontrivial determinant and from (6.1.14) we find the vectors z_1, which we substitute into (6.1.13).

By successive approximations, the right sides of Eq. (6.1.1) are represented in the form

$$F(x, t) = F(\tilde{x}, t) + F^*(X, \tau), \tag{6.1.15}$$

where $\tilde{x}$ and X denote the first and the second sums in the right side of (6.1.9), respectively. Relation (6.1.15), which we used to derive (6.1.11) and (6.1.12), defines the function F^*:

$$F^*(X, \tau) = F\left(\tilde{x}(\mu\tau) + X(\tau), \mu\tau\right) - F\left(\tilde{x}(\mu\tau), \mu\tau\right). \tag{6.1.16}$$

The equation for $Z_1(\tau)$ appears to be rather cumbersome:

$$\frac{dZ_1}{d\tau} = \left(\frac{\partial F}{\partial y}\right)_{00} \left[Y_1(\tau) + y_1(0) + \tau y_0'(0)\right]$$

$$+ \left(\frac{\partial F}{\partial z}\right)_{00} \left[Z_1(\tau) + z_1(0) + \tau z_0'(0)\right] + \tau \left(\frac{\partial F}{\partial \tau}\right)_{00}$$

$$- \left(\frac{\partial F}{\partial y}\right)_0 \left[y_1(0) + \tau y_0'(0)\right]$$

$$- \left(\frac{\partial F}{\partial z}\right)_0 \left[z_1(0) + \tau z_0'(0)\right] - \tau \left(\frac{\partial F}{\partial \tau}\right)_0, \qquad (6.1.17)$$

$$Z_1(0) = -z_1(0).$$

Here the subscript "00" means that the corresponding derivatives are evaluated at $y = y^0$, $z = z_0(0) + Z_0(\tau)$, and $t = 0$.

The monograph [60] contains a proof that formal series (6.1.9) are asymptotic expansions of exact solutions as $\mu \to 0$ $(\mu > 0)$.

Problem (6.1.1) permits a generalization to the case where the right sides of system (6.1.1) depend regularly on μ.

Example 1

Consider the following example:

$$\frac{dy}{dt} = z, \quad \mu \frac{dz}{dt} = y^2 - z^2, \quad y(0) = 1, \quad z(0) = 0.$$

The roots of the equation $F(y_0, z_0, t) = 0$ are $z = y$ and $z = -y$. Since $\partial F/\partial z = -2z$, then the root $z = y$ is stable for $y > 0$, and the root $z = -y$ is stable for $y < 0$. The initial point $y = 1$, $z = 0$ belongs to the domain of the stable root $z = y$. Thus, the unperturbed problem has the form

$$\frac{dy_0}{dt} = y_0, \quad y_0(0) = 1.$$

The solution of the equation is $y_0(t) = e^t$, and, therefore, $z_0(t) = e^t$. To find $Z_0(\tau)$ we use Eq. (6.1.11):

$$\frac{dZ_0}{d\tau} = -2Z_0(\tau) - Z_0^2(\tau), \quad Z_0(0) = -1,$$

which has solution $Z_0(\tau) = \tanh \tau - 1$. Hence the functions

$$y(t) = e^t, \quad z(t) = e^t + \tanh \tau - 1$$

approximate the solution with an accuracy of order $O(\mu)$ in any finite interval $0 \le t \le T_0 < \infty$.

The function $Y_1(\tau)$ can be found from relation (6.1.12)

$$Y_1(\tau) = \ln 2 + \ln (\cosh \tau) - \tau.$$

So, to obtain y_1 and z_1 in series (6.1.9) we have

$$\frac{dy_1}{dt} = z_1, \quad \frac{dz_0}{dt} = 2y_0y_1 - 2z_0z_1,$$

or

$$1 = 2y_1 - 2z_1, \quad y_1(0) = -Y_1(0) = -\ln 2.$$

This implies that

$$y_1(t) = \left(-\frac{1}{2} - \ln 2\right) e^t + \frac{1}{2}, \quad z_1(t) = \left(-\frac{1}{2} - 2\ln 2\right) e^t.$$

To find $Z_1(\tau)$ from (6.1.17) we write the equation

$$\frac{dZ_1}{d\tau} = 2\ln (\cosh \tau) - 2\tanh \tau \left(Z_1(\tau) - \frac{1}{2} - \ln 2 + \tau\right) - 1,$$

$$Z_1(0) = \frac{1}{2} + \ln 2,$$

with solution

$$Z_1(\tau) = \frac{1}{2} + \ln 2 - \tau + \frac{2}{\cosh^2 \tau} \int_0^\tau \cosh^2 x \ln (\cosh x) \, dx.$$

Therefore, combining these results we get

$$y(t, \mu) = e^t - t + \mu \left[\frac{1}{2} - \left(\frac{1}{2} + \ln 2\right) e^t + \ln 2 + \ln (\cosh \tau)\right] + O\left(\mu^2\right),$$

$$z(t, \mu) = e^t + \tanh \tau - 1 - t + \mu \left(\frac{1}{2} + \ln 2\right) (1 - e^t)$$

$$+ \mu \frac{2}{\cosh^2 \tau} \int_0^{t/\mu} \cosh^2 x \ln (\cosh x) \, dx + O\left(\mu^2\right).$$

The numerical (solid lines) and the asymptotic (dashed lines) values of the functions $y(t)$ (bold lines) and $z(t)$ (thin lines) are plotted for $\mu = 0.6$ in Fig. 6.1a and for $\mu = 0.2$ in Fig. 6.1b.

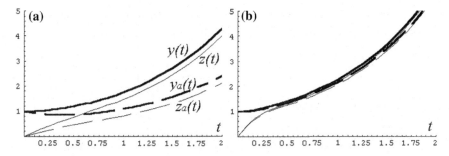

Fig. 6.1 Numerical values (*solid lines*) and asymptotic values (*dashed lines*) of $y(t)$ (*bold lines*) and $z(t)$ (*thin lines*) for $\mu = 0.6$ (*left*) and for $\mu = 0.2$ (*right*)

Example 2

We consider one more example, the reduced Van der Pol equation for a pendulum with a small mass:

$$\mu \frac{d^2x}{dt^2} + \left(1 - x^2\right)\frac{dx}{dt} + x = 0, \quad x(0) = a, \quad \frac{dx(0)}{dt} = 0, \quad a \neq 1. \quad (6.1.18)$$

As before, we seek a solution of equation (6.1.18) in the form of series (6.1.9):

$$x(t, \mu) = x_0(t) + \mu\left[X_1(\tau) + x_1(t)\right] + O\left(\mu^2\right).$$

For $\mu = 0$ we get a separable equation

$$\left(1 - x_0^2\right)\frac{dx_0}{dt} + x_0 = 0, \quad x_0(0) = a. \quad (6.1.19)$$

To find $x_0(t)$ one gets

$$t = \frac{x_0^2 - a^2}{2} - \ln\frac{|x_0|}{|a|}. \quad (6.1.20)$$

Write the equation for the boundary layer function $X_1(\tau)$ as

$$\frac{d^2X_1}{d\tau^2} + \left(1 - a^2\right)\frac{dX_1}{d\tau} = 0,$$

with solution

$$X_1(\tau) = C + C_1 e^{-(1-a^2)\tau}.$$

Since the function $X_1(\tau)$ should satisfy condition (6.1.10), then $C = 0$. The constant C_1 is evaluated from the second initial condition (6.1.18):

$$\frac{dX_1(0)}{d\tau} = -\frac{dx_0(0)}{dt} = -\frac{a}{1-a^2}, \quad C_1 = \frac{a}{(1-a^2)^2}.$$

To obtain x_1, we solve the equation

$$\frac{dx_1}{dt}\left(1 - x_0^2\right) + x_1\left(1 - 2x_0 \frac{dx_0}{dt}\right) + \frac{d^2x_0}{dt^2} = 0, \quad x_1(0) = -X_1(0).$$

By relations (6.1.19) and (6.1.20) this equation reduces to

$$\frac{dx_1}{dt} + x_1 \frac{1+x_0^2}{(1-x_0^2)^2} + \frac{x_0(1+x_0^2)}{(1-x_0^2)^4} = 0, \quad x_1(0) = -\frac{a}{(1-a^2)^2}. \tag{6.1.21}$$

We note that for this example it is hard to construct the analytic solution, i.e. to find an explicit function $x_0(t)$ from relation (6.1.20), but separating the boundary layer functions makes it easier to solve the problem numerically. The "slow" part of the solution can be found numerically from relations (6.1.20) and (6.1.21) while "the boundary effect" is obtained analytically.

6.1.3 Exercises

Find the exact and asymptotic solutions for the equations
6.1.1.

$$\mu \frac{d^2x}{dt^2} + \left(\frac{dx}{dt}\right)^2 - 1 = 0, \quad x(0) = x^0, \quad \frac{dx(0)}{dt} = 0. \tag{6.1.22}$$

6.1.2.

$$\mu \frac{d^2x}{dt^2} + x\frac{dx}{dt} = 0, \quad x(0) = x^0, \quad \frac{dx(0)}{dt} = \dot{x}^0, \quad x^0 > 0. \tag{6.1.23}$$

Find the main terms of the asymptotic solution of the equations
6.1.3.

$$\mu \frac{d^2x}{dt^2} + \frac{dx}{dt} + cx + kx^2 = 0, \quad x(0) = x^0, \quad \frac{dx(0)}{dt} = \dot{x}^0, \tag{6.1.24}$$

$$x^0 > 0, \quad c > 0, \quad k > 0.$$

6.1.4.

$$\mu \frac{d^2x}{dt^2} + \frac{dx}{dt} + cx + kx^3 = 0, \quad x(0) = x^0, \quad \frac{dx(0)}{dt} = \dot{x}^0,$$

$$x^0 > 0, \quad c > 0, \quad k > 0.$$

6.1.5.

$$\mu \frac{d^2x}{dt^2} + \frac{dx}{dt} x - Ax = 0, \quad x(0) = x^0, \quad \frac{dx(0)}{dt} = \dot{x}^0, \quad x^0 > 0. \qquad (6.1.25)$$

6.1.6.

$$\mu \frac{d^2x}{dt^2} + x \frac{dx}{dt} - A = 0, \quad x(0) = x^0, \quad \frac{dx(0)}{dt} = \dot{x}^0, \quad x^0 > 0. \qquad (6.1.26)$$

6.2 Perturbation of Nonlinear Boundary Value Problems with a Small Parameter

6.2.1 Introduction

We consider boundary value problems the solutions of which are representable as sums of slowly varying functions and functions of boundary layer type. Besides boundary layers in neighborhoods of the ends of the intervals of integration, *internal boundary layers* may also exist.

Consider the boundary value problem:

$$\frac{dy}{dt} = f(y, z, t), \quad \mu \frac{dz}{dt} = F(y, z, t), \quad t \in [0, 1], \qquad (6.2.1)$$

where y and z are m- and n-vectors, respectively. Let $m + n$ boundary conditions of the general type be introduced in the form

$$l_i (y(0), y(1), z(0), z(1)) = 0, \quad i = 1, \ldots, m + n. \qquad (6.2.2)$$

Again, the unperturbed system of equations has the form (6.1.2). However, the problem of introducing proper boundary conditions is nontrivial.

Let $z_0(y_0, t)$ be an isolated solution of equation (6.1.3). In a neighborhood of this solution we consider the adjoint system (6.1.6), where y and t are considered as parameters.

When solving a boundary value problem it is convenient to define stability in a more general way than stability used in the third condition of the Tikhonov theorem:

if all roots of the auxiliary characteristic equation (6.1.7) have negative real parts, then the root $z(y, t)$ is *stable to the right*. If all roots of Eq. (6.1.7) have positive real parts then the root $z(y, t)$ is said to be *stable to the left*.

Let Eq. (6.1.7) have n_1 roots with negative real parts and n_2 with positive real parts ($n_1 + n_2 = n$) for all t and y_0 from some domain containing a solution of equation (6.1.3). In this case, we seek a solution of the problem in a form similar to (6.1.9), but with two boundary layers:

$$x(t, \mu) = \sum_{k=0}^{\infty} x_k(t)\mu^k + \sum_{k=0}^{\infty} X_k^{(0)}(\tau_0)\mu^k + \sum_{k=0}^{\infty} X_k^{(1)}(\tau_1)\mu^k, \qquad (6.2.3)$$

$$X_k^{(0)}(\tau_0) \to 0 \quad \text{as} \quad \tau_0 \to \infty,$$
$$X_k^{(1)}(\tau_1) \to 0 \quad \text{as} \quad \tau_1 \to -\infty, \qquad (6.2.4)$$

where $\tau_0 = t/\mu$ and $\tau_1 = (t-1)/\mu$.

Here we discuss only the question of boundary conditions for the unperturbed problem (6.1.2). In the zeroth approximation, conditions, (6.2.2) may be written as

$$l_i \left[y_0(0), y_0(1), z_0(0) + Z_0^{(0)}(0), z_0(1) + Z_0^{(1)}(1) \right] = 0. \qquad (6.2.5)$$

By (6.2.4), the family of functions Z_0^0 and $Z_0^{(1)}$ contain n_1 and n_2 constants, respectively. Eliminating these functions from conditions (6.2.5) leads to m relations which do not contain boundary layer functions and are used as boundary conditions for the unperturbed problem (6.1.2).

Example 1

Consider boundary value problem:

$$\frac{dy}{dt} = z, \quad y(1) = 0,$$
$$\mu \frac{dz}{dt} = -(z - y - 1)(z - y)(z - y + 1), \quad z(0) = z^0, \qquad (6.2.6)$$

where y and z are scalar functions. In this case, equation $F(y, z, t) = 0$ has three distinct solutions: (1) $z = y + 1$, (2) $z = y$, and (3) $z = y - 1$, each of which may solve the boundary value problem (6.2.6). These solutions correspond to slowly varying functions of the zeroth approximation and the adjoint equation (6.1.6):

(1) $y_0 = e^{t-1} - 1$, $\quad z_0 = e^{t-1}$, $\quad \dfrac{dZ}{d\tau} = -Z(Z+1)(Z+2)$,

(2) $y_0 = 0$, $\quad z_0 = 0$, $\quad \dfrac{dZ}{d\tau} = -(Z-1)Z(Z+1)$,

(3) $y_0 = 1 - e^{t-1}$, $\quad z_0 = -e^{t-1}$, $\quad \dfrac{dZ}{d\tau} = -(Z-2)(Z-1)Z$. $\quad (6.2.7)$

(1) The root $z = y + 1$ of the equation $F = 0$ is stable to the right. The domain of attraction of the root $Z = 0$ of system (6.2.7) is $-1 < Z(0) < \infty$ or, by virtue of (6.1.11), problem (6.2.6) has a solution which converges to y_0, z_0 as $\mu \to 0$ and $t \geq \varepsilon > 0$ for z^0 satisfying the inequality

$$-1 + e^{-1} < z^0 < \infty. \tag{6.2.8}$$

(2) The root $z = y$ is stable to the left. Therefore, the boundary value problem (6.2.6) does not have solutions of the form (6.2.3) converging to $y_0 = z_0 \equiv 0$ as $\mu \to 0$, if $z^0 \neq 0$.

(3) The root $z = y - 1$ is stable to the right and its attraction domain is

$$-\infty < z^0 < 1 - e^{-1}. \tag{6.2.9}$$

Summing up the considered cases, we see that if condition $z^0 > 1 - e^{-1}$ is satisfied, then problem (6.2.6) has a unique solution converging to the functions y_0, z_0 (see (6.2.7) case 1) as $\mu \to 0$, as plotted in Fig. 6.2a.

For $-1 + e^{-1} < z^0 < 1 - e^{-1}$, problem (6.2.6) has two solutions, one corresponding to case (1) and the other to case (3) (see Fig. 6.2b).

For $z^0 < -1 + e^{-1}$, we again get a unique solution corresponding to case (3) (see Fig. 6.2c).

Example 2

Sometimes, in constructing a solution of a boundary value problem one should use not one, but several roots of the unperturbed equation $F(y, z, t) = 0$ and in the transition from a left-stable root to a right-stable root, an internal boundary layer may appear. Consider the model example

$$\frac{dy}{dt} = z, \quad \mu \frac{dz}{dt} = 1 - z^2, \quad y(0) = a, \quad y(1) = b, \quad |a - b| < 1, \tag{6.2.10}$$

which is discussed in detail in [60].

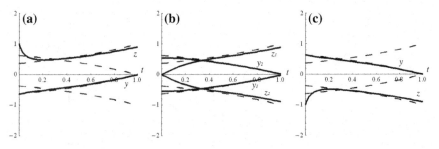

Fig. 6.2 Solution of problem (6.2.6)

It is easy, to find an exact solution for the problem

$$y = \mu \ln \left(\frac{e^{(t+b+1)/\mu} - e^{(t+a)/\mu} - e^{(b+1-t)/\mu} + e^{(2+a-t)/\mu}}{e^{2/\mu} - 1} \right),$$

$$z = \frac{e^{(2t+b-a+1)/\mu} - e^{2t/\mu} - e^{2/\mu} + e^{(b-a+t)/\mu}}{e^{(2t+b-a+1)/\mu} - e^{2t/\mu} + e^{2/\mu} - e^{(b-a+t)/\mu}}. \tag{6.2.11}$$

We seek an asymptotic solution. Equation $F = 0$ has two solutions: (1) $z = -1$ and (2) $z = 1$. The first solution is stable to the left and the second solution is stable to the right. Since both boundary conditions in (6.2.10) are imposed on the function y, both solutions,

$$\begin{aligned}
&(1) \quad z = -1, \quad y = -t + C, \quad C = a, \quad t < (a - b + 1)/2, \\
&(2) \quad z = 1, \quad y = t + C, \quad C = b - 1, \quad t > (a - b + 1)/2,
\end{aligned} \tag{6.2.12}$$

are used in the construction of a solution for problem (6.2.10). The lines $y = -t + a$ and $y = t + b - 1$ intersect at the internal point $t = (a - b + 1)/2$ and, in its neighborhood, an internal boundary layer appears as a smoothing jump of functions (6.2.12) at $t = (a - b + 1)/2$ (compare with the exact solution (6.2.11) and see Fig. 6.3 for $a = b = 0$).

Remark We note that the internal boundary layer may have a more complex form. The quasilinear problem

$$\mu y'' = g_1(y, t)y' + g_2(y, t), \quad y(t_0) = a, \quad y(t_1) = b. \tag{6.2.13}$$

was studied in [60]. It is shown that this problem can have a solution close to a discontinuous solution if the points (t_0, a) and (t_1, b) in the (t, y) plane are separated by a curve, on which the function $g_1(y, t)$ alternates.

A solution close to a discontinuous solution can be constructed as a solution with the following initial condition at the point $t_* \in (t_0, t_1)$:

Fig. 6.3 Boundary layers as smoothing jumps

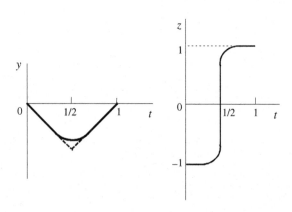

$$y(t_*, \mu) = y_0 + \mu y_1 + \cdots ,$$

$$z(t_*, \mu) = \frac{z_{-1}}{\mu} + z_0 + \mu z_1 + \cdots ,$$

Initially, the coefficients z_i, y_i and the value of t_* are unknown.

The solution of the problem has the form:

for $0 \le t \le t_*$,

$$z(t, \mu) = z_0^{(0)}(t) + \mu z_1^{(0)}(t) + \cdots + \frac{1}{\mu} Z_{-1}^{(0)}(\tau) + Z_0^{(0)}(\tau) + \cdots ,$$

$$y(t, \mu) = y_0^{(0)}(t) + \mu y_1^{(0)}(t) + \cdots + Y_0^{(0)}(\tau) + \mu Y_1^{(0)}(\tau) + \cdots ;$$

and for $t_* \le t \le t_1$,

$$z(t, \mu) = z_0^{(1)}(t) + \mu z_1^{(1)}(t) + \cdots + \frac{1}{\mu} Z_{-1}^{(1)}(\tau) + Z_0^{(1)}(\tau) + \cdots ,$$

$$y(t, \mu) = y_0^{(1)}(t) + \mu y_1^{(1)}(t) + \cdots + Y_0^{(1)}(\tau) + \mu Y_1^{(1)}(\tau) + \cdots .$$

In the zeroth approximation,

$$y_0^{(0)}(t_0) = a, \quad y_0^{(1)}(t_1) = b.$$

These conditions provide the solutions $\left(z_0^{(0)}(t), y_0^{(0)}(t) \right)$ and $\left(y_0^{(1)}(t), z_0^{(1)}(t) \right)$. For each of these solutions, the following condition must by satisfied:

$$z_{-1} + \int_{y_0}^{y_{0i}(t_*)} g_1(y, t_*) \, dy = 0, \quad i = 1, 2.$$

Subtracting one equation ($i = 1$) from the other ($i = 2$) we come to an equation for evaluating the point of a jump discontinuity ($t = t_*$) for the solution of problem (6.2.13):

$$\int_{y_{01}(t_*)}^{y_{02}(t_*)} g_1(y, t_*) \, dy = 0, \tag{6.2.14}$$

where $y_{01}(t)$ and and $y_{02}(t)$ are solutions of the unperturbed equation stable to the left and stable to the right, respectively.

Example 3

Consider the problem:

$$\mu y'' = -yy' + \alpha y, \quad y(0) = a, \quad y(1) = b, \quad a < 0, \quad \alpha, b > 0.$$

The unperturbed equation $y(y' - \alpha) = 0$ has solution $y' = \alpha$. Therefore the solution satisfying the left boundary condition has the form $y_{01} = a + \alpha t$ and the solution satisfying the right boundary condition has the form $y_{02} = b + \alpha(t - 1)$. For that, the following conditions must be satisfied

$$\begin{aligned}
(1) \quad & g_1(y_{01}, t) = -y_{01} = -a - \alpha t > 0, && 0 \leq t \leq t_* \\
(2) \quad & g_1(y_{02}, t) = -y_{02} = -b - \alpha(t - 1) < 0, && t_* \leq t \leq 1.
\end{aligned}$$

As seen from (6.2.14), the dependence of the point of jump at a and b has the form

$$t_* = \frac{1}{2} - \frac{b + a}{2\alpha}.$$

It is assumed that $|a + b| < \alpha < |b - a|$.

6.2.2 Exercises

Find the main terms in the asymptotic solution of the boundary value problem

6.2.1.

$$\mu \frac{d^2 y}{dt^2} + \frac{dy}{dt} + y + ky^2 = 0, \quad \frac{dy(0)}{dt} = \dot{y}^0, \quad y(1) = y^1. \tag{6.2.15}$$

6.2.2.

$$\mu \frac{d^2 y}{dt^2} + \frac{dy}{dt} + y + ky^3 = 0, \quad \frac{dy(0)}{dt} = \dot{y}^0, \quad y(1) = y^1.$$

6.2.3.

$$\mu \frac{d^2 x}{dt^2} + \left(\frac{dx}{dt}\right)^2 - \frac{dx}{dt} - 2 = 0, \quad x(0) = 0, \quad x(1) = 0. \tag{6.2.16}$$

6.2.4.

$$\mu \frac{d^2 x}{dt^2} + \frac{dx}{dt} x = 0, \quad x(0) = x^0, \quad x(1) = x^1, \quad x^0, x^1 > 0. \tag{6.2.17}$$

6.3 Bifurcation of Solutions of Nonlinear Equations

The basis of the bifurcation theory for solutions of nonlinear equations was laid in works by H. Poincaré, A.M. Lyapunov and E. Schmidt. In those papers it was shown that the problem on the number and behavior of solutions of integral or differential equations can be reduced to studying systems of implicit analytical functions called *ramification equation*.

The idea of the Lyapunov–Schmidt method is the decomposition of the equation under consideration into two equations: one in a finite dimensional subspace of dimension n and the other in its infinite dimensional orthogonal complement. This method, its different versions and developments are considered in [57].

Below, we briefly discuss one of the ways to solve the bifurcation problem on the example of the bending of a clamped uniform beam caused by an axial compressive load.

6.3.1 Statement of the Problem

The angle of rotation, $\theta(x)$, of the tangent to a beam satisfies the following equilibrium equation and boundary conditions:

$$\theta_{xx} + \Lambda \sin \theta = 0, \quad 0 \leq x \leq 1, \quad \theta_x(0) = \theta_x(1) = 0. \tag{6.3.1}$$

The value of Λ in Eq. (6.3.1) is proportional to the axial load. The function $\theta_0(x) \equiv 0$ is the solution of equation (6.3.1) for all values of Λ. This requires

(1) to find the values of the parameter Λ (for example, λ_0), for which the number of solutions of the equation changes;
(2) to find the number of solutions in a neighborhood of $\Lambda = \lambda_0$;
(3) to study the behavior of these solutions in a neighborhood of $\Lambda = \lambda_0$.

We linearize the problem $\theta_0(x) = 0$:

$$\theta_{xx} + \Lambda\theta = 0, \quad \theta_x(0) = \theta_x(1) = 0. \tag{6.3.2}$$

The eigenvalues $\Lambda_i = (\pi i)^2$ and corresponding eigenfunctions $\theta_i(x) = \cos(\pi i x)$ of this problem provide a family of functions orthogonal over the interval $[0, 1]$:

$$\int_0^1 \theta_i(x)\theta_j(x)\, dx = 0, \quad \text{for } i \neq j, \quad \text{and} \quad \int_0^1 \theta_i{}^2(x)\, dx = \frac{1}{2}.$$

If the function $\theta(x)$ is continuously differentiable over $[0, 1]$ and satisfies the boundary conditions $\theta_x(0) = \theta_x(1) = 0$, then it can be expanded in a uniformly convergent series of eigenfunctions of the linear problem (6.3.2) over $[0, 1]$:

$$\theta(x) = \sum_{n=1}^{\infty} c_n\theta_n(x), \quad c_n = 2\int_0^1 \theta(t)\theta_n(t)\, dt.$$

6.3.2 Solution of Nonlinear Problems

Consider the solution of the problem in a neighborhood of the first eigenvalue $\Lambda_1 = \pi^2$. Let $\Lambda = \pi^2 + \lambda$. We represent Eq. (6.3.1) in the form

$$B\theta = \theta_{xx} + \pi^2 \theta = -\theta\lambda - (\pi^2 + \lambda) \sum_{i=2}^{\infty} \theta^{2i-1} \frac{(-1)^{i+1}}{(2i-1)!}. \tag{6.3.3}$$

It is known that the solution of the non-homogeneous equation $B\theta = h(x)$ exists if the necessary and sufficient condition

$$\int_0^1 h(x)\theta_0(x)\,dx = 0, \tag{6.3.4}$$

holds. Here θ_0 is the solution of the homogeneous equation $B\theta = 0$.

We seek a solution θ of Eq. (6.3.3) in the form of a series in powers of the small parameter ξ:

$$\theta = \xi \cos(\pi x) + \sum_{i=1}^{\infty} \xi^{2i+1} \theta_{2i+1}, \quad \xi \ll 1. \tag{6.3.5}$$

This solution is expanded in odd powers of ξ since the operator B is an odd function. Substituting solution (6.3.5) in Eq. (6.3.3) we write only the main terms of the equation, taking the smallness of ξ and λ into account. Thus, we get a sequence of boundary value problems for the evaluation of θ_i:

$$\xi^3 B\theta_3 = \frac{\pi^2}{6} \xi^3 \cos^3(\pi x) - \lambda\xi \cos(\pi x), \quad (\theta_3)_x(0) = (\theta_3)_x(1) = 0. \tag{6.3.6}$$

The existence condition (6.3.4) for problem (6.3.6) provides the ramification equation

$$\frac{\pi^2}{8} \xi^3 - \lambda\xi = 0, \tag{6.3.7}$$

which has three solutions for $\lambda > 0$: one trivial ($\xi \equiv 0$) and two nontrivial solutions $\xi = \pm 2^{3/2} \pi^{-1} \sqrt{\lambda}$.

A particular solution of problem (6.3.6) is

$$\theta_3 = -\frac{\cos(3\pi x)}{24 \times 8}. \tag{6.3.8}$$

The general solution of problem (6.3.6) is a sum of the particular solution (6.3.8) and the solution of the homogeneous problem:

$$\theta_3 = -\frac{\cos(3\pi x)}{24 \times 8} + C \cos(\pi x),$$

Fig. 6.4 Value of ξ versus Λ in a neighborhood of the critical value

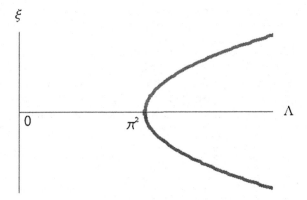

where C is a constant that can be found from the solvability condition for the following problem:

$$\xi^5 B\theta_5 = -\frac{\pi^2}{120}\xi^5\cos^5(\pi x) + \frac{\lambda\xi^3}{6}\cos^3(\pi x) - \lambda\xi^3\theta_3 + \frac{\pi^2\xi^5}{2}\cos^2(\pi x),$$

$$(\theta_5)_x(0) = (\theta_5)_x(1) = 0.$$

Hence, the angle of rotation of the tangent to the beam under loads close to the critical load is defined as

$$\theta \simeq \xi\cos(\pi x) - \frac{\xi^3}{24 \times 8}\cos(3\pi x) - \frac{5\xi^3}{16 \times 8}\cos(\pi x) + \cdots.$$

The graph for the value of ξ versus the parameter Λ in a neighborhood of the critical value $\Lambda = \pi^2$ is plotted in Fig. 6.4.

As it follows from the ramification equation (6.3.7) for small values of λ (in a neighborhood of the critical value of the axial load) there exist three solutions of equation (6.3.1), i.e. three equilibrium states.

We note that the bifurcation of solutions in neighborhoods of the other eigenvalues can be studied in a similar manner.

6.3.3 Exercises

To find the ramification equation and the solution of the boundary value problems in a neighborhood of the first eigenvalue.

6.3.1. $y_{xx} + \Lambda y = -y^3$, $y(0) = y(1) = 0$.

6.3.2. $y_{xx} + \Lambda y = y^2$, $y(0) = y(1) = 0$.

6.3.3. Find the ramification equation and the solution of Exercise 6.3.2 in a neighborhood of the second and third eigenvalues.

6.3.4. Find the ramification equation for the problem of axisymmetric buckling of a clamped circular plate under a uniform compressive load applied on the edge of the plate. The angle of rotation of the normal to the neutral plate surface, $\theta = -w'_r$, where w is the plate deflection and r is the plate radius, and the membrane radial force, $T_r - 1$, satisfy the nonlinear Karman equations:

$$\frac{d^2\theta}{dr^2} + \frac{1}{r}\frac{d\theta}{dr} - \frac{\theta}{r^2} + \Lambda(1 - T_r)\theta = 0, \quad 0 \le r \le 1,$$

$$\frac{1}{r^3}\frac{d}{dr}\left(r^3\frac{dT_r}{dr}\right) + \frac{\theta^2}{2} = 0, \quad \theta(1) = T_r(1) = 0,$$

and $\theta'(0) = T'_r(0) = 0$ from the symmetry conditions. The value of Λ is proportional to the applied load.

6.4 Answers and Solutions

6.1.1. The exact solution of equation (6.1.22) is $x = -t + \mu \ln((e^{2t/\mu} + 1)/2) + x^0$. To construct the asymptotic solution, we write Eq. (6.1.22) as a system:

$$\mu\frac{dz}{dt} = -z^2 + 1, \quad \frac{dx}{dt} = z, \quad x(0) = x^0, \quad z(0) = 0. \tag{6.4.1}$$

For $\mu = 0$, the roots of the equation $F(x_0, z_0, t) = -z_0^2 + 1 = 0$ are $z = 1$ and $z = -1$. Since $\partial F/\partial z = -2z$, then the root $z_0 = 1$ is stable. Hence, the unperturbed problem has solution $x_0 = t + x^0$, $z_0(t) = 1$. To find $Z_0(\tau)$ we use Eq. (6.1.11):

$$\frac{dZ_0}{d\tau} = -2Z_0(\tau) - Z_0{}^2(\tau), \quad Z_0(0) = -1,$$

which has solution $Z_0(\tau) = -2/\left(e^{2\tau} + 1\right)$.

From relation (6.1.12) one can obtain the function

$$X_1(\tau) = -2\tau + \ln\left(e^{2\tau} + 1\right).$$

Hence, the functions

$$x(t) \simeq t + x^0, \quad z(t) \simeq 1 - 2\left(e^{2t/\mu} + 1\right)^{-1}$$

approximate the solution of system (6.4.1) to an accuracy of order $O(\mu)$ in any finite interval $0 \leq t \leq T_0 < \infty$. In other words, the difference between the exact solution and the zeroth approximation has order μ. With the same accuracy, the solution of equation (6.1.22) can be represented in the form

$$x(t) = t + x^0 + \mu \left[-2t/\mu + \ln \left(e^{2t/\mu} + 1 \right) \right] + O(\mu)$$

$$= -t + \mu \ln \left(e^{2t/\mu} + 1 \right) + x^0 + O(\mu).$$

Therefore, to satisfy both the equation and the boundary conditions to an accuracy of order $O(\mu)$, the smooth part of the solution should be obtained to an accuracy of order $O(\mu)$ and the boundary layer to an accuracy of order $O\left(\mu^2\right)$: $x(t) = x_0(t) + \mu X_1(t) + O(\mu)$. For evaluating x_1 we have

$$\frac{dx_1}{dt} = z_1 = 0, \quad x_1(0) = -X_1(0) = -\ln 2,$$

whence $x_1(t) = -\ln 2$.

Therefore the solution fond to an accuracy of order $O(\mu^2)$ coincides with the exact solution.

6.1.2. The exact solution of equation (6.1.23) is

$$x = a \frac{(a + x^0)e^{at/\mu} + x^0 - a}{(a + x^0)e^{at/\mu} - x^0 + a}, \quad \text{where} \quad a = \sqrt{x^{02} + 2\mu \dot{x}^0}.$$

To obtain the asymptotic solution we represent Eq. (6.1.23) in the form of the system

$$\mu \, dx/dt = -zx, \quad dx/dt = z, \quad x(0) = x^0, \quad z(0) = \dot{x}^0.$$

For $\mu = 0$, Eq. (6.2.1), $F = 0$, has root $z_0 = 0$, which is stable for $x^0 > 0$ since $\partial F/\partial z = -x$. The unperturbed problem has solution $x_0 = x^0$, $z_0(t) = 0$. To find $Z_0(\tau)$ we find the solution of equation (6.1.11)

$$dZ_0/d\tau = -x^0 Z_0(\tau), \quad Z_0(0) = \dot{x}^0,$$

that is, $Z_0(\tau) = \dot{x}^0 e^{-x^0 \tau}$. From equality (6.1.12) one can find the function $X_1(\tau) = -\dot{x}^0 e^{-x^0 \tau}/x^0$.

The solution of equation (6.1.23) can be written as

$$x(t) = x^0 - \mu \left(\dot{x}^0/x^0 \right) e^{-x^0 t/\mu} + O(\mu).$$

The function x_1 is evaluated from the equation

$$dx_1/dt = z_1 = 0, \quad x_1(0) = -X_1(0) = \dot{x}^0/x^0.$$

Therefore, $x_1(t) = \dot{x}^0/x^0$ or $x(t) = x^0 + \mu(\dot{x}^0/x^0)\left(1 - e^{-x^0 t/\mu}\right) + O\left(\mu^2\right)$.

6.1.3. To construct the asymptotic solution we write Eq. (6.1.24) as a system:

$$\mu\frac{dz}{dt} = -z - y - ky^2, \quad \frac{dy}{dt} = z, \quad y(0) = x^0, \quad z(0) = \dot{x}^0.$$

For $\mu = 0$, we have $z_0 = -cy_0 - ky_0^2$. Therefore, the unperturbed problem

$$dy_0/dt = -cy_0 - ky_0^2, \qquad y(0) = x^0$$

has solution $y_0 = (cx^0)[(c + kx^0)e^{ct} - kx^0]^{-1}$. To obtain $Z_0(\tau)$ we write

$$dZ_0/d\tau = -Z_0(\tau), \qquad Z_0(0) = \dot{x}^0 + cx^0 + kx_0^2,$$

whence it follows that $Z_0(\tau) = \left(\dot{x}^0 + cx^0 + kx^{0^2}\right) e^{-\tau}$. From equality (6.1.12) one can find the function $Y_1(\tau) = -\left(\dot{x}^0 + cx^0 + kx^{0^2}\right) e^{-\tau}$. Thus, the solution of equation (6.1.24) has the form

$$x(t) = \frac{cx^0}{(c + kx^0)e^{ct} - kx^0} - \mu\left(\dot{x}^0 + cx^0 + kx^{0^2}\right) e^{-t/\mu} + O(\mu).$$

6.1.4.

$$x(t) = \sqrt{\frac{cx^{0^2}}{(c + kx^{0^2})e^{2ct} - kx^{0^2}}} - \mu\left(\dot{x}^0 + cx^0 + kx^{0^3}\right) e^{-t/\mu} + O(\mu).$$

6.1.5. To construct the asymptotic solution we write Eq. (6.1.25) as the system:

$$\mu\frac{dz}{dt} = (A - z)y, \quad \frac{dy}{dt} = z, \quad y(0) = x^0, \quad z(0) = \dot{x}^0.$$

For $\mu = 0$, we have $z_0 = A$, and the unperturbed system

$$\frac{dy_0}{dt} = A, \qquad y(0) = x^0$$

has solution $y_0 = At + x^0$. To obtain $Z_0(\tau)$ we write

$$dZ_0/d\tau = -x^0 Z_0(\tau), \quad Z_0(0) = \dot{x}^0 - A$$

or $Z_0(\tau) = \left(\dot{x}^0 - A\right) e^{-x^0\tau}$.

From equality (6.1.12) we obtain the function $Y_1(\tau) = -\dfrac{\dot{x}^0 - A}{x^0} e^{-x^0\tau}$. The solution of equation (6.1.25) is represented in the form

$$x(t) = At + x^0 - \mu \frac{\dot{x}^0 - A}{x^0} e^{-x^0 t/\mu} + O(\mu).$$

To find x_1 we solve the equations

$$\frac{dy_1}{dt} = z_1, \quad 0 = -\left(At + x^0\right) z_1, \quad x_1(0) = -X_1(0) = \frac{\dot{x}^0 - A}{x^0},$$

from where $x_1(t) = \left(\dot{x}^0 - A\right)/x^0$. Thus the solution of equation (6.1.25) is

$$x(t) = At + x^0 + \mu \frac{\dot{x}^0 - A}{x^0} \left(1 - e^{-x^0 t/\mu}\right) + O\left(\mu^2\right).$$

6.1.6. Equation (6.1.26) is equivalent to the system:

$$\mu \frac{dz}{dt} = A - zy, \quad \frac{dy}{dt} = z, \quad y(0) = x^0, \quad z(0) = \dot{x}^0.$$

For $\mu = 0$, we have $A - z_0 y_0 = 0$. The unperturbed problem

$$dy_0/dt = A/y, \quad y(0) = x^0,$$

has solution $y_0 = \sqrt{2At + x^{0^2}}$, $z_0 = A/\sqrt{2At + x^{0^2}}$. To find $Z_0(\tau)$, we solve the problem

$$dZ_0/d\tau = -x^0 Z_0(\tau), \quad Z_0(0) = \dot{x}^0 - A/x_0,$$

or $Z_0(\tau) = \left(\dot{x}^0 - A/x^0\right) e^{-x^0\tau}$. From equality (6.1.12) we get

$$Y_1(\tau) = \left(A - \dot{x}^0 x^0\right) e^{-x^0\tau} / \left(x^0\right)^2.$$

The solution of equation (6.1.26) can be represented in the form

$$x(t) = \sqrt{2At + (x^0)^2} + \mu \frac{A - \dot{x}^0 x^0}{(x^0)^2} e^{-x^0 t/\mu} + O(\mu).$$

6.2.1. To construct the asymptotic solution we write Eq. (6.2.15) as a system:

$$\mu \frac{dz}{dt} = -z - y - ky^2, \quad \frac{dy}{dt} = z, \quad y(1) = y^1, \quad z(0) = \dot{y}^0.$$

For $\mu = 0$, we find the root $z_0 = -y_0 - ky_0^2$. The unperturbed problem

$$dy_0/dt = -y_0 - ky_0^2, \quad y(1) = y^1,$$

has solution $y_0 = y^1/[(1 + ky^1)e^{t-1} - ky^1]$. To obtain $Z_0(\tau)$, where $\tau = t/\mu$, we get the equation

$$dZ_0/d\tau = -Z_0(\tau), \quad Z_0(0) = \dot{y}^0 + y_0(0) + ky_0^2(0),$$

or $Z_0(\tau) = \left[\dot{y}^0 + y^1 e(1 + ky^1)(1 + ky^1 - ky^1 e)^{-2}\right] e^{-\tau}$.

The root $z = -y - ky^2$ of the equation $F = 0$ is stable to the right and the attraction domain of the root $Z = 0$ of Eq. (6.2.17) is the entire real axis. From equality (6.1.12) one finds the function

$$Y_1(\tau) = -\left[\dot{y}^0 + y^1 e(1 + ky^1)(1 + ky^1 - ky^1 e)^{-2}\right] e^{-\tau}.$$

Thus, the solution of problem (6.2.15) to an accuracy of order $o(\mu)$ is

$$y = \frac{y^1}{(1 + ky^1)\, e^{t-1} - ky^1} - \mu \left[\dot{y}^0 + \frac{y^1 e(1 + ky^1)}{(1 + ky^1 - ky^1 e)^2}\right] e^{-t/\mu}.$$

6.2.2.

$$y(t) = \sqrt{\frac{(y^1)^2}{(1 + k(y^1)^2)\, e^{2t-2} - k(y^1)^2}}$$

$$- \mu \left[\dot{y}^0 + \frac{1 + ky^{1^2}}{1 + ky^{1^2} - k(ey^1)^2} \sqrt{\frac{(ey^1)^2}{1 + k(y^1)^2 - k(ey^1)^2}}\right].$$

6.2.3. To construct the asymptotic solution we write Eq. (6.2.16) as the system:

$$\mu \frac{dz}{dt} = -z^2 + z + 2, \quad \frac{dx}{dt} = z, \quad x(0) = 0, \quad x(1) = 0.$$

For $\mu = 0$, we have two roots, $z = -1$ and $z = 2$. The first root is stable to the left and the second root is stable to the right. Since both boundary conditions are introduced for the function x we have two solutions,

$z = -1,$	$x = -t + C,$	$x(0) = 0$	$C = 0,$	$x = -t,$
$z = 2,$	$x = 2t + C_1,$	$x(1) = 0,$	$C_1 = -2,$	$x = 2t - 2,$

Fig. 6.5 Internal boundary
layer at $t_0 = 2/3$

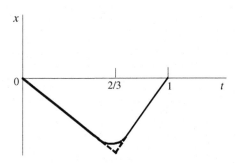

which are parts of the solution of the problem. The straight lines $x = -t$ and
$x = 2t - 2$ intersect at the interior point $t_0 = 2/3$ and in a neighborhood of that
point an internal boundary layer appears (Fig. 6.5).

6.2.4. To construct the asymptotic solution we write Eq. (6.2.17) as the system

$$\mu \frac{dz}{dt} = -zx, \quad \frac{dx}{dt} = z, \quad x(0) = x^0, \quad x(1) = x^1.$$

For $\mu = 0$, the root of the system is $z_0 = 0$. Since $\partial F/\partial z = -x$, then for $x^0, x^1 > 0$
that root is stable to the right. Thus, the unperturbed problem has the solution $x_0 = x^1$,
$z_0(t) = 0$, in a neighborhood of the edge $x = 0$ and the solution has the boundary
layer $x(t) = x^1 + X(\tau)$, $\tau = t/\mu$.
  To evaluate the function $X(\tau)$ we solve the equation

$$dX/d\tau = -x^1 X(\tau), \quad X(0) = \left(x^0 - x^1\right),$$

as $X(\tau) \to 0$ and $\tau \to \infty$, which has the solution $X(\tau) = \left(x^0 - x^1\right) e^{-x^1 \tau}$. The
solution of the initial value problem has the form

$$x(t) \simeq x^1 + \left(x^0 - x^1\right) e^{-x^1 t/\mu} + O(\mu).$$

  In Fig. 6.6, the exact and asymptotic solutions of equation (6.2.17) are plotted for
$x^0 = 3$, $x^1 = 1$, $\mu = 0.1$ (a), and for $x^0 = 1$, $x^1 = 5$, $\mu = 0.1$ (b).

6.3.1. The linearized problem $y_{xx} + \Lambda y = 0$, $y(0) = y(1) = 0$, has the eigenvalues
$\Lambda_i = (\pi i)^2$ and the corresponding eigenfunctions $y(x) = \sin(\pi i x)$, which form a
set of orthogonal functions over the interval $[0, 1]$.
  Consider the solution of the problem in a neighborhood of the first eigenvalue,
$\Lambda_1 = \pi^2$. Let $\Lambda = \pi^2 + \lambda$. We represent the given equation in the form $By =
y_{xx} + \pi^2 y = -\lambda y - y^3$. We seek the solution of the problem, y, in a series in odd
powers of the small parameter ξ: $y = \xi \sin(\pi x) + \sum_{i=1}^{\infty} \xi^{2i+1} y_{2i+1}$, for $\xi \ll 1$.
Substituting this solution into the given equation and taking the smallness of ξ and

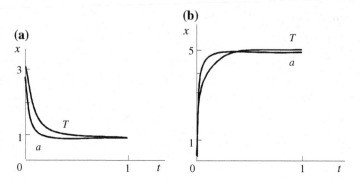

Fig. 6.6 Exact and asymptotic solutions: **a** for $x^0 = 3, x^1 = 1, \mu = 0.1$, and **b** for $x^0 = 1, x^1 = 5$, $\mu = 0.1$

λ into account, we get a sequence of boundary value problems for evaluating the functions y_i. In particular, for the evaluation of y_3 we come to the problem

$$\xi^3 By_3 = -\xi^3 \sin^3(\pi x) - \lambda\xi \sin(\pi x), \quad y_3(0) = y_3(1) = 0. \tag{6.4.2}$$

Condition (6.3.4) for the boundary problem $By = h(x)$ provides the ramification equation

$$-\frac{3}{4}\xi^3 - \lambda\xi = 0,$$

which, for $\lambda < 0$, has three solutions: the trivial solution ($\xi \equiv 0$) and two other solutions $\xi = \pm\sqrt{-\lambda/3}$. The particular solution of the homogeneous problem: $y_3 = -(1/32\pi^2) \sin(3\pi x) + C \sin(\pi x)$, where the constant C can be obtained from the solvability condition for the following problem:

$$\xi^5 By_5 = -3\xi^5 \sin^2(\pi x)y_3 - \lambda\xi^3 y_3, \quad y_5(0) = y_5(1) = 0.$$

Thus,

$$y \simeq \xi \sin(\pi x) - \frac{\xi^3}{32\pi^2} \sin(3\pi x) - \frac{\xi^3}{64\pi^2} \sin(\pi x) + \cdots .$$

In Fig. 6.7, the graph for the value of ξ versus the parameter Λ is plotted in a neighborhood of the critical value $\Lambda = \pi^2$.

6.3.2. The linearized problem has the same form as in Exercise 6.3.1. One seeks the solution of the problem

$$By = y_{xx} + \pi^2 y = -\lambda y + y^2 \tag{6.4.3}$$

Fig. 6.7 ξ versus Λ in a neighborhood of the critical value $\Lambda = \pi^2$

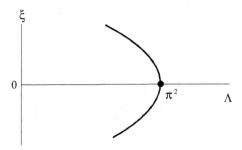

in the form $y = \xi \sin(\pi x) + \sum_{i=2}^{\infty} \xi^i y_i$, $\xi \ll 1$. Substituting this expression for y in Eq. (6.4.3) and keeping only the main terms, one gets:

$$\xi^2 B y_2 = \xi^2 \sin^2(\pi x) - \lambda \xi \sin(\pi x), \quad y_2(0) = y_2(1) = 0.$$

The existence conditions for the solution of this problem yields the ramification equation $-\lambda \xi / 2 + 4 \xi^2 / (3\pi) = 0$, which has the nontrivial solution, $\xi = 3\pi \lambda / 8$, for both $\lambda < 0$ and $\lambda > 0$. So, we obtain

$$y_2 = \frac{4x - 2}{3\pi^2} \cos(\pi x) + \frac{1}{2\pi^2} + \frac{1}{6\pi^2} \cos(2\pi x) + C \sin(\pi x),$$

where the constant C is evaluated from the solvability condition for the problem:

$$B y_3 = [2 \sin(\pi x) - 8/(3\pi)] y_2, \quad y_3(0) = y_3(1) = 0.$$

Therefore,

$$y \simeq \left[\xi + \frac{2\xi^2}{\pi^3} - \frac{5\xi^2}{16\pi} \right] \sin(\pi x)$$
$$+ \xi^2 \left[\frac{4x - 2}{3\pi^2} \cos(\pi x) + \frac{1}{2\pi^2} + \frac{\cos(2\pi x)}{6\pi^2} \right] + O\left(\xi^3\right).$$

In Fig. 6.8 the dependence of ξ versus the parameter Λ is plotted in a neighborhood of the critical value $\Lambda = \pi^2$.

6.3.3. In a neighborhood of the second eigenvalue, the problem can be represented in the form $B_1 y = y_{xx} + 4\pi^2 y = -\lambda y + y^2$. We seek a solution of the problem as $y = \xi \sin(2\pi x) + \sum_{i=2}^{\infty} \xi^i y_i$, $\xi \ll 1$. Substituting this solution into the equation and keeping only the main terms, we get

$$\xi^2 B_1 y_2 = \xi^2 \sin^2(2\pi x) - \lambda \xi \sin(2\pi x), \quad y_2(0) = y_2(1) = 0.$$

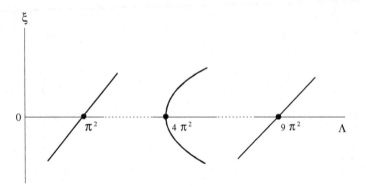

Fig. 6.8 ξ versus Λ in neighborhoods of three critical values

The solvability condition for this equation, $\int_0^1 h(x) \sin(2\pi x)\, dx = 0$, provides $\lambda = 0$. This means that, in this case, $\lambda \ll \xi$ and

$$y_2 = C \sin(2\pi x) + \frac{\cos(4\pi x)}{24\pi^2} - \frac{\cos(2\pi x)}{6\pi^2} + \frac{1}{8\pi^2}.$$

For the next approximation we have the equation

$$\xi^3 B_1 y_3 = 2y_2 \xi^3 \sin(2\pi x) - \lambda \xi \sin(2\pi x), \quad y_3(0) = y_3(1) = 0.$$

The solvability condition for this equation provides the ramification equation $\lambda = 5\xi^2/(24\pi^2)$, where the constant C in the expression for y_2 is found from the solvability condition for the following problem:

$$\xi^4 B_1 y_4 = \xi^4 y_2^2 + 2\xi^4 \sin^2(\pi x) y_3 - \lambda \xi^2 y_2, \quad y_4(0) = y_4(1) = 0.$$

The ramification equation has a nontrivial solution for $\lambda > 0$.

In a neighborhood of the third eigenvalue, we have

$$B_1 y = y_{xx} + 9\pi^2 y = -\lambda y + y^2.$$

The solution is similar to the solution of the problem in a neighborhood of the first eigenvalue. In this case the ramification equation is $\lambda \xi - 8\xi^2/(9\pi) = 0$. The nontrivial solution in a neighborhood of the third eigenvalue has the form:

$$y \simeq \xi \sin(3\pi x) + \xi^2 \left[\frac{4x-2}{27\pi^2} \cos(3\pi x) + \frac{1}{18\pi^2} + \frac{\cos(6\pi x)}{54\pi^2} \right]$$
$$+ C_1 \xi^2 \sin(3\pi x) + O\left(\xi^3\right),$$

where the constant C_1 is evaluated from the solvability condition for the problem for y_3.

The dependence of the variable ξ on the parameter Λ in a neighborhood of the critical values $\Lambda = 4\pi^2$ and $\Lambda = 9\pi^2$ is plotted in Fig. 6.8.

6.3.4. Consider the linearized problem

$$\frac{d^2\theta}{dr^2} + \frac{1}{r}\frac{d\theta}{dr} - \frac{\theta}{r^2} + \Lambda\theta = 0, \quad 0 \le r \le 1,$$

$$T_r \equiv 0, \quad \theta(1) = \theta(0) = 0.$$

For $\Lambda = \Lambda_n$ ($n = 1, 2, \ldots$), the problem has nontrivial solutions $\theta_n = C J_1(\sqrt{\Lambda_n}\,r)$ and $J_1(\sqrt{\Lambda_n}) = 0$, i.e. the eigenvalue Λ_n is such that $\sqrt{\Lambda_n}$ is the nth root of the Bessel function J_1. The eigenfunctions form a set of orthogonal functions

$$\int_0^1 J_1\left(\sqrt{\Lambda_i}\,r\right) J_1\left(\sqrt{\Lambda_j}\,r\right) r\, dr = 0, \quad i \ne j.$$

We find the solution of this problem in neighborhoods of the eigenvalues $\Lambda = \Lambda_n + \delta$:

$$\boldsymbol{B}\theta \equiv \frac{d^2\theta}{dr^2} + \frac{1}{r}\frac{d\theta}{dr} - \frac{\theta}{r^2} + \Lambda_n\theta = (\Lambda_n + \delta)T_r\theta - \delta\theta.$$

Since, from the second equation of the problem, $T_r = \int_r^1 \frac{1}{2\tau^3}\int_0^\tau \theta^2 s\, ds\, d\tau$, then the equation can be represented in terms of the angle θ as

$$\boldsymbol{B}\theta = (\Lambda_n + \delta)\theta \int_r^1 \frac{1}{2\tau^3}\int_0^\tau \theta^2 s\, ds\, d\tau - \delta\theta.$$

We seek the solution of this problem in the form of a series:

$$\theta = \mu J_1\left(\sqrt{\Lambda_n}\,r\right) + \mu^3\theta_{n3}(r) + \mu^5\theta_{n5}(r) + \cdots$$

Substituting this solution into the given equation and keeping only the main terms, since μ and δ are small, we get a sequence of eigenvalue problems for θ_{ni}. For θ_{n3}, we have

$$\mu^3 \boldsymbol{B}\theta_{n3} = -\delta\mu J_1\left(\sqrt{\Lambda_n}\,r\right)$$

$$+ \mu^3\Lambda_n J_1\left(\sqrt{\Lambda_n}\,r\right)\int_r^1 \frac{1}{2\tau^3}\int_0^\tau J_1\left(\sqrt{\Lambda_n}\,r\right)^2 s\, ds\, d\tau.$$

The existence condition for the solution of the problem

$$\boldsymbol{B}\theta = f, \quad \theta(0) = 0, \quad \theta(1) = 0,$$

is $\int_0^1 f J_1 \left(\sqrt{\Lambda_n} \, r \right) r \, dr = 0$. It provides the ramification equation $\delta = \Lambda_n A \mu^2$, where

$$A = \frac{\int_0^1 r J_1^2 \left(\sqrt{\Lambda_n} \, r \right) F_1(r) \, dr}{\int_0^1 J_1^2 \left(\sqrt{\Lambda_n} \, r \right) r \, dr}, \quad F_1(r) = \frac{1}{2} \int_r^1 \frac{1}{\tau^3} \int_0^\tau s J_1^2 \left(\sqrt{\Lambda_n} \, s \right) ds \, d\tau.$$

For $\Lambda_1 = 14.682$ ($\sqrt{\Lambda_1} = 3.832$), for example, $A = 0.0369$, and the ramification equation has the form $\delta = 0.537 \mu^2$. For $\Lambda_2 = 49.219$ ($\sqrt{\Lambda_2} = 7.016$) we obtain $\delta = 1.606 \mu^2$. For $\Lambda_3 = 103.500$ ($\sqrt{\Lambda_3} = 10.174$) we get $\delta = 2.911 \mu^2$.

Bibliography

1. M. Abramowitz, I.A. Stegun, *Handbook of Mathematical Functions with Formulas, Graphs, and Mathematical Tables* (Dover Publications, New York, 1972), 1046 pp
2. I.V. Andrianov, J. Awrejcewicz, L.I. Manevitch, *Asymptotical Mechanics of Thin-walled Structures* (Springer, Berlin, New York, 2004), 535 pp
3. I.V. Andrianov, L.I. Manevitch, *With Help from Michiel Hazewinkel, Asymptotology: Ideas, Methods, and Applications* (Kluwer Academic Publishers, Boston, 2003), 252 pp
4. J. Awrejcewicz, I.V. Andrianov, L.I. Manevitch, *Asymptotic Approaches in Nonlinear Dynamics* (Springer, Berlin, Heidelberg, 1998), 310 pp
5. V.M. Babich, V.S. *Buldyrev, Short-wavelength Diffraction Theory: Asymptotic Methods*, translated by E.F. Kuester (Springer, Berlin, New York, 1991), 445 pp
6. G.A. Baker Jr, P. Graves-Morris, *Pade Approximants*, 2nd edn. (Cambridge University Press, Cambridge, New York, 1996), 746 pp
7. N.S. Bakhvalov, G.P. Panasenko, Homogenization: Averaging Processes in Periodic Media. Mathematical Problems in the Mechanics of Composite Materials (Nauka, Moscow, 1984); English Translated in Mathematics and Its Applications (Soviet Series), vol. 36 (Kluwer Academic Publishers Group, Dordrecht, 1989)
8. R.G. Barantsev, *Asymptotic versus classical mathematics, in Topics in Mathematical Analysis* (World Scientific, Singapore, 1989), 49–64 pp
9. L. Barbu, G. Morosanu, *Singularly Perturbed Boundary-Value Problems, International Series of Numerical Mathematics*, vol. 156 (Birkhauser, Basel, Boston, Berlin, 2007), 236 pp
10. S.M. Bauer, S.B. Filippov, A.L. Smirnov, P.E. Tovstik, Asymptotic methods in mechanics with applications to thin shells and plates, in *Asymptotic Methods in Mechanics, CRM Proceedings and Lecture Notes* (American Mathematical Society, Providence, 1993), pp. 3–141
11. C.M. Bender, S.A. Orszag, *Advanced Mathematical Methods for Scientists and Engineers: Asymptotic Methods and Perturbation Theory* (Springer, Berlin, New York, 1995), 593 pp
12. N. Bleistein, R.A. Handelsman, *Asymptotic Expansions of Integrals* (Dover Publications, New York, 2010), 448 pp
13. N.N. Bogolyubov, Y.A. Mitropolsky, *Asymptotic Methods in the Theory of Non-linear Oscillations*, 2nd edn. (Gordon and Breach Science Publishers, New York, 1961), 537 pp
14. N.G. de Bruijn, *Asymptotic Methods in Analysis* (Dover Publications, New York, 1981), 200 pp
15. K.W. Chang, F.A. Howes, *Nonlinear Singular Perturbation Phenomena, Theory and Applications* (Springer, New York, Berlin, Heidelberg, Tokyo, 1987)

© Springer International Publishing Switzerland 2015
S.M. Bauer et al., *Asymptotic Methods in Mechanics of Solids*,
International Series of Numerical Mathematics 167,
DOI 10.1007/978-3-319-18311-4

16. P.G. Ciarlet, *Introduction to Linear Shell Theory, Series on Applied Mathematics Paris* (France, Dauthier-Villars, 1998), 184 pp

17. E.T. Copson, *Asymptotic Expansions* (Cambridge University Press, Cambridge, 2004), 132 pp

18. L.H. Donnell, *Beams, Plates and Shells* (McGraw-Hill, New York, 1976), 453 pp

19. W. Eckhaus, *Asymptotic Analysis of Singular Perturbations* (Elsevier, Amsterdam, 2011), 287 pp

20. A. Erdèlyi, *Asymptotic Expansions* (Dover Publications, New York, 2010), 128 pp

21. M.A. Evgrafov, Asymptotic Estimates and Entire Functions, translated by Allen L. Shields (Gordon and Breach, New York, 1961), 181 pp

22. M.V. Fedoruk, Asymptotics: Integrals and Series (Nauka, Moscow, 1987), 544 pp (in Russian)

23. M.V. Fedoruk, *Asymptotic Analysis: Linear Ordinary Differential Equations*, translated from the Russian by Andrew Rodick (Springer, Berlin, New York, 1993), 363 pp

24. M.V. Fedoruk, *Saddle Point Method, in Encyclopedia of Mathematics* (Springer, New York, 2001)

25. S.F. Feshchenko, N.I. Shkil', L.D. Nikolenko, *Asymptotic Methods in the Theory of Linear Differential Equations*, translated by Scripta Technica (American Elsevier Publishing Company, New York, 1967), 270 pp

26. S.B. Filippov, Theory of Joint and Reinforced Shells (SPb., Sankt-Petersburg: SPb University Press, 1999), 196 pp (in Russian)

27. N. Fröman, P.O. Fröman, J.W.K.B. Approximation, *Contributions to the Theory* (North-Holland Publishing Company, Amsterdam, 1965), 138 pp

28. A.L. Gol'denveizer, *Theory of Elastic Thin Shells, translation from the Russian*, ed. by G. Herrmann (Pergamon Press, New York, 1961), 658 pp

29. A.L. Gol'denveizer, Asymptotic method in theory of shells. Adv. Mech. **5**(1/2), 137–182 (1982)

30. A.L. Goldenveizer, V.B. Lidsky, P.E. Tovstik, Free Vibrations of Thin Elastic Shells (Nauka, Moscow, 1979), 384 pp (in Russian)

31. E.I. Grigolyuk, V.V. Kabanov, Stability of Shells (Nauka, Moscow, 1978), 360 pp (in Russian)

32. J. Heading, *An Introduction to Phase-Integral Methods* (Methuen, London, 1962), 160 pp

33. E.J. Hinch, *Perturbation Methods* (Cambridge University Press, Cambridge, 1991), 160 pp

34. M.H. Holms, *Introduction to Perturbation Methods* (Springer, New York, 1995), 337 pp

35. E. Jahnke, F. Emde, F. Lösch, Tables of Higher Functions, 6th edn. revised by Friedrich Lösch (McGraw-Hill, New York, 1960), 318 pp

36. J.D. Kaplunov, L.Y. Kossovich, E.V. Nolde, *Dynamics of Thin Walled Elastic Bodies* (Academic Press, San Diego, Calif, 1998), 226 pp

37. J. Kevorkian, J.D. Cole, *Multiple Scale and Singular Perturbation Methods* (Springer, New York, 1996), 632 pp

38. J. Kevorkian, J.D. Cole, *Perturbation Methods in Applied Mathematics* (Springer, New York, 1981)

39. R.E. Langer, The asymptotic solutions of linear differential equations of second order with two turning points. Trans. Am. Math. Soc. **90**, 113–142 (1959)

40. A. Libai, J.G. Simmonds, *The Nonlinear Theory of Elastic Shells* (Cambridge University Press, Cambridge, 1998), 562 pp

41. C.C. Lin, A.L. Rabestein, On the asymptotic solutions of a class of ordinary differential equations of the fourth order. Trans. Amer. Math. Soc. **94**, 24–57 (1960)

42. S.A. Lomov, *Introduction to the General Theory of Singular Perturbations*, translated from the Russian by J.R. Schulenberger (American Mathematical Society, Providence, 1992), 375 pp

43. V.P. Maslov, Th*e Complex WKB Method for Nonlinear Equations I: Linear Theory*, translated from the Russian by M.A. Shishkova, A.B. Sossinsky (BirkhöUser, Basel, Boston, 1994), 300 pp

44. V.P. Maslov, V.E. Nazaikinskii, *Asymptotic of Operator and Pseudo-Differential Equations* (New York, London, Consultant Bureau, 1988), 313 pp

45. D.R. Merkin, *Introduction to the Theory of Stability* (Springer, Berlin, Heidelberg, New York, 1996), 344 pp

46. D.R. Merkin, F.F. Afagh, S.M. Bauer, A.L. Smirnov, *Problems in Theory of Stability* (St. Petersburg University Press, St. Petersburg, 2000), 113 pp
47. G.I. Mikhasev, P.E. Tovstik, Localized Vibrations and Waves in Thin Shells. Asymtotic Methods (Moscow, Nauka, 2009), 290 pp (in Russian)
48. J.A. Murdock, Perturbations: Theory and Methods (SIAM, 1999), 509 pp
49. A.H. Nayfeh, *Perturbation Methods, Wiley Classics*, Library edn. (Wiley, New York, 2000), 425 pp
50. A.H. Nayfeh, *Introduction to Perturbation Techniques* (Wiley, New York, 1981), 529 pp
51. S.A. Nazarov, Asymptotic Analysis of Thin Plates and Beams. V. 1. Downturn of Dimension and Integrated Estimations (Novosibirsk, Nauchnaya kniga, 2002), 408 pp (in Russian)
52. R.E. O'Malley, *Introduction to Singular Perturbations* (Academic Press, New York, London, 1974), 206 pp
53. F.W.J. Olver, *Asymptotics and Special Functions* (Mass, A.K. Peters Wellesley, 1997), 572 pp
54. E.Y. Riekstynsh, Asymptotic Expansions of Integrals, Zinatne, Riga, vol. I, 1974, p. 391; vol. II, 1977, p. 370; vol. III, 1981, p. 370 (in Russian)
55. E. Sanches-Palencia, *Non-homogeneous Media and Vibration Theory* (Springer, Berlin, Heidelberg, 1980), 398 pp
56. P.E. Tovstik, A.L. Smirnov, *Asymptotic Methods in the Buckling Theory of Elastic Shells, Series A*, vol. 4 (World Scientific, Singpore, 2002), 347 pp
57. B.R. Vainberg, *Asymptotic Methods in Equations of Mathematical Physics*, translated from the Russian by E. Primrose (Gordon and Breach Science Publishers, New York, 1989), 498 pp
58. M.M. Vainberg, V.A. Trenogin, *Theory of Branching of Solutions of Non-linear Equations*, translated by Israel Program for Scientific Translations (Noordhoff International Publications, Leyden, 1974), 485 pp
59. M. Van Dyke, *Perturbation Methods in Fluid Mechanics* (The Parabolic Press, Stanford, 1975), 271 pp
60. A.B. Vasil'eva, V.F. Butuzov, L.V. Kalachev, *The Boundary Function Method for Singular Perturbation Problems* (Society for Industrial and Applied Mathematics, Philadelphia, 1995), 221 pp
61. A.B. Vasil'eva, *Singularly Perturbed Equations in the Critical Case* (University of Winsconsin, Mathematics Research Center, Madison, 1980), 156 pp
62. M.I. Vishik, L.A. Lyusternik, Regular degeneration and boundary layer for linear differential equations with small parameter. Russ. Math. Surv. **12**(5), (77) 3–122 (1957)
63. M.I. Vishik, L.A. Lyusternik, The solution of some perturbation problems for matrices and selfadjoint or non-selfadjoint differential equations. Russ. Math. Surv. **15**(3), (93), 1–73 (1960)
64. W.R. Wasow, Linear Turning Point Theory, Applied Mathematical Sciences, vol. 54. (Springer, New York), 246 pp
65. W.R. Wasow, Asymptotic Expansions for Ordinary Differential Equations (Dover, New York 1987), 374 (1965)
66. R. Wong, *Asymptotic Approximation of Integrals, Classics in Applied Mathematics*, vol. 34 (Society for Industrial and Applied Mathematics, Philadelphia, 2001), 543 pp

Index

© Springer International Publishing Switzerland 2015
S.M. Bauer et al., *Asymptotic Methods in Mechanics of Solids*,
International Series of Numerical Mathematics 167,
DOI 10.1007/978-3-319-18311-4

Printed in the United States
By Bookmasters